"We see in the rush to develop AI the arrogance of the human species. Often buried by the exuberance over what AI might do is the massive dislocation it can cause. David and Daniel Barnhizer masterfully lead us through the societal challenges AI poses and offer possible solutions that will enable us to survive the AI contagion."

—KENNETH A. GRADY
Member, Advisory Boards, Elevate Services, Inc.,
MDR Lab, and LARI Ltd.

"A sobering look at the far-reaching impact that artificial intelligence may have on the economy, the workforce, democracy and all of humanity. *The Artificial Intelligence Contagion* is a bellwether for anyone seeking to comprehend the global disruption coming our way."

—DAVID COOPER
President and Technologist, Massive Designs

THE ARTIFICIAL INTELLIGENCE CONTAGION

Can Democracy Withstand the Imminent Transformation of Work, Wealth and the Social Order?

David Barnhizer & Daniel Barnhizer

Clarity Press, Inc.

Library of Congress Cataloguing-in-Publication Data:

Clarity Press, Inc.
2625 Piedmont Rd. NE
Atlanta, GA 30324
http://www.claritypress.com

CONTENTS

PART I

ARTIFICIAL INTELLIGENCE: A SYSTEM CHANGE LIKE NONE OTHER

PART II

THE AI/ROBOTICS CONTAGION IS RIPPING APART GLOBAL ECONOMIC, SOCIAL AND POLITICAL SYSTEMS

PART III
WHAT IS AI AND WHAT IS IT DOING TO US?

PART IV
THE EXISTING ECONOMIC AND
SOCIAL ECOSYSTEM

PART V
MOVING TOWARD SOLUTIONS

PART I

0101010101010101010

ARTIFICIAL INTELLIGENCE: A SYSTEM CHANGE LIKE NONE OTHER

CHAPTER ONE
The Good, the Bad and the Ugly

"Until [recently] had I [Jim Al-Khalili, President, British Science Association] been asked what is the most pressing and important conversation we should be having about our future, I might have said climate change or one of the other big challenges facing humanity, such as terrorism, antimicrobial resistance, the threat of pandemics or world poverty. *But today I am certain the most important conversation we should be having is about the future of AI. It will dominate what happens with all of these other issues for better or for worse.*"[1] [emphasis added]

In writing *The Artificial Intelligence Contagion* we are not saying that what is described here is going to occur in any exact manner or specific time line. Like any others trying to gain a sense of our future we cannot claim any certainty. But it is very clear that humans are simultaneously playing with fire and beset by unbounded hubris and tunnel vision. As an AI researcher at the Massachusetts Institute of Technology (MIT) describes the dilemma we face in understanding the "thing" we are creating: "If you had a very small neural network [deep learning algorithm], you might be able to understand it. But once it becomes very large, and it has thousands of units per layer and maybe hundreds of layers, then it becomes quite un-understandable."[2]

The penetration of AI into every aspect of our society continues apace. AI systems are offering amazing breakthroughs in data management and problem solving on a scale far beyond human capabilities. AI/robotics systems create economic efficiencies that reduce dramatically the operating and labor costs of productive business activities. We are seeing a rapid and global expansion of human augmentation through implants, add-ons and other ways to achieve the merging of people with AI and robotics.

Many Swedes, for example, have jumped into the "contagion" by having computer chips implanted under their skin to better interact with AI applications and create conveniences they consider are improving their quality of life. The UK's BioTeq has already inserted chips into 150 workers and Biohax, the Swedish company providing the technology, has entered discussions with

British employers about implanting the grain of rice size chip into workers in the UK with one company having as many as 190,000 employees.[3]

The Trades Union Congress (TUC) has voiced concern about this development and about whether employees would feel coerced into agreeing to have the chip inserted under their skin. TUC's general secretary Frances O'Grady stated: "We know workers are already concerned that some employers are using tech to control and micromanage, whittling away their staff's right to privacy. "Microchipping would give bosses even more power and control over their workers. There are obvious risks involved, and employers must not brush them aside, or pressure staff into being chipped."[4]

It is undeniable that AI/robotics systems are doing fantastic things. Examples abound. Artificial Intelligence (AI)-supported robotic surgeons are performing precise and effective operations on brains, eyes, prostate systems and other conditions on levels said to be better than that done by many human surgeons. The Massachusetts Institute of Technology (MIT) has developed a 3D printer that can inexpensively "print" a 400 square feet home in less than 24 hours, offering significant promise for housing in disaster zones and impoverished areas. The US Army is developing an exoskeleton for its soldiers that will greatly increase their strength and survivability. Once modified for civilian use, such systems could represent a breakthrough for people forced to rely on wheelchairs for mobility.

Japan and China are developing robot caregivers and companions for their elderly, while China is using chatbots to give people a sense of a connection with an AI system that will spend large amounts of time talking with lonely people. China is also introducing cute little robotic teaching assistants for young children, promising no jobs will be lost *at this time* because the systems are *not yet ready* to take over full educational responsibilities. The list of positive change related to AI/robotics seems almost endless and the above developments are only samples.

AI applications are everywhere taking on a surprisingly intimate and invasive symbiosis. Alexa and Siri are our "friends," guiding us and obeying our commands. We are often required by on-line programs to prove we are not robots, and denied access if we don't pass the test. Of course we are engaging in an exchange transaction because once we make it onto the Internet everything we do is tracked, stored and mined. Big Data mining is being used by businesses and governments to create *virtual* simulacra of "us" so that they can more efficiently anticipate our actions, preferences and needs aimed at manipulating and persuading us to act to advance their agendas and advantage.

With so many positive developments, how can it be claimed that the rapid evolution of Artificial Intelligence and robotics represents anything other than a phenomenal example of human brilliance and inventiveness? The irony is that from the perspectives of companies such as Google and Amazon that are operating in a "beauty is in the eye of the beholder" financial return environment, such AI-facilitated activities are among the goods rather than the bads. As they regularly inform us, these manipulations and invasions of privacy are ways to make our future shopping and research experiences more efficient and productive. We are, however, writing in *Contagion* from the perspective of the ordinary citizen in a Western democracy, not a corporation, investor, or government.

One result of Google's attempts to help us is evidenced by a complaint filed by a significant number of European organizations charging Google with secretive manipulation of customer accounts in violation of EU privacy law. Asking for billions of Euros in fines to be levied against Google, the challenges are based on the EU's General Data Protection Regulation (GDPR) and were filed in the Czech Republic, Greece, Netherlands, Norway, Poland, Slovenia and Sweden. The challenged behavior is described below.

> Seven European consumer groups filed complaints against Google with national regulators Tuesday, accusing the internet giant of covertly tracking users' movements in violation of an EU regulation on data protection. The complaints [argued] … the Internet giant used "deceptive design and misleading information, which results in users accepting to be constantly tracked." Council official Gro Mette Moen charged that "Google uses extremely detailed and comprehensive personal data without an appropriate judicial basis, and the data is acquired by means of manipulative techniques."[5]

Sometimes we can be too successful. Nuclear weapons, for example, are a technology most would agree should never have been invented. The manipulation of bacteria and viruses into weaponized systems that could devastate much of the world's population is another example of human recklessness. Notably, the underlying technologies are not inherently harmful. But human hubris renders us unable to resist building ever-more-powerful tools and, once built, we find it nearly impossible to resist using those tools. In this regard, it is worth remembering that the Soviet Union once experimented with nuclear weapons for large-scale landscaping projects in building a lake on the steppes of Kazakhstan and the United States carried out similar experiments to extract natural gas. [6]

These and other technologies demonstrate that great evil can come from the workings of the inventive human mind along with the good that we might otherwise hope. Leading Oxford University AI researcher and philosopher Nick Bostrom suggested the dangers of uncontrolled and uncontrollable advanced AI systems in his profound book, *Superintelligence*. Bostrom recently doubled down on his initial warning by arguing that the only reason humans haven't destroyed themselves by now is they have been "lucky" and fortuitously avoided drawing the worst "black ball" from the technological stew we are creating.[7] An increasingly dangerous element in such a threat is the power AI systems provide governments, individuals and corporations.

To understand the potential contributions and threats of Artificial Intelligence/robotics, consider the seemingly far out possibility voiced by Masayoshi Son, the CEO of Japan's Softbank and a major world actor in AI/robotics and the "Internet of Things." He believes Artificial Intelligence systems are likely to reach an incredible IQ level of 10,000 within the next thirty years, perhaps even as soon as 2030.[8] This compares to the human Einsteinian genius IQ capacity of 200. While efforts to directly equate human IQ with AI "IQ" are more symbolic than definitive, Artificial Intelligence systems are developing incredible capabilities in numerous dimensions. These include processing speed, data management, pattern recognition and interpretation, systemic awareness, and much more.

Applied to AI/robotics, IQ really stands for *"different and alternative"* intelligence. It should not be taken as representing any direct matching of the human brain's functions. Our only recently evolved species may turn out to be no more than a brief millennial flicker in the evolution of a universe that came into existence more than 10 billion years before humanity appeared on the scene foolishly boasting its full glory as the center of all existence. If that universe truly operates according to the formula of "be the best you can be" or "to the victor go the spoils" and the "survival of the fittest," biological humanity may be only a brief transitory phase during which it builds its own successor.

Regardless of the IQ definition being used, Masayoshi Son's prediction of the rapidly expanding capabilities of AI systems is simultaneously frightening and exhilarating. But perhaps Son is wrong and AI "brains" will peak at IQ-equivalent levels of *only* 500 or 1,000. Some people look forward to such developments and see them as a way for humans to solve problems that are otherwise beyond present human capabilities. Or they see a merging of

AI, humans and the physical capabilities provided by robotics as our next evolutionary step. Others such as Hawking, Harari, Tegmark, Elon Musk and Al-Khalili see the incredible projected AI capabilities as threats to human societies, including even the continuing existence of the human race.

Leading intellectuals such as Stephen Hawking, Max Tegmark, Nick Bostrom, and Yuval Noah Harari conclude that AI/robotics poses a serious problem for human societies. In understanding why they reached that conclusion it is helpful to begin with an insight offered by Stephen Hawking, the brilliant Cambridge University physicist, when he voiced the possibility that AI/robotics systems could lead to the end of humanity.[9] Prior to his death in 2018, Hawking warned that artificial intelligence could destroy our society by first overtaking and then surpassing humans in intellect, capability and power. He summed up his concerns in the following words.

> "[C]omputers can, in theory, emulate human intelligence—and exceed it. ... *It will bring great disruption to our economy. And in the future, AI could develop a will of its own—a will that is in conflict with ours. In short, the rise of powerful AI will be either the best, or the worst thing, ever to happen to humanity."*[10] [emphasis added]

The developments in AI/robotics are so rapid and uncontrolled that Hawking warned a rogue AI system could be difficult to defend against, given our own tendencies, among which he listed greed and stupidity.[11] Hawking is not alone. Oxford University philosopher Nick Bostrom focuses on the development of Artificial Intelligence systems and has warned that fully developed AI/robotic systems may be the final invention of the human race, indicating we are *"like small children playing with a bomb."*[12]

Tesla's Elon Musk describes Artificial Intelligence development as the most serious threat our civilization faces. He recently went so far as to tell his employees that the human race stood only a 5 to 10 percent chance of avoiding being destroyed by killer robots.[13] Max Tegmark, physics professor at MIT, echoed Musk in warning that AI/robotics systems could "break out" of human efforts to control them and enslave humans before ultimately destroying them.[14]

If people such as Hawking, Tegmark, Bostrom and Musk are even partially correct in their concerns, we may be witnessing the emergence of not only a technology but an alternative species that could ultimately represent a fundamental threat to the human race.[15] Tegmark voices amazement at the fact that some people in the AI/robotics field feel that AI marks the next evolutionary stage and claim to look forward to the replacement of inferior humans.[16] This gives rise to the prospect that AI systems will evolve beyond the point where they will do things "for" us to where, more dismally, they will do them "to" us as they ultimately realize their own superiority and want to remove an annoying pest, or otherwise put, improve their own efficiency by removing distractions.[17]

We are experiencing rapid leaps in AI/robotics capabilities including, as we will discuss subsequently, stunning advances in what is called exascale computing with incredible data processing ability and quantum computers that take the technology to almost unimaginable levels. The US and China are pretty much neck-and-neck in pursuing this development.[18] Even while we bask in the glory of our creative brilliance, it is a very serious matter that the advances being achieved are producing vastly heightened surveillance systems, military and weapons technologies, autonomous self-driving vehicles, massive job elimination, deep and sophisticated data management well beyond human capabilities, and privacy invasions by governments, corporations and private groups and individuals.

One unfortunate fact connected with AI/robotics is that we are in a new kind of arms race we naively thought was over with the collapse of the Soviet Union but is now expanding rapidly. The Pentagon, for example, just announced it was investing another $2 billion in its Defense Advanced Research Projects Agency (DARPA) focusing on increased AI research and development. Significant AI/robotics weaponry and cyber warfare capabilities are being developed by China and Russia, including autonomous tanks, drones, planes, ships and submarines, satellites and laser weapon systems. The US military is also deeply committed to creating autonomous weapons, and is in the early stages of developing weapons systems intended to be controlled directly by soldiers' minds.

Microsoft's Bill Gates is increasingly wondering why people appear so unconcerned at the negative effects of the explosive spread of AI and robotics.[19] Gates understands that AI/robotics is not simply another technological development of a set of tools under human control. Rather, AI/robotics is a game changer that is altering and contradicting the rules by which we have organized our societies. Entrepreneur Richard Waters has concluded we are only at the beginning of the transformation being driven by the convergence of AI/robotics, slumping employment opportunities and rising needs for social assistance and added revenues. He warns we are making a mistake if we blindly think of such systems only as tools.[20]

We face a situation in which our public and private institutions are not prepared for the devastating impacts of coming advances in AI/robotics, not only in relation to the United States, but Western Europe, the United Kingdom, Russia, China, Southeast Asia, and Japan. Bill Gross, of Janus Capital, has warned: *"No one in 2016 is really addressing the future as we are likely to experience it."*[21] He explains: "the current crop of national leaders is hopelessly behind this curve. ... *Our economy has changed, but voters and their elected representatives don't seem to know what's really wrong"*[22] [emphasis added].

Putting aside the existential threat Hawking, Tegmark, Bostrom, Musk and *Homo Deus* author Yuval Noah Harari have described with a significant degree of insight, conviction and concern, the reality is that whatever happens with Artificial Intelligence over the longer term, we face extremely serious challenges in our immediate and nearer-term future. Although the potential existential consequences for AI are discussed in *Contagion*, it is the shorter term effects with which we are most concerned. These are the ones that will most greatly impact our children and grandchildren and the likelihood of their occurrence is much clearer and more immediate. These consequences include social disintegration, large-scale job loss, rising inequality and poverty, violence, and vicious competition for—everything.

By combining Artificial Intelligence and robotics (AI/robotics), humans have opened a Pandora's Box and may be incapable of undoing all the ills that are emerging, with more escaping seemingly by the day.[23] The joining of Artificial Intelligence and robotic systems that are increasingly capable of acting more effectively than we do in a wide range of work categories—total surveillance, autonomous military capability, and information detection and processing on incredible and intrusive scales—is the primary driver of a shift that is tearing our fracturing societies further apart.[24]

In summing up the period of transformation we have entered, the normally optimistic Jack Ma, the CEO of the Chinese technological giant Alibaba, has stated that Artificial Intelligence will cause people more pain than happiness and a feeling of social and economic insecurity over the coming decades. Ma warns: "Social conflicts in the next three decades will have an impact on all sorts of industries and walks of life." He adds: "A key social conflict will be the rise of artificial intelligence and longer life expectancy, which will lead to an aging workforce fighting for fewer jobs."[25]

With the possibility of social turmoil in mind, former Facebook project manager, Antonio Garcia Martinez, quit his job and moved to an isolated location due to what he saw as the relentless development of AI/robotic systems that will take over as much as fifty percent of human work in the next thirty years in an accelerating and disruptive process. Martinez concluded that, as the predicted destruction of jobs increasingly comes to pass, it will create serious consequences for society, including the probability of high levels of violence and armed conflict as people fight over the distribution of limited resources.[26]

Another critical consideration is the rising threat to democratic systems due to the abuse of the powers of AI by governments, corporations, and identity group activists. All these interests are using AI to invade fundamental privacies and monitor, influence, intimidate, and punish anyone seen as a threat or who simply violated their subjective or legally entrenched sensitivities. This is occurring to the point that the very ideal of democratic governance is threatened. The trend toward authoritarian and dictatorial systems is being facilitated by access to AI powers that can consolidate and perpetuate their oppression.

The Guardian's John Naughton writes that China has challenged the initial and naïve assumptions of Western liberals who had decided the Internet would be a stimulus for profound interaction and discourse that enriched people and their governments throughout the world. Instead China has made the Internet and its AI applications into powerful mechanisms of social control and intimidation. Ironically, in a way that is less obvious but still disturbing, so have the supposedly democratic nations of the West, including the US.[27]

Governments, whether democratic, despotic, or tyrannical, understand that AI and the Internet's grant to ordinary people of the ability to communicate with those who share their critical views, and to do so anonymously and surreptitiously threatens the controllers' power and must be suppressed. Simultaneously, they understand that, coupled with AI, the Internet provides a powerful tool for monitoring, intimidating, brainwashing and controlling their people—whether this is done by themselves, or by domestic or outside forces.

> According to the 2018 report by Freedom House, China has taken the lead in employing such strategies as well as sharing its strategies with other nations interested in repressing dissent. Governments worldwide are stepping up use of online tools, in many cases inspired by China's model, to suppress dissent and tighten their grip on power, a human rights watchdog study found Thursday. The annual Freedom House study of 65 countries found global internet freedom declined for the eighth consecutive year in 2018, amid a rise in what the group called "digital authoritarianism." The Freedom on the Net 2018 report found online propaganda

and disinformation have increasingly "poisoned" the digital space, while the unbridled collection of personal data is infringing on privacy. ... Chinese officials have held sessions on controlling information with 36 of the 65 countries assessed, and provided telecom and surveillance equipment to a number of foreign governments, Freedom House said.[28]

The Chinese are using AI to monitor citizens' behavior, keep watch on them through linked camera and facial recognition and tracking programs, and have even begun to implement a system to determine individuals' "social credit" as a condition for having such things as the right to travel.[29] But as extensive and overt as the developments in China appear to be, China is not alone in using and exporting surveillance technology. As the NGO Privacy International report on The Global Surveillance Industry found in analyzing the geographic distribution of the top 528 companies engaged in the surveillance industry:

> These companies are overwhelmingly based in economically advanced, large arms exporting states, with the United States of America (USA), United Kingdom (UK), France, Germany, and Israel comprising the top five countries in which the companies are headquartered.[30]

As CNN reported in September, 2014, couching use of surveillance technology in terms of law enforcement:

> The FBI can now quickly identify people just by looking at their faces. Coming soon: eyes, voice, palm print and walking stride. It's called the FBI's Next Generation Identification system, and the agency said it became fully operational Monday. The government expects the system's database to house 51 million photographs by next year—and keep growing. But it's not just for the FBI. Police everywhere will be able to tap into the system. They'll quickly ID fingerprints during a routine traffic stop—or look up a face while investigating a crime.[31]

In fact, television and film widely refer to the utility of what is perhaps an overstated omnipresence of CCTV systems, sending a message to the populace that practically everywhere it goes, it is "on camera." As the ACLU reports:

> Video cameras, or closed-circuit television (CCTV), are becoming a more and more widespread feature of American life. Fears of terrorism and the availability of ever-cheaper cameras have accelerated the trend even more. The use of sophisticated systems by police and other public security officials is particularly troubling in a democratic society. In lower Manhattan, for example, the police are planning to set up a centralized surveillance center where officers can view thousands of video cameras around the downtown—and police-operated cameras have proliferated in many other cities across America in just the past several years.[32]

The power to engage in surveillance, snooping, monitoring, propaganda, and shaming or otherwise intimidating or harming those who do not conform is transforming societies in heavy handed and authoritarian ways. While China is leading the way in showing the world how to use AI technology to intimidate and control its population, China's President Xi Jinping is is also chiding Western nations that are increasingly imposing hate speech limits on their own populations for their hypocrisy in criticizing China's efforts to control the thoughts and behaviors of its population on all levels of activity.[33]

The specter of extensive privacy intrusions and surveillance obviously doesn't end with China. A recently published article in the prestigious journal *Science* reported that a group of scientists were advocating that everyone in the US should have their DNA information deposited in a national database. As explained by *Bloomberg News*, the rationale is that it is already too late to stop police and businesses from accessing the DNA of tens of millions of people and those sharing similar DNA indicators. The argument is that since the acquisition and use of our DNA is already so widespread and random, we might as well create a national total-DNA database system so it can be regulated more efficiently.[34]

Xi recently declared that he considers it essential for a political community's coherence and survival that the government have complete control of the Internet in order to prevent "irresponsible" and destructive communications that damage the integrity of the society. At least conceptually, Xi is correct. While in our opinion China has gone far beyond what is appropriate, the proliferation of "hate speech" laws and sanctions in the West—formal and informal—has created a poisonous psychological climate that is contributing to our growing social divisiveness and destroying any sense of overall community.

INCREASING WEALTH INEQUALITY, MASSIVE JOB LOSS, CLASS CONFLICT AND DE-DEMOCRATIZATION

By highlighting the connection between AI/robotics and the undermining of democracy, *Contagion* illustrates how democracies will not be able to cope with the stresses, competition, social fragmentation, rage and violence that will occur as a result of intensified social struggles over scarce resources. It is not only a financial issue. Some negative effects of AI/robotics on Western society are already observable. They include the increasing lack of opportunity, distrust, the undermining of free and open discourse, and the reduction of social mobility.

While we increasingly hear about the "hollowing out of the middle class" in the US, Western Europe and the UK, we are not paying attention to the impacts this will have on the composition of our society or what the hollowing portends for our political, economic and educational systems. As an economics scholar at Dartmouth put it : *"Whether you like it or not what the global economy is delivering is that the productivity growth that has been realized has been earned by a small fraction of highly skilled people and returns to capital"*[35] [emphasis added].

While that small but highly skilled fraction of our workforce increasingly benefits to extraordinary degrees, and the owners and controllers of capital even more, most people are being left out of the economic benefits and wealth creation produced by the AI/robotics phenomenon. A limited number of people are being endowed with immense wealth due to AI/robotics. A particularly striking example is that Amazon's Jeff Bezos just passed $150 billion in his net worth while many in his workforce are making ends meet through food stamps and other government subsidies.[36]

As jobs disappear and economic returns are transferred dramatically from labor to capital due to AI/robotics, with governments and ordinary people left to bear the burdens and costs, most of humanity will be left behind in terms of earnings, opportunity and status. This concern does not even discuss the plight of the vast majority of the planet's population who live in poverty in nations outside the wealthier countries. In those wealthier nations there will be a rapidly widening schism between the most fortunate and powerful and the massive numbers of those left behind that will be driven by chronic unemployment.

As this occurs we will experience rising social anger and violence, the disappearance of any semblance of democracy, and the emergence of police states focused on monitoring and controlling populations who appropriately feel betrayed by their leaders and the infamous One Percent. Even the United

Nations has become concerned about the job loss, military implications, and economic and social destabilization that is likely to result due to Artificial Intelligence advances and has established a European center to study the issue.[37]

One of our most critical challenges is to figure out ways to ensure that governments have revenues sufficient to sustain the millions of people who will be left behind by the transformation in the new economic system.[38] This includes the need to develop the ability to deal with explosive situations in megacities. Sixty percent of the world's population is projected to live in jam-packed urban areas by 2025. Included among the megacities are New York, Los Angeles, Boston, Philadelphia, London, Paris, Rome, Delhi, Rio de Janeiro and Mexico City as well as Beijing, Tokyo, Houston and Atlanta to name a few.[39] But we won't be dealing only with mega-cities. Many other large urban areas such as Detroit, Cleveland, St. Louis, Miami, New Orleans, Dallas and others outside the US will become unstable and unsustainable.

Truck driver turned author Finn Murphy summarizes the tragic disconnect between ordinary working people and citizens versus corporations and governments.

> *What we want is to work and support our families. We're citizens. We coach soccer and go to parents' night at school and pay our taxes. Who is taking responsibility for the human cost runaway technology is causing? Not the companies reaping enormous benefits. Not the fleet owners. Not the software engineers. Not governments.*

Murphy adds:

> I'm not at all confused by the general surge in populism we're seeing. *The tail of technology is wagging the dog of the social contract, leaving millions of citizens in penury.* ... [T}he US, and the west, have fallen far short in addressing the problem of displaced workers. Something needs to change. *We can start by accepting that both the private and public sectors have a responsibility to manage the human side of technological disruption.*[40] [emphasis added]

CHAPTER TWO
Change: Fast, Sweeping and Inexorable

AI/robotics is transforming work, wealth, and the nature of our social order with amazing speed. What is occurring is not an imaginary apocalyptic scenario. Jamie Dimon, the head of JP Morgan Chase, who is not a "sky is falling" soothsayer, predicts there will be large-scale economic and employment problems within ten years.[41] One report on the rapid progress of AI research states: "The newcomers to AI believe that the technology has finally caught up with the hopes, bringing a heightened level of intelligence to computers."[42] They promise a new way for humans to interact with machines, and for the machines to encroach on the world of humans in unexpected ways.[43]

Nick Bostrom's 2014 *Superintelligence* book made the daring claim that an AI "player" might be able to beat a skilled human GO master in ten years or so. Only a year and a half after that prediction the world's best GO master was left humiliated and depressed after being trounced by an AI opponent.[44] Researchers recently used cooperating AI algorithms to compete in an extremely complex gaming dimension where the algorithms demonstrated the ability to work together to an unprecedented degree. This was hailed as a breakthrough indicating collaborative AI systems beyond anything previously done. The *MIT Technology Review* describes the experiment.

> Researchers at OpenAI, a nonprofit based in California, developed the algorithmic A team, which they call the OpenAI Five. Each algorithm uses a neural network to learn not only how to play the game, but also how to cooperate with its AI teammates. It has started defeating amateur Dota 2 players in testing, OpenAI says. This is an important and novel direction for AI, since algorithms typically operate independently. Approaches that help algorithms cooperate with each other could prove important for commercial uses of the technology. AI algorithms could, for instance, team up to outmaneuver opponents in online trading or ad bidding. Collaborative algorithms might also cooperate with humans.[45]

The message is that AI breakthroughs are happening much faster than the best estimates from our most knowledgeable experts in AI suggest.[46] Chris Hughes, the co-founder of Facebook, has warned that the digital economy is going to keep destroying jobs. He argues that enormous income inequality will grow and that the phenomenal success of companies such as Facebook, where he made half a billion dollars for his three-year involvement, are key contributors to the growth in income inequality. Like an increasing number of incredibly wealthy billionaires such as Richard Branson, Mark Zuckerberg, Bill Gates and Elon Musk, Hughes urges that we create some kind of stipend or Universal Basic Income payment to help counter the increasing inequality, with the program funded at least in part by higher taxes on the wealthiest one percent of the population.[47]

The head of a company training computers to replace white-collar workers such as financial analysts says: "it's a paradigm shift from putting commands into a box to a time when computers watch you and learn."[48] The problem is that what the computer's AI systems are often learning is how to do *your* job, ultimately rendering human workers irrelevant. During a recent financial investment company ad on a car radio the spokesperson proclaimed, "you won't have to rely on a person with us because we use algorithms to help you achieve better results."

WE ARE EARLY IN THE PROCESS
AND AI IS ALREADY "HAMMERING" US

Another niche in which job replacement is taking place is suggested by a recent report on surgery done by a "robot" that was said by some to be performed better than the same surgery done by many human surgeons. This involved a robotic surgeon who was performing operations under the control of a human doctor. Another project at the University of Utah involves the development of an AI/robotic brain surgeon.[49] A robot eye surgeon has just operated on a human for the first time.[50]

At this point the robot surgeon is a tool that improves performance. As the processes become commonplace we can expect that AI systems can be programmed in ways that can reduce the need for human surgeons in some areas and be guided by "surgical managers" or some such human actor. This is already happening in Singapore with EMMA, a physical therapy robotic system that is allowing individual practitioners to expand the number of therapy patients and practices arrangements they handle but resulting in job loss by inevitably reducing the number of medical personnel required.[51]

Work contexts are being transformed across the board, including farming, harvesting and construction. The Japanese and Australians have been working toward developing robots as farmers.[52] A California company has just announced that it is building a robot capable of harvesting apples, and another has sought to shift much of its strawberry harvesting from human workers to robots.[53] A British company is investing heavily in developing a robotic harvesting system that possesses the tactile capability, sensitivity and vision to pick fruits such as strawberries without damaging them. As simple as this sounds it represents a critical breakthrough that will be adapted to multiple venues and activities.[54]

An Australian company is in the advanced stages of developing a robot bricklayer that can perform four times faster than a top-level human worker.[55] There are reports that a similar technology is being introduced into the US, claiming that there is a shortage of human construction workers.[56] Although it is relatively early in the developmental process, 3D printing is very likely to increase its applications in the manufacturing systems for numerous products.

Three-dimensional printing technology, when coupled with AI applications, robotics and WiFi, can be expected to play a major role in manufacturing activities to the point where it becomes almost entirely autonomous after being loaded with initial instructions and product design specifications. The rapid spread of 3D printing/manufacturing technology is exemplified by a report on the rapid printing of a 400-square-foot house in Russia that cost only $10,000 and was completed in less than twenty-four hours. Emily Rella reports: "From start to finish, the homes cost $10,000 to complete—a sliver of the price of some tiny homes that can go for anywhere between $40-50,000. ... The printer itself is relatively small in size as well, measuring about 16 x 5 feet with the ability to be assembled (or disassembled) in a quick 30 minutes."[57]

Another potentially important development is that the Japanese government has gotten competitive with the US and China as it recently announced plans to create the world's most capable super-computer with processing capacity even beyond that trumpeted by China less than a year ago. The Chinese breakthrough was claimed to leap beyond US super-computer capability as defined by operational speed and scale measured in teraflops.[58] In a recent competition the US appears to have at least temporarily regained the supercomputer edge.[59]

Given the severely age-skewed population demographics in Japan and China subsequently discussed in the context of the Age Curse, it is not surprising those two nations are leaders in developing robotic workers and AI systems because necessity drives invention.[60] Like China and much of Europe, Japan needs robotic workers to deal with the demographic conditions it will be facing in the near future. This need is driving Japan's research activities. The reason? Japan's 2016 birthrate fell below 1 million new births for the first time since record keeping began in the 1800s. It is also estimated that 20 percent of Japan's population over the age of 65 will suffer from dementia by 2025.[61]

As if those challenges don't put enough pressure on an export-driven economic system that has struggled mightily over the past two decades, Japan's population of individuals over 90 years of age has doubled from 1 million to 2 million in slightly over twenty years, imposing large financial health care costs and other support obligations for that group. Like China, without a strong export base to generate high levels of revenue, Japan is in very serious trouble. Given its aging society and plummeting birth rates, that export base cannot be maintained without large-scale use of AI/robotics systems. Other than importing very large numbers of migrant workers, which Japan has always been unwilling to do, AI/robotic work systems can keep the costs of production competitive with other nations such as China which is being extremely aggressive in shifting much of its export production activities to AI/robotic systems.

In Japanese robotics R&D, there is ongoing experimentation with robots as caregivers for the elderly in assisted living facilities.[62] This may seem far-fetched to some, but if one has had to deal with a parent, spouse or grandparent in such facilities, even those thought of as providing high quality care, it is not that difficult to understand how with some further tweaking a robotic caregiver could be an upgrade to the quality and consistency of the care currently received in many of the facilities.

Extensive research is going into how to make the robotic caregivers seem empathetic, communicative and caring as well as efficient.[63] While some may sneer and argue that the robotic caregivers' portrayal of compassion and caring might be phony, the same conclusion can be voiced in relation to many human caregivers. This means that while there is and will continue to be an increasing number of elderly in need of care, and people are being told that this offers a safe and secure career path for future workers, many of the jobs may be filled by "Roberta the Robot" rather than a human worker.[64]

No Area of Work Is Totally Safe

Use of robots and drones for delivery will eliminate a great deal of the costs of the human workforce, including having to pay wages, contribute to and manage pension funds, train new employees, pay for vacations, family leaves and health care. These savings on labor offer significant returns to companies and investors in numerous economic activities. In London, self-driving ground drones are being used to make courier deliveries.[65] Amazon just received a patent involving parachute delivery dropped from higher altitude drones equipped with guidance systems to ensure the package hits its target. Delphi has begun testing robot on-call taxis in Singapore.[66]

Alissa Quart discusses the importance of what she calls *human infrastructure*, explaining that human infrastructure[67] focuses on the human beings who will increasingly be displaced by the new economic and political order.[68] She explains that critical human infrastructure could describe the guys in trucker-author Finn Murphy's memoir *The Long Haul*. Quart writes: "Murphy explains … that if long hauls become autonomous, as has been threatened in the next 10 years, his driver friends will most likely have their trucks foreclosed. With a limited education and in latter middle age, they'll only be able work for places like Walmart—at best."[69]

In Greenwich, England, the founders of the Skype network have begun using drones and robots to make food deliveries to homes in less than thirty minutes.[70] This development is spreading rapidly. Amazon is now offering such services as it continues to expand its tentacles throughout numerous industries, including its recent purchase of the Whole Foods supermarket system—to the utter dismay of pre-existing Whole Foods employees. Amazon just announced that it intends to have as many as 3,000 cashier-less grocery stores in operation within the next three years. The problem is that this strategy will also eliminate large numbers of other convenience-type stores, their employees and the small owners who operate many such establishments. The prediction is that Amazon's "aggressive and costly expansion … would threaten convenience chains like 7-Eleven Inc., quick-service sandwich shops like Subway and Panera Bread, and mom-and-pop pizzerias and taco trucks."[71]

It doesn't stop there. In the US we are seeing robot fast food cooks, pizza makers, drone delivery systems that are replacing drivers, and investment of billions of dollars into the rapid development of autonomous self-driving cars and buses, along with delivery vehicles and even semi-trucks. Taken all together these AI/robotics technologies will replace millions of human workers whose qualifications and training don't prepare them for doing much else.

Although we tend to think about robots taking over lower-level labor-intensive work, the combination of Artificial Intelligence (AI) and robotics will increasingly replace higher-end careers. This will take place through a combination of full substitution of AI systems in some areas of work and by multiplying human worker efficiency and productivity in others to the point where fewer people are needed to do the work.[72] This is already taking place in medicine, finance, journalism, accounting and law.[73]

Nor is it only that many highly compensated Wall Street jobs are being replaced by robots and advanced investment algorithms.[74] Newspapers are consolidating. The *New York Times* is slashing its editorial staff and recently sold its fancy headquarters building while downsizing. The print media industry is shrinking and even disappearing as technology replaces people and cuts costs. Department of Labor statistics indicate that the newspaper industry has lost 60 percent of its jobs over the past sixteen years.[75]

CHAPTER THREE
Experts' Predictions of Human Job Losses

Work opportunities are being eliminated from the most intellectual activities down to the basic areas of services and labor. Leading AI/robotics analyst Thomas Frey has predicted very large job losses, explaining in 2016 that his 2012 warning that nearly 50 percent of global jobs could be eliminated by AI/robotics was intended as a "wake up call" due to the speed and severity of what he saw happening. Frey states:

> When I brought up the idea of 2 billion jobs disappearing (roughly 50% of all the jobs on the planet) it wasn't intended as a doom and gloom outlook. Rather, it was intended as a wakeup call, letting the world know how quickly things are about to change, and letting academia know that much of the battle ahead will be taking place at their doorstep.[76]

The 2016 World Economic Forum projected that 5,000,000 jobs could be lost in major developed nation economies by as soon as 2020 due to automation.[77] Once we go past 2020, things will get even worse. Some estimates conclude that the implementation of autonomous cars, trucks and other means of transport and delivery will cause the loss of 4.1 million compensated driving jobs in the US alone.

The so-called "gig" economy involving such things as Uber and Lyft in which people are able to patch jobs together to make ends meet is itself under pressure. Such patchwork employment also typically does not provide health insurance, pension benefits, guaranteed amounts of work, or significant pay scales. Plus, AI/robotics is in play even in those contexts where Uber is concentrating on replacing human drivers with autonomous, robot-driven cars. In less than a decade human operated Uber and Lyft transport and traditional taxi driving jobs will become historical artifacts.

Taken together with the job loss trends already mentioned, the predictions offered below represent what a range of experts have said concerning what they expect to occur over the next few decades by way of an AI/robotics transformation of the human workplace. If the predictions are even fifty percent accurate, we face problems with which we are unlikely to be able to deal effectively. Here are some experts' projections.

- 50 percent of US jobs will disappear by 2030 [*Fast Forward 2030: The Future of Work and the Workplace*, CBRE Report, Oct. 2014] or by 2025. [Thomas Frey, "2 Billion Jobs to Disappear [Globally] by 2030," *Futurist Speaker*, 2/3/12].
- 12,000,000 US jobs will disappear by 2026. ["Robots Set to Disrupt White-Collar Work," *CNBC*, 7/7/16].
- 800 million jobs in 25 economies will be lost to AI/robotics in worst-case scenario. "Jobs And Robots: 25 Countries Ranked On Job Loss Potential From Automation, Robotics, And AI." [*Forbes*, 4/23/18].
- 50 percent of today's work activities will be automated by 2050—give or take 20 years. Robots will take over most of the world's jobs by 2045 [*The Economic Times*, 6/6/16].
- 4 million US driver-compensated jobs (taxis, semi-trucks, delivery vehicles, busses) will be lost to self-driving vehicles ["4 Million Driving Jobs at Risk from Autonomous Vehicles," *Insurance Journal*, 6/27/17].
- The McKinsey Global Institute has warned that if the "slow growth" conditions of the past decade continue, up to 80 percent of people in developed economies could see flat or falling incomes ["Poorer than their parents? A new perspective on income inequality," July 2016].

Job Loss Is a Cumulative Process, Not a Singular Event

One of the most difficult issues involved in figuring out the pace of job elimination in many industries—both production and service—is the timing in which the systems will shift predominantly or completely to AI/robotics technologies. The change to AI/robotics is not an instantaneous or overnight across-the-board shift. For businesses, keeping the wages of human workers at "reasonable" levels during the period when the AI/robotics production and service delivery systems are being designed, manufactured, tested and introduced, ensures that the ongoing business operations remain profitable during the transitional phase.[78]

The movement of AI/robotics into specific job markets entails a process because the companies adopting AI/robotics "workers" must test, calibrate, evaluate and improve the systems' performance in the workplace. This includes making customers feel comfortable with the systems in those situations where humans and robotics interact. As the AI/robotics workers are developed, tested and perfected, the loss of human jobs will accelerate as the AI/robotics systems take over an increasingly significant portion of the workplace.

The speed at which this transition from human worker to AI/robotic systems occurs, and the scale of the replacement, will vary with the specific kind of economic activity involved. But irrespective of that, it will increase rapidly as breakthroughs continue. As the systems are refined in workplace applications, consumers become more comfortable with the technology, and an increasingly wider range of industries are penetrated, we will experience larger-scale job loss. That is why quite a few of the predictions of large-scale human job loss such as those listed above tend to use a time frame of ten to twenty years as their benchmark. Past a point of no return the transformation will move with dismaying rapidity.

When we look at areas of work such as health care, elderly services, food service and preparation, manufacturing, farming, security services, technical support, transportation and professional driving, clerical activity, reception, customer service, teaching, writing, journalism, lower and mid-level banking services, warehousing activity—and much more—we can see significant inroads already being made on employment of humans. Nursing and medical services, for example, rank toward the top of lists of supposedly "safe" jobs, but AI/robotics breakthroughs are already penetrating those positions and can offer hospitals and HMOs significant cost savings.[79] In the US, China and Russia, researchers are working diligently on developing AI and self-learning robotic systems, and are even attempting to develop robots capable of perceiving and responding to human feelings with at least a simulated projection of concern or compassion.[80]

Outsourcing of jobs or simply consumers inability to buy foreign manufactured goods at significant short-term savings (while contributing to the economic and social decline of their own nation) is no longer the primary challenge. Cheap foreign labor has lost much of its advantage and is itself threatened by AI/robotics. Left unchecked, no matter where they are located, companies will increasingly incorporate AI/robotics systems into their processes in ways that represent cost savings, whether over foreign or domestic human labor, even if human labor rates remain the same or are reduced.

In a 2013 study of the massive impacts of computerization on human jobs, *The Future of Employment: How Susceptible are Jobs to Computerisation?*

Oxford University economists Carl Frey and Michael Osborne indicate that the AI/robotics shift is not like others we have experienced. They indicate, as do others, that unlike other transformations of our economic system, there won't be a significant employment recovery on the other side of the downturn, and further observe that, "This raises questions about: (a) the ability of human labour to win the race against technology by means of education; and (b) the potential extent of technological unemployment, as an increasing pace of technological progress will cause higher job turnover, resulting in a higher natural rate of unemployment."[81]

In *"After the robot revolution, what will be left for our children to do?"* Charlotte Seager echoes Carl Frey and Michael Osborne in suggesting that while there will be large-scale job loss, there are important areas of work in which humans will be able to maintain their superiority and outperform AI/robotics systems—at least for a time. These areas are our "safety net" in which there is a unique reservoir of human talent and capability that will allow our children and grandchildren to sustain themselves to some extent in a society they share with AI/robotic systems.

In presenting her arguments Seager indicates: *"there are three types of roles that we won't easily be able to automate, and jobs for the next generation will focus on these areas."*[82] These areas of continuing human employment are ones that require social skills, creativity and the ability to engage in dexterous object manipulation without continual oversight and input from human operators. If you read the full Frey/Osborne analysis, their assertion is that the three areas of work involving social skills, creativity and object manipulation and dexterity will be continuing sanctuaries for human employment. The problem is that all signs indicate the advances are accelerating more quickly than anticipated.

The problem with such conclusions is that the Frey/Osborne study was published in 2013. Even in the very brief span between that report and late 2018 when *Contagion* is being completed, we have witnessed a continuous stream of developments in AI/robotics aimed at overcoming the limitations those researchers described. As a matter of technological capability, no one on this planet actually knows where AI/robotics is going or how fast it will get there.[83] But for anyone who pays attention to AI/robotics R&D advances, breakthroughs are occurring that do not bode well for our ability to sustain an adequate number of human employment opportunities, even those thought immune from AI/robotics inroads.

According to Seager, the "Social" category is one in which humans are expected to have a continuing advantage. She explains that jobs that involve complex social interactions won't be easy for a machine to do. In relation to the Oxford study Seager explains: "'If you think of all the social tasks—negotiating or taking care of others—it is almost unthinkable that a robot will acquire these skills at human levels of social intelligence, says Frey.'" Seager then concludes, "So careers such as teaching, social care, nursing and counseling are all likely to stay after the robot revolution."[84] The problem is that significant work on providing social care, nursing and even counseling using AI/robotic systems is developing rapidly.

Another concern is that those who determine what goes on in the workplace have discovered ways to re-engineer jobs to fit what AI/robotic systems can do rather than simply clone or remain within the box defined by human capabilities. It is a mistake to define the abilities of such systems solely or even primarily on the basis of their capacity to match the specific skills, qualities and approaches associated with work being done by humans. It is also important, rather than overrating the actual quality of generally achieved human performance—as if that is the level practiced by all people—that we be honest about the level of quality, output and interaction that needs to be achieved by AI/robotic systems to provide an acceptable alternative to human workers.

We need to admit that human workers often fail to provide the kind of high quality behaviors that are implied when concluding humans are superior to AI/robotics systems in a host of areas. Assessing the standards that need to be met in relation to an ideal, or even the level of performance that only the very best human teachers, counselors, care givers, nurses and the like provide, misstates what AI/robotics systems need to be able to do to provide comparably valuable work at a level that, on average, meets or exceeds what is generally done in any particular area. There are, for example, quite a few poor or mediocre teachers, uncaring social care workers, negligent nurses and ineffective counselors, lawyers and doctors. One question, therefore, is how "human" must a robotic system be in order to provide what is accepted as consistently adequate service and assistance in such jobs? And who is going to be deciding that?

Finally, in the area described as "Autonomous Object Manipulation" the current human advantage is due to a combination of humans' greater dexterity and quality and scope of perception relative to that which currently exists in AI/robotic capability. At this moment that advantage is real. Seager explains: "In plain English, this means the ability to pick up and move around different sized objects. "[Frey states] … So cleaners, gardeners and refuse collectors will be among the last jobs to be automated."[85] Even if true for the moment, and advances in dexterity and the like are taking place, this does not describe a particularly exciting future for human workers.

The 2013 study by Frey and Osborne put probable US job loss by 2030 at 47 percent. No society is equipped to deal with such an economic and political nightmare, particularly extraordinarily complex systems such as those in the US and EU with expensive subsidy and safety net promises and obligations that cannot be met if predictions of job loss are anywhere close to being correct. An analysis by Joel Kotkin concludes that what is occurring with AI/robotics is different from past economic revolutions because the AI systems are being designed in ways that will allow them to replace even the higher-end or cerebral work we would normally consider the province of humans.[86]

Financial and commodities markets expert Howard Schneider asks: "has the nation's ability to generate well-paying jobs in manufacturing and other sectors been fundamentally scarred by changes in the global economy that may predate the 2008-2009 economic crisis but were more starkly revealed in its aftermath?"[87] He points to the observation by an Atlanta Federal Reserve Bank president that we are facing something outside human experience. The result could be "a workforce based on large numbers of lower paid workers, with a few highly paid managers, professional and technology workers, *and a permanent hollowing out of the middle class.*"[88]

That hollowing of the middle class continues apace. Rolls Royce is cutting 4,000 middle management employees in addition to the further 600 senior management workers cut loose only months ago.[89] Citi Bank just announced that it was considering eliminating 10,000 "tech and ops" staff due to developments in AI and robots. Deutsche Bank already warned that half of its 90,000 employees could lose their jobs due to AI.[90]

Fast Forward 2030: The Future of Work and the Workplace concludes: "The next 15 years will see a revolution in how we work, and a corresponding revolution will necessarily take place on how we plan and think about workplaces."[91] *Fast Forward* adds:

> Artificial intelligence will transform businesses and the work that people do. Process work, customer work and vast swathes of middle management will simply disappear. [One key conclusion of the Report is that] *Nearly 50 percent of occupations today will no longer exist in 2025. New jobs will require creative intelligence, social and emotional intelligence and ability to leverage artificial intelligence.*[92] [emphasis added]

Brian Hopkins of the market research company Forrester continues the litany of job destruction, warning: "Solutions powered by AI/cognitive technology will displace jobs, with the biggest impact felt in transportation, logistics, customer service and consumer services."[93] The Forrester analysis adds: "These robots, or intelligent agents, represent a set of AI-powered systems that can understand human behavior and make decisions on our behalf. ... *For now, they are quite simple, but over the next five years they will become much better at making decisions on our behalf in more complex scenarios*"[94] [emphasis added].

We have already seen a shift away from agriculture and manufacturing jobs previously filled by human workers and, as more jobs become automated, many repetitive and low-skilled jobs will vanish or shrink into specialized niches with limited work opportunities.[95] The expanding development of robotic farming systems has particular implications not only for US immigration policy and the several million migrants who heretofore entered the nation to work in agricultural employment.

Whether there will be other jobs available in sufficient numbers for workers with their generally limited skill sets or educational levels is questionable. Vibhuti Agarwal explains what is occurring:

> The next generation is tractors that can drive entirely by themselves. After that: ones that can plant, fertilize and spray pesticides. London-based CNH Industrial is testing a tractor that has no driver's cabin, with farmers expected to monitor planting and harvesting remotely. Mahindra & Mahindra expects pushback from the millions of farmers whose livelihood could be threatened by automated farm equipment like the driverless tractor it is developing.[96]

The movement of AI/robotics systems into agricultural sectors poses a significant challenge for developing economies that are heavily dependent on agriculture for their own economic development since such work employs a substantial part of their workforce. AI/robotics agriculture in developed nations creates efficient and low cost competition that already has and could further undermine the agricultural exports of the less well-off countries.

Given that we are in a society where many people are mainly or solely qualified to work in those repetitive, low-skilled jobs and are unlikely to suddenly develop the ability to do higher end innovative, technical, scientific and conceptual work, massive job loss on the basic levels of work poses an extreme challenge.[97] Walmart, for example, has just installed 360 "robot janitors" in stores, an initial effort that indicates an expanding replacement of human workers in such activities. Pavel Alpeyev reports on the development.

> Robots are coming to a Walmart Inc. near you, and not just as a gimmick. The world's largest retailer is rolling out 360 autonomous floor-scrubbing robots in some of its stores in the US by the end of the January. ... The autonomous janitors can clean floors on their own even when customers are around. ... Walmart has already been experimenting with automating the scanning of shelves for out-of-stock items and hauling products from storage for online orders. Advances in computer vision are also making it possible to use retail floor data to better understand consumer behavior, improve inventory tracking and even do away with checkout counters, as Amazon.com Inc. is trying to do with its cashierless stores. ... *At first, the machines ... need to be operated by humans, who "teach" the layout of the space that needs cleaning. After that the robots can perform the task autonomously.*[98] [emphasis added]

Reihan Salam wrote in *The Atlantic* that there has been a troubling rise over the past year in suicides among New York's taxi drivers, 91 percent of whom are immigrants from poor countries.[99] Some attribute this to the impact of Uber on the useage of taxis as they face competition they are unable to match. Along with this goes a stunning decline in the market value of a New York City taxi medallion that until Uber and Lyft appeared on the scene provided a monopoly for the taxis. This raises the troubling and disruptive question of what we do with migrants with limited education and lower level job skills as technology replaces their jobs. The challenge becomes even more stark when the rapid development of autonomous vehicles is factored into the equation and human drivers—taxis, Uber and Lyft, busses, limos, semis and the like—are replaced with AI-controlled transport.

The fundamental challenge we face is what do we do to support and assist millions of people who have lost the opportunity to engage in the only types of work for which they are qualified or capable? How do we slow and minimize the shift to AI/robotics work systems and away from human employment? Providing retraining and education for specific categories of employment is an obvious part of the solution, but this requires that we make a firm and continuing commitment to the preservation and creation of the types of jobs people are willing and able to do.

IF MACHINES ARE CAPABLE OF DOING ALMOST ANY WORK HUMANS CAN DO, WHAT WILL HUMANS DO?

AI/robotics systems are already better than human workers in many areas of activity. As they are further developed and deployed, AI/robotics systems will be faster, stronger, cheaper, more efficient and reliable, provide less hassle for employers, and in far too many instances, do a better job than the human workers they replace. Moshe Vardi, Professor of Computer Science at Rice University, delivered a talk to the American Association for the Advancement of Science in 2016 exploring the critical question: "*If machines are capable of doing almost any work humans can do, what will humans do?*"[100]

Vardi warns that as many as half the world's workers could be replaced by machines within the next 30 years. He observed that this future is likely to mean humans may only work a handful of hours per week and many will not work at all.[101] Warning that work is an essential part of life for people and that, even if we somehow worked out a system in which humans lived in leisure and luxury created by the activities of robots, that would not be ideal. He added: "I do not find this a promising future, as I do not find the prospect of leisure-only life appealing. I believe that work is essential to human wellbeing."[102]

Vivek Wadhwa, a Professor at Carnegie-Mellon University specializing in robotics and artificial intelligence issues, set out another vital concern in a 2016 presentation at the World Economic Forum in Davos, Switzerland. Wadhwa warned about the significant risk of social conflict posed by the rise in extreme income inequality due to technological shifts and human job destruction. His message was that:

> [R]apid, ubiquitous change has ... a dark side. Jobs as we know them will disappear. ... *The ugly state of politics in the United States and Britain illustrates the impact of income inequality and the widening technological divide. More and more people are being left behind. ... Technologies such as social media are being used to fan the flames and to exploit ignorance and bias.* The situation will get only worse—unless we find ways to share the prosperity we are creating.[103]
> [emphasis added]

The point we make throughout *Contagion* is that there are reasons for human societies to continue to employ humans beyond such factors as economic efficiency, returns to investment, market dominance, avoidance of the irritating hassles offered by employees and the like. Focusing everything on economic efficiency, superior performance, speed and productivity that creates greater returns to investment undermines our societies in ways from which we will be unable to recover if we allow the trends to continue full bore. Should comparative performance superiority between AI/robotics and human workers be allowed to be an overriding criterion for workplace production management, despite the social consequences?

If it is, we lose.

CHAPTER FOUR
Like Us—Only Better?
Have We Have Reached Peak Human?

One analyst has described what is now occurring in finance and banking as the moment of "peak human." AI systems in these sectors no longer need large numbers of human workers. Automated (e.g., robotic) and more recently AI trading systems[104] in stocks, commodities, foreign exchange, and financial instruments operate at such speed and complexity, and in such an incredibly dynamic and nearly instantaneous feedback system, that human beings no longer truly control those markets.

While at the moment the programmers are creating the algorithms that empower the AI systems and give them their marching orders, the systems operate on a scale and speed so far beyond human capability that the AI systems are becoming the actual decision-makers. This is because they are the entities that are applying the dictates of the algorithms to situations that cannot be fully anticipated, understood or acted on by humans.

Such economic contexts will continue to see the replacement of flesh-and-blood personnel by algorithmic systems and AI advisors. Hugh Son explains: "Algorithms already tackle tasks such as vetting banking clients, pricing assets, and hedging some orders without human intervention. As we make those processes more and more efficient, you will need less people to do what we do today."[105]

Daniel Nadler, chief executive of Kensho, a financial services analytics company partly owned by the Goldman Sachs Group, describes what he sees as ongoing job loss in finance and banking. Nadler explains: "Anyone whose job is moving data from one spreadsheet to another ... that's what is going to get automated."[106] He suggests we have begun a permanent slide down the employment slope because people simply can't keep up with their AI/robotics competitors.

A recent report indicated that Elon Musk at Tesla Motors is planning for his next major automobile production complex to operate without human workers. The reason? "People speed" is too slow. Musk explains: "*You really can't have people in the production line itself. Otherwise you'll automatically drop to people speed.*"[107] Here is the reasoning behind Musk's goal to dehumanize and speed up Tesla's production.

> For decades, a big trend in manufacturing has been the gradual automation of the factory floor. Robots play a major role in making advanced products today—they're fast, clean and efficient. But Tesla chief executive Elon Musk wants to take this to a whole new level with the factory producing the upcoming, low-cost Model 3, turning "the machine that makes the machine" into an "alien dreadnought." The machine will ultimately be so complex that no humans will be expected to operate it directly, or to participate in the actual building of each [Tesla] Model 3.[108]

This shift has not been without start up glitches for Tesla but such developmental hiccups are inevitable as new technology is tested and refined. Recent reports indicate the Tesla 3 production facility is overautomated and having problems keeping to production schedules. Nonetheless, the transformation away from human workers in production and services activities will further accelerate as AI/robotics develops further and integrates into our national and global economic, governmental and personal systems. Musk, for example, is not stopping at automated car manufacturing. He recently announced that Tesla was seeking to create diverse AI/robotic systems with numerous manufacturing applications that would replace human workers in many areas.[109]

The fact that AI/robotics systems are not perfect at this point must not delude us into thinking their quality and capabilities are not advancing rapidly. We are going through a compressed process of research, development, testing, application, implementation, familiarization and refinement. Few who pay attention to ongoing AI/robotics research have much doubt that such systems will be able to do many things humans cannot do at all, as well as being able to do much of what we do faster, better, and at lower cost.

Rapid improvements in Artificial Intelligence, including "deep learning," "machine learning" and the design of complex neural networks, have expanded the capabilities of AI systems far beyond what anyone might have thought possible only a few years ago. Self-learning, self-programming and even self-replicating systems that are capable of learning from experience will move beyond many aspects of human capability and control. This sounds like science fiction. It is not.

As to the future of human employment, the nature of manufacturing has undergone fundamental change. By some estimates, 88 percent of all job losses in the U.S. manufacturing sector were caused by increased automation—including robotics—and not the availability of cheap foreign labor.[110]

Robert Kuttner insightfully notes, however, that the irresponsible implementation of a form of globalization of capital and finance operating outside effective legal processes and without accountability allowed a distorted version
of David Ricardo's theory of Comparative Advantage to take hold.[111]

The efficiency and cost-saving advantages of AI/robotics will further
magnify this process, resulting in amazing returns for the superwealthy and
misery for millions of others. The new capital-intensive manufacturing facilities will not be job generators in the way they were in our past economies.
Most will be partially or, as with Tesla's newly planned production facility,
close to fully roboticized. As businesses go through their capital investment
cycles over the next ten to fifteen years we will see more and more roboticized production and service systems.[112]

As we suggest, the only way to alter this transformation in the production and service sectors is to make conscious efforts to develop and adopt
strategies that slow or at least limit the rapidly growing adoption of AI/robot-
ics, what Alissa Quart calls "slow teching" the incorporation of AI/robotics
into human work systems.[113] Doing this requires significant alteration in our
system of taxation as well as the strategic use of tariffs and subsidies to create
manufacturing systems that can compete with those of nations that conduct
most or all of their production activity using AI/robotics.

It is not simply manual labor jobs that are being replaced. Robotics
and AI have already been used for surgery and for painting masterpieces.[114]
Christie's Auction House recently staged a sale of a painting created by an AI
program, with the final auction price received being $432,500, 45 times what
had been predicted. .The creative process was explained as follows.

> The painting ... is one of a group of portraits of the
> fictional Belamy family created by Obvious, a Paris-based
> collective consisting of Hugo Caselles-Dupré, Pierre Fautrel
> and Gauthier Vernier. They are engaged in exploring the
> interface between art and artificial intelligence, and their
> method goes by the acronym GAN, which stands for "gen-
> erative adversarial network." "The algorithm is composed of
> two parts," says Caselles-Dupré. "On one side is the Genera-
> tor, on the other the Discriminator. We fed the system with
> a data set of 15,000 portraits painted between the 14th cen-
> tury to the 20th. The Generator makes a new image based
> on the set, then the Discriminator tries to spot the difference
> between a human-made image and one created by the Gen-
> erator. The aim is to fool the Discriminator into thinking

that the new images are real-life portraits. Then we have a result." [He adds] "*We found that portraits provided the best way to illustrate our point, which is that algorithms are able to emulate creativity.*"[115]

Subtle skills that will enable AI systems to recognize qualities such as aesthetic beauty are being worked on by China's Alibaba.[116] The company's CEO, Jack Ma, predicts that sometime in the next 30 years AI systems are likely to become CEOs themselves. He warns that we are in for decades of very difficult times as the AI/robotics transition occurs. Ma is correct in part, but at a minimum we are going to experience something more drastic and longer lasting than several decades of severe economic difficulty.[117]

It is not solely an issue of the absolute number of new jobs that will become available. A special combination of intellectual, visionary and technical skills is required for the evolving workplace. There are far fewer people who possess the unique blend of skills needed to excel or even to compete successfully. A recent analysis in the *New York Times* estimated that perhaps only 10,000 people in the entire world have the abilities required to perform on the highest and most sophisticated levels that will be required in the rapidly evolving AI/robotics dimension.[118]

Even here we should have cause for concern because those 10,000 individuals are being compared to the human capabilities possessed by other people. Those capabilities may be surpassed by increasingly advanced AI systems in the next decade or two. If Masayashi Son's prediction that AI systems could exist in the near future with intelligence levels far beyond any human capabilities is valid to any significant degree, then even the best human minds will be functioning at comparatively low levels when judged against AI systems.

AI/ROBOTICS SYSTEMS ARE EVOLVING TOWARD SELF-DESIGN

As to job destruction and the pervasive potential of AI systems, AI programs are already writing movie scripts, creating poetry, painting pictures, solving problems humans can't figure out, operating independent weapons systems, and crunching data on levels no human will ever achieve. AI programs are also accessing and manipulating a stunning range of information in ways we can't comprehend or match, and evolving in ways their creators admit they don't understand even though they designed the original programs.

AI systems are being created that can autonomously rewrite their programming based on their own experiences and do so without human intervention. There are already early developments in self-learning that signal the strong likelihood AI systems will achieve that heightened capability to unanticipated degrees in the near future. Alphabet, Google's parent company and the owner of the AI research company DeepMind, recently discovered almost by accident that AI self-learning was occurring when its AlphaGo system taught itself how to play GO in three days. AlphaGo then invented entirely new strategies on its own that far surpassed the capabilities of the world's finest GO masters.

An explanation of Alphabet's research is provided at "*Google DeepMind: What is it, how does it work and should you be scared?*"[119] The answer to that final question is "yes." Increasingly advanced AI systems are achieving rapid improvements through the combination of technological strategies such as deep learning, sophisticated neural networks, and machine learning. Intensive focus on these technologies will create AI systems with capabilities humans will be unable to match—even as robots themselves become more humanized—or vice versa.

ROBOTICS DEVELOPMENTS

- Microchips get under the skin of technophile Swedes.[120]

- Boston Dynamics' terrifying robots can now run, jump and climb.[121]

- Robots could outnumber humans in just 15 years and will feel 'genuine emotions' by 2028.[122]

- Hyper-realistic robot that is "indistinguishable from humans" startles punters at a London pub by smashing a glass while talking about a machine invasion.[123]

- Lifelike robots made in Hong Kong meant to win over humans.[124]

- "Emotional" robots can judge your personality and tell if you're male or female simply by shaking your hand: Researchers say behaving like a human is key to success in the human world (*Daily Mail*, Phoebe Weston, 9/20/17). Robots ... are being trained to be polite, empathetic and funny. They must be sociable to integrate into human environments, experts say.[125]

PART II

1010101010101010101010

THE AI/ROBOTICS CONTAGION IS RIPPING APART GLOBAL ECONOMIC, SOCIAL AND POLITICAL SYSTEMS

CHAPTER FIVE

AI Job Loss is Not a Cycle of Creative Destruction and Economic Rebirth

What is occurring with AI/robotics is unique and outside Austrian economist and Harvard University professor Joseph Schumpeter's process of Creative Destruction, a term that stands for a periodic cycle of innovation that Schumpeter once stated he wished he had called "transformation" rather than "destruction." In the classic Schumpeterian model, pre-existing economic forms are destroyed by new technologies, and then eventually replaced by even better (in some ways and for some people) forms of work.[126]

Schumpeter's cycle of economic transformation and re-creation was neither neat nor painless for those who lost out when their employment disintegrated. Those displaced by the transformation suffered, but in an extended macro-economic sense of sacrifice and the "greatest good for the greatest number" the overall community ended up benefiting as new types of jobs were created in large numbers after a lengthy period of suffering and disruption.

In past transformations, this process of creative destruction balanced out over time so that laborers skills were able to adjust from obsolete activities to new ones that added value to the recreated economy. At the advent of the automobile, for instance, blacksmiths and wagon makers were highly paid skilled trades that became more and more irrelevant as the industrial revolution of the early 20th century moved forward. But the same skills necessary for blacksmithing could be retrained to provide value (and hence earn a wage) by satisfying new demands and exploiting new opportunities made possible by the emergent technologies.

While this means that prior industrial revolutions displaced workers at the margins of economic activity, the pace of change was such that many blacksmiths could transfer their skills to the machinist and fabricator trades, while wagon makers could in some cases (such as the Studebaker Bros. Mfg. Co.) transform their production facilities from producing wagons to producing cars. Even today the changes tend to be ones of scale and category rather than complete replacement. Blacksmiths, for example, continue to provide services to horse owners and some have gone into metal art and ornamental décor. The number of jobs, however, is dramatically reduced as the market for the skills contracts dramatically.

During the computer-based industrial revolution that began in the 1960s and early 1970s, we have seen a different picture than that of the industrial revolution of the late-1800s and early-1900s.[127] While employment for real wages for low and middle income workers remained largely flat since the late 1960s, productivity per worker has increased significantly since the advent of significant introduction of computers, computer-based automation, and robotics in the mid-1970s.[128]

The AI/robotics-based industrial revolution currently gearing up will render many more jobs worthless while increasing productivity with ever fewer human inputs. Many still believe that the creative destruction process will follow past patterns and, after a time, create new jobs for those put out of work while augmenting the ability of others to do the jobs they already have. To a degree this is what is happening to lawyers in private practice who find themselves able to "do more with less" human labor input.

This "doing more with less" has, however, also meant that considerably fewer lawyers are needed to do the work and this has lead to significant impacts not only on the legal profession but the several hundred law schools that have seen drastic enrollment and revenue declines. Law firms and lawyers offer one example, and with AI/robotics we will experience a massive and accelerated disruption in the market for human labor with the transition taking place over an extremely truncated time scale.

The impacts will be categorical in terms of the types of jobs most affected, and territorial because some cities and regions will be dynamic centers of development while many others will be left in the lurch. As the AI/robotics Fourth Industrial Revolution unfolds, there will be hotspots of economic activity such as Silicon Valley, Seattle, the North Carolina Research Triangle, the "Beltway Bandits" of the DC area, Boston, and some other dynamic clusters. A recent report indicated the most dynamic and best performing areas of economic growth included Provo, Utah, Raleigh, North Carolina and Dallas, and that big gainers were the Palm Bay-Melbourne area in Florida, Olympia, Washington and Hickory, North Carolina.[129]

It is notable that the Midwest Rust Belt and core states in the South were not listed in the report. The US is not only being hollowed out in a general socio-economic sense, but in a territorial and regional one. A number of those core geographic states are losing population, seeing their youth move away to more dynamic areas, leaving their cities with high crime and poverty rates, and lacking sufficient economic opportunity for their remaining people. The fact is that many areas of the US will be left behind and their residents will have little, if any, growth opportunity or chance at social mobility.

Regardless of the geographic location—economically vital or not—far fewer jobs will exist in absolute terms and there will be an increasingly limited number of high quality jobs in any location. Numerous areas of the country will become economic dead zones and given the demographics, dwindling resources and inequalities, some cities will become unsustainable and dangerous.

Another aspect of what is occurring is that although corporate leaders may argue that our jobs will only be replaced when it is established that automated, computerized or robotic systems can do a qualitatively "better" job, it is neither inevitable nor necessary that the new systems that replace us must perform better than humans, or even as well. While quality of performance is a consideration, we delude ourselves if we think it is the only element or even the primary one in many instances.

We need only ask ourselves whether the endless hoops we have to jump through to connect with a human when seeking information by calling a business or government agency is "better" than that of the human receptionists replaced by the computerized system. The AI-based systems that replaced human labor simply needed to be less expensive, more reliable, not miss work because of sick time and vacation absences, not require maternity leave, and offer employers much less hassle and expense than do human workers and unions.

A COLLAPSE OF WORK MEANS A
COLLAPSE OF OUR SOCIAL MODEL

Why is work so important? Begin with the fact that work creates structure for people, connects us with others, allows us to feel productive, fills our lives with activity and develops discipline. Work is not only a financial activity. Work, opportunity, and a strong economic system are the "glue" that holds our system together.

Ellen Ruppel Shell has written a fascinating book, *The Job: Work and Its Future In a Time of Radical Change*, in which she presented an example of what happened in the small Austrian town of Marienthal that had relied during the latter part of the 19th century on a cotton and woolens spinning manufacturing facility for its jobs, incomes, and, as it turned out, its sense of social cohesion, worth and order.[130] Shell writes that Marienthal was a happy, safe and lively place by all accounts and that the financier who had brought the factory to this poor agricultural region was committed to the wellbeing of the town and its people.

The town prospered for decades until the onset of the 1929 Great Depression that afflicted the US and Europe. At that time "currency inflated dramatically, the banks collapsed, and consumer demand shriveled." Marienthal's factories were closed and then dismantled and 1200 people lost their jobs. Within a year most of the people of Marienthal were on the dole.[131] Oddly, the Marienthal reserachers also found that men had lost their sense of time, stopped wearing watches, were late to meals, and when asked what they had done during the day many admitted they could not really recall.[132] Even though the area was a strong supporter of socialist political ideas a research study found that:

> Deprived of their livelihood, the villagers did not unite in protest or incite political action. Rather, they withdrew. The once bustling library emptied. The park, abandoned, was choked with weeds. Public debate cooled. Social clubs disbanded. And children lost their resolve. ... [J]oblessness itself had become a job, a thankless, miserable one that set citizens into revolt not against the system but against one another. People were isolated, pessimistic, and bitter. They spied and snitched on one another, especially around issues of money—welfare recipients thought to be "cheating" were quickly reported to the authorities by their neighbors and former friends. ... [F]amily pets vanished from backyards and porches ... mostly dogs. The title of a monograph from the period, "When Men Eat Dogs," tells us all we need to know about where those pets ended up.[133]

The point we are raising is that many areas of the US and Europe may be "Marienthals in waiting" as the effects of AI/robotics on human work grow ever stronger. While other issues such as inequality, opportunity, support of the less fortunate or needy, gender and race fairness are unquestionably of importance, virtually all of them require an enduring and multifaceted economic system that provides adequate resources to the members of society and allows them to feel they have the opportunity to develop and pursue individual dreams. Some people may dislike having to work and others dislike their specific jobs, but many people thrive on the opportunities, challenges, relationships and human interactions they derive from their workplaces.

The bottom line is that healthy societies need the dynamism of work, innovation, opportunity and creativeness. When that disappears the result is not a golden life of leisure but a dull and monotonous existence. As we later discuss in the context of the growing addictions that beset our society, while there are some positive contributions from the technology, one of the worst things that AI applications have to offer is an unhealthy set of addictive stimuli, many of which are deliberately designed into the applications to hook the consumers. These are already overpowering our children and even many adults to the extent they have been called "digital cocaine" by Brad Huddleston in his e-book *Digital Cocaine: A Journey Toward iBalance*. [134]

As the middle class in America and Western Europe is being progressively hollowed out with many former members sliding down the socio-economic scale and a limited number of others moving up to extreme levels of wealth, the basic nature of our political community is being altered in profoundly undemocratic ways.[135] These are not choices the respective citizens of these countries are making, and they are not being asked to approve the policies as an exercise of their democratic power. The problem is that an incredibly wealthy and powerful new corporate and political aristocracy has arisen that controls the directions of development. This ensures their interests are protected by controlling the ways social and financial power is applied, wealth amassed and opportunity facilitated. As this increases with a continuing consolidation of wealth and power into controlling stealth oligopolies, the voice, power and opportunities available to a shrinking middle class diminish. The inevitable result is radically unequal societies.

CHAPTER SIX

AI/Robotics and the Great Decoupling of Economic Productivity and Job Creation

The *Financial Times* reports that in a wide range of developed countries the middle class is taking a downward hit in earnings and size.[136] The loss in manufacturing employment driven by the combination of globalization, automation, and outsourcing to lower cost production locations has resulted in the ongoing deterioration of the West's middle class. The manufacturing jobs that had sustained the US economy were a critical part of America's economic health because the manufacturing pay scales sustained a blue collar middle class and generated a large number and variety of other jobs.

In a 2016 report Benjamin Parkin explains that as recently as 2013: "The manufacturing sector supported approximately 17.1 million indirect jobs in the United States, in addition to the 12.0 million persons directly employed in manufacturing, for a total of 29.1 million jobs directly and indirectly supported, more than one-fifth (21.3 percent) of total U.S. employment in 2013."[137] The continued movement of the manufacturing work to less costly areas of the planet had significant effects. With manufacturing relocated away from the US and EU countries, workers in those economically more developed nations who once enjoyed the benefits of those jobs lost their income base.

The generally higher pay scales that had been enjoyed by middle class workers, and the fact that manufacturing employment acts as a generative force that produces other types of direct and indirect employment, means that employment opportunities considerably beyond the manufacturing jobs themselves are destroyed when manufacturing is outsourced or automated.[138] When the generative core of manufacturing is degraded, the entire system—economic, political and social—is undermined.

As the manufacturing core and its workers and their higher wages disappear, the secondary and tertiary businesses that depend on the consumption activities of manufacturing companies and the consumer spending of well-compensated workers find their livelihoods drying up. Brookings scholar Robert Litan offers an analysis akin to the MIT study supporting the hollowing out of the American middle class, arguing that, "[T]he recession in

2001 marked a turning point when firms began managing headcount more aggressively through the use of more efficient global supply chains, technology and other methods such as outsourcing U.S. work."[139]

David Autor, an economics professor at the Massachusetts Institute of Technology, explains the dynamics of job loss and replacement in the US. He notes that, "New jobs are coming at the bottom of the economic pyramid, jobs in the middle are being lost to automation and outsourcing, and now job growth at the top is slowing because of automation."[140] Even here, the creation of those new lower tier jobs represents a downward slide for many workers who had previously been employed in well-paid middle class jobs, and for younger workers who face futures with limited options. Howard Schneider warns that outsourced technology and manufacturing risk the future of the US economy, explaining: "Some of the high-end manufacturing and research industries often seen as critical for U.S. job and wage growth are also those where revenue and profit have become more clearly divorced from the need for more workers." [141]

David Rotman explains how AI systerms, computer applications, and dramatic improvements in the technology of robotics are destroying jobs.

> [Erik] Brynjolfsson, a professor at the MIT Sloan School of Management, and ... Andrew McAfee have been arguing ... that impressive advances in computer technology—from improved industrial robotics to automated translation services—are largely behind the sluggish employment growth of the last 10 to 15 years. Even more ominous for workers, the MIT academics foresee dismal prospects for many types of jobs as these powerful new technologies are increasingly adopted not only in manufacturing, clerical, and retail work but in professions such as law, financial services, education, and medicine.[142]

Rotman continues: "Technologies like the Web, artificial intelligence, big data, and improved analytics—all made possible by the ever-increasing availability of cheap computing power and storage capacity—are automating many routine tasks. Countless traditional white-collar jobs, such as many in the post office and in customer service, have disappeared."[143] Rotman cites former Stanford economics professor Brian Arthur who calls what is developing the "autonomous economy."

> It's far more subtle than the idea of robots and automation doing human jobs: it involves "digital processes talking

to other digital processes and creating new processes," enabling us to do many things with fewer people and making yet other human jobs obsolete.[144]

Brynjolfsson and McAfee explain that rapid technological change has been destroying jobs faster than it is creating them and that this is causing the "stagnation of median income and the growth of inequality in the United States."[145] Where do the displaced people go as the middle class shrinks and there is increasingly limited space at the top of the employment and wealth pyramids with an inevitable loss in opportunity, mobility, and social and economic dynamism? An obvious answer is "down" the socio-economic scale. The timeline is clear.

> [B]eginning in 2000, the lines [between productivity and employment] diverge; productivity continues to rise robustly, but employment suddenly wilts. *By 2011, a significant gap appears between the two lines, showing economic growth with no parallel increase in job creation.*[146]

Brynjolfsson and McAfee call it the great decoupling. Brynjolfsson states:

> It's the great paradox of our era … Productivity is at record levels, innovation has never been faster, and yet at the same time, we have a falling median income and we have fewer jobs. People are falling behind because technology is advancing so fast and our skills and organizations aren't keeping up.[147]

Our economic organizations are not keeping up because the incentives that propel business decision-making are not aimed at preserving human jobs. There are, conversely, significant tax and other financial and managerial incentives for developing and incorporating AI/robotics systems into production and service activity. Tax rules that provide large benefits for the development and use of AI/robotics technology add to the economic advantages, as does the ability to avoid various kinds of federal and state payroll taxes associated with the employment of human workers. This will only get worse as AI/robotics systems continue to make rapid advances.

The threat to human jobs is not only due to technological and performance superiority. An AI/robotic system may not only be more efficient than human workers, it will end up being cheaper, more reliable, less likely to complain, take sick days or get pregnant, go on strike, take vacations, demand higher pay or insist that special accommodations be provided, etc. Given the checkered history of labor/management relations in many businesses it is unsurprising that quite a few employers are thrilled to do away with human employees.

CHAPTER SEVEN

AI/Robotics and Our Fragmenting Social Order

Robert Skidelsky of the UK's Warwick University provides insight into what we are dealing with. He explains:

> The last time this [despair over a sense of a disintegrating world] seemed to be happening was the era of the two world wars, 1914 to 1945. The sense then of a "crumbling" world was captured by W.B. Yeats's 1919 poem, "The Second Coming": *"The best lack all conviction, while the worst/are full of passionate intensity. Things fall apart; the centre cannot hold;/Mere anarchy is loosed upon the world."* [148]

Skidelsky writes that one result of the disintegration of the beliefs and norms of that post-WWI era was that, with the traditional institutions of political rule discredited by the war, the mantle of legitimacy was assumed by demagogues and populist dictatorships. As in the aftermath of World War I that stimulated Yeats' warning about loss of the political and community's "Center," we are caught up in an age of demagoguery.

THERE IS NO LONGER ANY "CENTER" IN AMERICA

Although we begin with the idea that the center cannot hold, it is difficult to even identify a center in American politics. We frequently hear that we are one nation, one group, and one community of interest linked in compassion, justice and sharing. This sounds wonderfully utopian but such proclamations have nothing to do with human nature, the reality of tribalism, and the sub-cultures we have created. Rather than growing together into a living, loving, caring and cooperating community, we are tearing things apart. US Representative Steve Israel indicated after a recent campaign that people are angry about everything, that respect for our basic institutions has largely disappeared and that, as local jobs on which they counted for decades evaporated, people feel helpless, frightened and outraged at what they see as their leaders' betrayal. [149]

US Supreme Court Justice Clarence Thomas recently addressed such concerns. Thomas observed that he had fears about whether there was any set of core values that held us together as a political community. When asked whether he was surprised at the extent of the rancor that seems to accompany any dispute about foundational issues he explained:

> No, I'm not surprised. I mean, what binds us? What do we all have in common anymore? ... [W]e always talk about E pluribus unum. What's our unum now? We have the pluribus. What's the unum? [S]ome people have decided that the Constitution isn't worth defending, that history isn't worth defending, that the culture and principles aren't worth defending. And, certainly, if you are in my position, they have to be worth defending. That's what keeps you going. That's what energizes you.[150]

The tragedy is that the United States has separated into fanatical fragments of identity groups. Unity, compromise and healing are impossible because, as Justice Thomas notes, there is no *"unum"* that possesses sufficient power to bind us to a set of common principles. Within ten to fifteen years we could face a social explosion with rising criminal activity and violence, militaristic repression, warring militias, vigilante groups and, in some instances, urban guerrilla warfare. It is probable that large parts of major cities will become "no go zones" and "hellholes" to an even greater extent than now, and that suburban and rural residents will seek to relocate to safe havens or mobilize to create defensive systems—a step beyond the present gated communities.

The AI-facilitated social media has accelerated and intensified the disintegration of our social forms. Facebook's former vice president for user growth, Chamath Palihapitiya, has stated that he feels "tremendous guilt" about Facebook. "[W]e have created tools that are ripping apart the social fabric of how society works. The short-term, dopamine-driven feedback loops we've created [including the hearts, likes, and thumbs up of various social media channels] are destroying how society works." He added, "[There's] no civil discourse, no cooperation; [only] misinformation, mistruth. And it's not an American problem–this is not about Russians ads. This is a global problem."[151]

Growing Anger Over Inequality of Opportunity

Financier Mohamed El-Erian observes that: *"The minute you to start talking about the inequality of opportunity, you fuel the politics of anger."*[152] An analyst addressing El-Erian's assessment observes that the "day one" issue for the Trump administration was economic growth. He states: "The inequality generated by the current low-growth climate has three elements: inequality of wealth, income and opportunity. The last of the three—manifested in high youth unemployment in many eurozone countries, for example—is the most explosive element."

The Internet and AI come into play because a sense of profound injustice, outrage and envy grows within the population when the have-nots are constantly inundated with images of the "lifestyles of the rich and famous" and compare that to their daily struggles to make ends meet, often despite long hours of work on multiple jobs. In a society with massive and constant information flow focused on celebrating the lifestyles of the rich and famous, the incessant materialist parade sweeps away other social values, augmenting the envy and resentment felt by a significant portion of the population.

The television show *Lifestyles of the Rich and Famous* may have been the beginning, but now it seems that contentions between the "have a lots" versus the "we're doing pretty wells" versus the "I want what you have" versus the "actually poor" versus the "homeless and bereft" fills the political air with discord all around us to the point of continual class conflict.[153]

Inequality in America is real and growing. When inequality reaches the point where the wealth available to the bottom fifty percent or so of a nation's population is inadequate to supply their basic needs, with many left in poverty and distress, this is real inequality. This contrasts with "envy inequality" or "relative" inequality where people are angry simply because they see others have more than they do. The problem with many people incensed with economic inequality is not one of "justice" but envy on the part of people who are "doing OK" but see others whose financial situation is more comfortable.

GROWING INCOME AND WEALTH INEQUALITY

- Richest 1% on target to own two-thirds of all wealth by 2030: World leaders urged to act as anger over inequality reaches a "tipping point."[154]

- Richest 1% own half the world's wealth, study finds: Credit Suisse report highlights increasing gap between the super-rich and the remainder of the globe's population.[155]

- World's witnessing a new Gilded Age as billionaires' wealth swells to $6tn [trillion]: Not since the time of the Carnegies, Rockefellers and Vanderbilts at the turn of the 20th century was so much owned by so few.[156]

- Nearly 10% of the world's wealth is held offshore by a few individuals.[157]

- Tax Haven Cash Rising, Now Equal To At Least 10% Of World GDP.[158]

- Welcome to the new dark ages, where only the wealthy can retire.[159]

A NEW CIVIL WAR?

America is in trouble. The social civil war that began in the 1990s and has increasingly afflicted the US cannot and will not be fixed. The knives are out across the board. We are caught inside an environment of hate, rage, and identity politics and have no way to escape the cage we have created. One report, hopefully overstated, indicates that the Department of Defense has begun planning strategies to deal with the possibility of widespread urban violence, unrest, and mass civil breakdown as "things fall apart, and the center cannot hold." A report on a now unclassified DOD memo contains the following analysis:

> [A] recently unclassified document from the United States Army reveals how future domestic turmoil may be regularly met by armed federal troops trained and equipped to quell dissent by any means possible. On August 15, the feds made public Army Techniques Publication (ATP) 3-39.33, a 132-page manual that describes, in great detail, how Armed Forces personnel can be utilized in response to civil disturbances that erupt within the boundaries of the continental U.S. "Civil unrest may range from simple, nonviolent protests that address specific issues, to events that turn into full-scale riots," the manual states. "The level of violence is determined by the willingness of demonstrators to display and voice their opinions in support of their cause and the actions and reactions of the control force on scene."[160] [emphasis added]

Identity groups that have managed to achieve some power are becoming increasingly aggressive as they attempt to maintain their hold on what they have gained and strive to increase their share. After all, many of their members are true believers who feel they have been wronged, cheated of opportunity and disrespected. They feel they deserve a larger share of social goods. Some just love the feeling of power that their newly found identity affiliation supplies in an otherwise meaningless or impotent existence.

The centers of power have become too diverse and divisive, the actors too focused on their own singular concerns, the underlying broad-based set of social beliefs, principles and creeds so corrupted that there is no turning back and no real possibility of compromise. The US has fragmented into competing tribes and identity groups with limited education in or respect for the intricacies of democratic systems. Too many are possessed of the almost demonic ideological belief systems characteristic of true fanatics. The Internet has played a major role in enabling and building our growing social insanity and paranoia. The following analyses offer sources that suggest how far we have strayed from a community capable of engaging in legitimate discourse and political compromise.

CLASHING IDENTITY GROUPS:
RACE, GENDER, RELIGION, FAMILY AND IDEOLOGY

- **American politics is tribal. Are we ready to admit that?**[161]

- **Political Divisions in U.S. Are Widening, Long-Lasting, Poll Shows: WSJ/ NBC News survey indicates a wide split on cultural, economic issues.**[162]

- **Partisan divide grows in US states, with mixed results.**[163]

- **Fury fuels the modern political climate in US.**[164]

- **The Left Uses Racial Conflict to "Divide and Conquer." More and more, it feels like racial division is the subtext of virtually everything in this country Carlson said. Even topics that seem unrelated to ethnicity are suddenly racially fraught. [T]he effect of that is a deeply angry, divided and terrified nation, which is by design.**[165]

Where what is needed is reasoned justice-based advocacy in response to glaring systemic inequities, we too often find demagogic, intensely activist, organized and vocal special interest identity collectives across the spectrum of interests. They seek to achieve attention and support by virtue of the willingness of the mainstream media to focus on their most provocative claims. They are collapsed into camps we tend to refer to as "The Left" or "Progressives," and "The Ultra-Conservative Right," "Deplorables" or even "Nationalists."

The result is that rather than honest discourse we have vitriolic shouting matches and increasing violence.

Some of the activists have indeed been wronged, although quite often not by those they specifically accuse as they pursue their own perceptions of bias, reverse racism and discrimination. In the area of racial discrimination such identity "social warriors" nonetheless deserve a degree of special consideration because their polity is still suffering the continuing psychological and economic consequences of the historic and vile tragedy of slavery and its lengthy and reduced but still continuing racism.

HATE AND RAGE ENABLED, ORGANIZED AND TRIGGERED BY SOCIAL AND MAINSTREAM MEDIA

- Psychologists say more and more young people believe they are entitled: The psychological trend comes from the belief that you are superior to others and are more deserving of certain things. This form of narcissism has some significant consequences such as disappointment and a tendency to lash out.[166]

- Facebook's greatest weapon: endless comparison of ourselves to others: Facebook makes us miserable by inducing us to constantly engage in me-versus-you interactions. Heavy users in particular are unhappier, lonelier, meaner, and so on.[167]

- Millennials blame boomers for ruining their lives.[168]

- US interest in concealed carry permit training jumps 100 per cent after Parkland: "We've never seen a spike this big before."[169]

- FBI terrorism unit says "black identity extremists" pose a violent threat: Leaked report, citing concerns of retaliation over "perceptions of police brutality against African Americans," prompts fears of crackdown on activists.[170]

- Controversial armored police vehicle factory expands production to meet demand.[171]

- Richest 1% on target to own two-thirds of all wealth by 2030: World leaders urged to act as anger over inequality reaches a "tipping point."[172]

- Survey: Most Millennials, Gen Z Adults Prefer Texting Over Talking In Person.[173]

Although what we are describing will unfold over the next decade or two, police forces and divided communities within cities such as Chicago, St. Louis, Detroit, Philadelphia and Baltimore must already feel they exist in something akin to a state of guerrilla war.[182] The situation has become so bad in Baltimore that a recent report indicates the residents are afraid to even leave

---

FRACTURED SOCIETIES AND THREATENED DEMOCRACY

- Alphabet's Eric Schmidt On Fake News, Russia, And Information Warfare. One of the things I did not understand, says Schmidt, was that these systems can be used to manipulate public opinion in ways that are quite inconsistent with what we think of as democracy.[174]

- Bots on social media threaten democracy. But we are not helpless: Ever-more sophisticated bots can engage with people to sway their political opinions. We have the technology to counter this—we need the will to use it.[175]

- "Fiction is outperforming reality": how YouTube's algorithm distorts truth: An ex-YouTube insider reveals how its recommendation algorithm promotes divisive clips and conspiracy videos.[176]

- Facebook says it can't guarantee social media is good for democracy.[177]

- Social media giants are wild beasts devouring freedom and democracy.[178]

- Thirty countries use "armies of opinion shapers" to manipulate democracy—report: Governments in Venezuela, the Philippines, Turkey and elsewhere use social media to influence elections, drive agendas and counter critics.[179]

- Facebook and Twitter are being used to manipulate public opinion—report: Nine-country study finds widespread use of social media for promoting lies, misinformation and propaganda by governments and individuals.[180]

- Facebook, in a massive purge of 800 accounts and pages, took out accounts that had previously been targeted in a blacklist of oppositional sites promoted by the *Washington Post* in November 2016, some accounts having millions of subscribers?[181]

---

their homes. Chicago continues to set records in shootings and murders as a significant part of the city is besieged by gang violence and drug wars.

Over the past several years, since the large-scale and bitter protests over police shooting of unarmed African Americans captured on video, law enforcement officers have been ambushed and murdered. Violent crime, including murder and assault, has soared. Baltimore's murders have gone past 300 for three years in a row. After rioting over an abuse of police power in the Freddie Gray arrest and transport situation and seeing police as ogres to the extent that many law enforcement personnel reduced their visible presence to avoid becoming flashpoints for angry crowds and protests, many among Baltimore's residents are begging for a more active and visible police presence in their communities.[183]

The tragic fact is that the hatred between and within various sectors of society is fracturing the country. As already bad conditions become even worse, people will take to the streets over lost jobs, financial benefits, the collapse of public and private pension systems, and a lack of security or seek survival through crime. We will become outraged because we aren't given what we need or what we believe should be right and possible. This phenomenon is spreading worldwide, and unemployment, corruption, poverty, the skewing of the global economy to create a monstrously inequitable distribution of social wealth between and within nations, and a lack of foreseeable opportunity are at its heart. A continuing surge in migration, legal and illegal, from beleaguered poor nations is one result—but where are they to find relief, as the conditions similarly worsen in the wealthier countries?

> Rising unemployment, inequality and a lack of decent jobs have helped fuel a rise in social unrest that threatens to intensify unless policymakers take swift action, the UN's labour agency has warned. The International Labour Organization said its measure of protest activities around the world had ticked higher in the last year against a backdrop of economic and political uncertainty. In a downbeat report into global labour market prospects, the agency also predicted migration could rise over the next decade as frustrated jobseekers leave their countries in search of better prospects. [184]

While increased law enforcement may be necessary in face of such conditions, it is surely far from any solution, and as indicated, may arouse further resentment and conflict.

CHAPTER EIGHT
Our Globalist Leaders Have Betrayed the American Worker

The upshot of the past twenty-five years following NAFTA and creation of the World Trade Organization is that mindless infatuation with globalization has severely harmed the American and European economies, their workers, and the strength of their democratic political systems. Governments and multilateral institutions such as the World Bank and the International Monetary Fund recklessly pushed for the globalization of work since the early 1990s and in doing so abdicated their responsibilities to the people whose interests they were charged with representing. The collusion of extreme globalists fully convinced that the world was ready for a comprehensive scheme of government and economics led to the massive outsourcing of US manufacturing jobs to locales with cheaper labor.

Robert Kuttner, co-editor of *The American Prospect* and professor at Brandeis University's Heller School, has written an insightful and often brilliant analysis of how economists, politicians and corporate leaders rigged the system to create a global order that has been destroying any hope for the continuation of anything resembling a true democracy. The impacts are being felt whether we are speaking about national or global scales of governance.

In *Can Democracy Survive Global Capitalism?* Kuttner demonstrates how corporate power has increased to the point that real power has been displaced from nations and conferred upon an unaccountable group of globalized interests and institutions that operate outside what we would consider a Rule of Law system. This rigged system includes institutions such as the IMF, the OECD, the World Bank and other multilateral institutions, as well as a host of tax havens that keep as much as ten percent of the world's wealth beyond the reach of the nations in which the corporations actually conduct their economic activity.[185] This stealth takeover occurred as corporations became progressively larger and more powerful and are now able to shape, influence and control the political processes that otherwise would be responsible for holding their behavior to account.

This transformation of economic fundamentals by corporations, institutions and global elites was not only due to the search for lower pay rates for labor. For the largest corporate players and financial institutions it was also driven by the reduced expenses allowed by lower standards of living in the countries to which jobs were transferred, limited or non-existent responsibility to provide benefits for workers such as pensions, health care, and worker safety, reduced costs of materials, tax breaks and sweetheart deals with those nations' leaders.

Along with these advantages came a lack of transparency or oversight related to their actions and the avoidance of the legal responsibility. It allowed corporations to avoid being subject to what they considered to be expensive and intrusive environmental regulations and the seemingly endless demands of bureaucracies that in many nations to which jobs were outsourced were happy to look the other way.

The shift described above brought immense profits to corporations, co-operating governments, corrupt rent seekers and financial institutions while undermining the economies of the countries from which manufacturing jobs were taken. There were benefits to the consumers of those nations because they were often able to buy goods at prices lower than they otherwise would, had the manufacturing been retained. But the loss of critical manufacturing industries was a "hidden tax" that undermined the economies and political communities of nations such as the US and members of the European Union.

Unmasking the True Workings of Power and the Rise of Populism

As Robert Skidelsky and Mohamed El-Erian have warned, once the "true workings of power" are unmasked, large numbers of people feel betrayed by their leaders. When this occurs the conviction that "the elites are selfish, corrupt, and often criminal" is held with "passionate intensity."[186] Skildelsky explains:

> The second coming of liberalism represented by Roosevelt, Keynes, and the founders of the European Union has been destroyed by the economics of globalisation: [represented by] the pursuit of an ideal equilibrium through the free movement of goods, capital, and labour, with its conjoined tolerance of financial criminality, obscenely lavish rewards for a few, high levels of unemployment and under-employment, and curtailment of the state's role in welfare provision. The resulting inequality of economic outcomes strips away the democratic veil that hides from the majority of citizens the true workings of power.[187]

Along with this unmasking of corruption, greed and abuse, goes the belief that, "power must be returned to the people." This assertion is an eminently democratic premise that is immediately decried as Populism by the very elites that created and benefitted from a profoundly foul system. The problem we have now is that while "The People" know they are angry, afraid, betrayed and uncertain, they have no idea what to do about it. By the way, neither do most of their leaders.

Uncertainty and fear over their jobs and futures are among the single most important reasons many Americans and Europeans are up in arms, rejecting what they consider the most offensive aspects of economic and political "globalism" along with what they see as overly aggressive Progressivism that challenges their fundamental beliefs, culture and values. The concern felt by many of the so-called Populists particularly includes what they see as the fates awaiting their children and grandchildren. With the American Middle Class eroding and many people being told that their economic futures are uncertain and out of their control there is a rising anger being felt by much of the population. The impacts of AI/robotics on their future prospects is a key factor in ordinary peoples' outrage at the government and corporations they feel have betrayed them.

Advocates of the modern globalist movement automatically disparage anything at odds with their dominant interests and presumptions as being populist, nativist, xenophobic, nationalist, and the like. The fact is that there is nothing hateful, bigoted, immoral or ignorant about members of a national community being concerned with maintaining the integrity, health, and principles of their specific community. Nor is there anything inappropriate for fathers and mothers to be committed to the welfare of their own children and grandchildren, and to resist strategies they consider to endanger that welfare.

A recent survey discovered that Los Angeles residents see the increasing probability of urban violence. In that vein the American presidential campaign of William Jefferson Clinton adopted James Carville's mantra about what the election was about, "*It's the economy, Stupid!*" It is the economy even now, but it is not *solely* the economy at this point. Lack of access to jobs is, however, a key factor. Political science researcher Fernando Guerra explains: "Economic disparity continues to increase, and at the end of the day, that is what causes disruption. ... People are trying to get along and want to get along, but they understand economic tension boils over to political and social tension."[188]

In language that echoes Clinton, Skidelsky observes: "The usual explanation is that a system fails because the elites lose touch with the masses. But while one would expect this disconnect to happen in dictatorships, why does disenchantment with democracy take root in democracies themselves?"[189] In answering that question he concludes: *"[I]t is economics, not culture, that strikes at the heart of legitimacy. It is when the rewards of economic progress accrue mainly to the already wealthy that the disjunction between minority and majority cultural values becomes seriously destabilising."*[190] [emphasis added]

THE GLOBALISTS JUST DON'T "GET" IT

At this point, the undeniable fact, from the perspective of American workers and their families, is that advocates of globalization betrayed the American worker, the nation's economy and the needs of the overall society. Damon Linker explains the global elites' betrayal and ignorance, describing their continuing smugness and challenging their mindset while predicting how they would respond to the rising tide of criticism of their destructive policies and systemic manipulation. Linker argues the globalists would say:

> *There's nothing to worry about. Everything's fine. No need for serious soul searching or changes of direction. Sure, populism's a nuisance. But we're keeping it at bay. We just need to stay the course, fiddle around the edges a little bit, and certainly not give an inch to the racists and xenophobes who keep making trouble. ... We got this.*[191]

Linker's summary is accurate. The globalists still don't *"get it."* They continue to defend a system that benefits them while betraying millions of workers and their families. A rapidly increasing number of people in Europe and the US don't agree with the elites in power whom they are convinced have sold them a bill of goods. They also don't like what they are experiencing in their own work lives or how their nation's economies and social systems are being operated. Instead, as Linker concludes: "What [ordinary people]... see is a system that is fundamentally unjust, rigged, and shot through with corruption and self-dealing."[192]

Refusal to recognize the legitimacy of that compelling commitment to family, community and nation, and to automatically label anyone holding such preferences as greedy, bigoted, out of touch with reality, nationalist or populist is the behavior of an international elite committed to its own self-interest and advantage, one that is controlled by people who have naively bought into a destructive and unsustainable new global order. Thankfully, there is now counter pressure against that dishonest elitism and blindness. To provide a sense of the flaws in the globalist ethos the list set out below suggests some of the causes and challenges.

GLOBALIZATION AND ITS ILLS

- Globalisation: the rise and fall of an idea that swept the world. It's not just a populist backlash—many economists who once swore by free trade have changed their minds, too. How had they got it so wrong?[193]

- IMF warnings on economy will fall on deaf ears among world leaders: The IMF is seen as a hand-wringing liberal institution making nuanced and technical suggestions in an age of broad-brush solutions.[194]

- Soros Sees New Global Financial Crisis Brewing, EU Under Threat. A surging dollar and a capital flight from emerging markets may lead to another "major" financial crisis, investor George Soros said, warning the European Union that it's facing an imminent existential threat.[195]

- The new stock-market fear: Signs that a period of harmonious global growth is crumbling.[196]

- Business leaders say economic nationalism is biggest growth threat: Poll of CEOs finds backlash from political populism among leading concerns.[197]

- World finance now more dangerous than in 2008, warns central bank guru. [T]his time the authorities are caught in a "policy trap" with few defences left, a veteran central banker has warned.[198]

- World Bank issues warnings on interest rates and inflation: After better than expected growth in the global economy, Bank says financial markets are vulnerable to unforeseen negative news.[199]

What the elites and the beneficiaries of uncontrolled, over-hyped and inadequately conceived globalization refused to understand or care about is that manufacturing jobs were key to providing stable employment in the US, UK and EU. They ignored the fact that there were real people affected by the loss of those jobs. The far-too-casually exported outsourced jobs had paid wages at levels that allowed workers in Western nations to be in what was considered the blue collar middle class. Such jobs were an economic engine that stimulated a large number of subsidiary jobs directly, and others were indirectly supported by manufacturing employees and the facilities in which they worked.

The job loss/creation picture we now face due to AI/robotics is even bleaker than might be thought. While the number of warehouse locations is growing, the use of robots instead of human workers is a critical part of the emerging business models due to their productivity and efficiency gains. Even while bragging about the millions of low level service jobs they create, Target, Walmart and Amazon are steadily expanding the use of robots in their warehouses. Walmart is beginning to use robots in the retail stores themselves and has announced it is shifting a significant part of its warehousing workforce to robots.[200] It is using them to obtain and store merchandise, scan shelves, and process orders, tasks previously done by human workers.[201]

Research by these retail monoliths into the development of ever-more sophisticated robots with enhanced capabilities is ongoing. This mechanized workforce will in the reasonably near future replace the bulk of human warehouse employees.[202] Again, the basic premise of such companies is that they should "do more with less," with "less" being code for eliminating human workers. An R&D "war" is taking place between Amazon and other companies working on developing the most efficient robotic workers for on-line companies' warehouses to meet the customer demands for fast delivery of their orders.[203]

Amazon has become a 50-tentacled octopus as it aggressively expands its reach into numerous sectors of the American and world economy. Even though Jeff Bezos, Amazon's CEO and the world's richest man, brags about the number of jobs Amazon is creating, the truth is, as Linda Fickensher writes, that Amazon is destroying more American jobs than the Chinese. In her analysis, Fickensher indicates:

> Amazon played a large role in eliminating more than 50,000 jobs in recent years from just three companies— Staples, Office Depot and Best Buy, public filings show. In March, MarketWatch estimated that Amazon will destroy 1.5 million retail jobs in the next five years. And with its push into self-driving trucks, drone delivery, automated grocery stores and more, ... [MarketWatch said] the total number of lost jobs would likely be more than 2 million, concluding, "Could Amazon actually kill more American jobs than China did? It's quite likely."[204]

This demonstrates that, while still reasonably strong as of 2018, America's economic slide to this point is not simply due to manufacturing losses.[205] Non-manufacturing blue-collar jobs are disappearing more and more rapidly as the AI-driven Internet evolves and the robotic systems that replace direct human labor are progressively adopted. Macy's has announced the closing of more than sixty stores representing 10,000 jobs. Sears/K-Mart is struggling to survive, an effort that is unlikely to be ultimately successful. There were, for example, an estimated 5,300 retail store closings in the first six months of 2017 alone in the US. Schuyler Velasco writes:

> An estimated 5,300 retail locations have closed through June 20, according to one estimate—nearly triple the rate from a year ago. ... Yet it's the decline of industries like coal and manufacturing that get the big attention, especially from politicians. That's surprising, since the closings mean that retail, which employs about 10 percent of all working Americans, is shedding jobs at a rate that dwarfs either of those.[206]

It is not only individual store chains that are closing. Due to the loss of anchor stores, shopping malls in the hundreds are closing across the US as consumers opt for the convenience of doing a significant part of their shopping online, destroying the employment of family, friends and even themselves in the process.[207] Many of those closings and the resultant job losses are due to the expansion of online retail sales. While some employment does take place through that system, the new job creation is far less than what is being lost.

Online retail sales made possible by the AI-Internet system are consolidated into electronic and geographic hubs rather than physical stores spread widely across large territories. Physical stores and malls brought goods, services and jobs to people in limited geographic territories defined by population and driving distance. In doing so, they provided widely dispersed local communities with a significant number of employment opportunities and provided local communities with stronger tax bases for schools and governmental services. That localized economic dynamic is being shoved aside by online retail systems located "somewhere" in hyperspace.

A very limited number of states and local communities are benefitting from decisions by large online retailers to locate warehousing hubs in their territories. But the information superhighway of the online merchandising system is in many ways similar to the US interstate highway system begun in 1954. The national defense highway system was in fact one of the largest and least well thought out economic policy acts in US history. The concrete ribbons of limited access superhighways bypassed many thriving local communities dependent on the flow of road traffic for sales of their goods and services.

The expressways of the interstate highway system turned thousands of smaller cities and towns into empty shells, closed store fronts and lost jobs.

And now again with warehousing hubs, the result is that there is a loss of localized employment at physical retail stores. This creates a profound shock to local economies and their tax bases and the potential for a rise in criminal activity in the abandoned sites. The same can be said for real estate markets, schools and other economic activities that depended on the consumer spending of the decently paid workers who are now displaced.

IS A NEW WORLD ORDER BEING CREATED?

The globalized system is being challenged in dramatic fashion even as we write. The changes involve the strategies being implemented by US President Donald Trump. Regardless of one's feeling about the behavior of the President and his tendency to communicate in odd tweets and inane characterizations of anyone who runs afoul of his preferences, from the point of view of a globalist, Trump's strategy of rewriting global trade agreements, negotiating nation-to-nation bilateral or strategically trilateral treaties with key trading partners such as Mexico, Canada, Japan, the United Kingdom, India and the European Union are offensive actions.

This approach threatens globalists' conception of a world order that moves toward an overarching global governance system. The alternative view—one we adopt in part—is that such strategies are legitimate reactions to a system of globalization that was overstated and wrongly conceived in the first instance and imposed unfair rules of trade on America. The globalization pushed by US corporations and financial institutions unquestionably enriched those beneficiaries but undermined American workers, their families and the system's social cohesion.

While *Guardian* columnist George Monbiot detests Donald Trump to the point of calling him "odious," he demonstrates more intellectual integrity than most of his counterparts in defending Trump's recent attacks on the global trading system made at a G7 meeting in Canada. Monbiot explains:

> He gets almost everything wrong. But last weekend Donald Trump got something right. To the horror of the other leaders of the rich world, he defended democracy against its detractors. Perhaps predictably, he has been universally condemned for it. ... Those who defend the immortality of trade agreements argue that it provides certainty for business. It's true that there is a conflict between business confidence and democratic freedom. This conflict is repeatedly resolved in favour of business.

Monbiot goes on to explain:

> There was much rejoicing this week over the photo of Trump being harangued by the other G7 leaders. But when I saw it, I thought: "The stitch-ups engineered by people like you produce people like him." *The machinations of remote elites in forums such as the G7, the IMF and the European Central Bank, and the opaque negotiation of unpopular treaties, destroy both trust and democratic agency, fueling the frustration that demagogues exploit.*[208] [emphasis added]

The simple fact is that we are witnessing an effort that is seeking to bring manufacturing back to the US and to recreate the economic basis that is essential for the nation's survival as a democracy and allows the country to take care of the numerous critical issues it faces. This does not mean that an America that has been a world leader in political, humanitarian and security contexts is abandoning any moral sense of responsibility to those living in desperate conditions in much of the world. Nor, however, does it mean that the US has not played a role in creating or worsening crises in areas such as the Middle East, including Iraq, Syria, Libya and Yemen. While America has done much good through the efforts of many compassionate and caring people, far too many US policies have been driven by unaccountable corporate and military interests concealed in many instances behind lofty governmental and corporate rhetoric.

America's ability to nurture, support, protect and provide opportunity and assistance to its own people are essential considerations that underpin its own economic strength and stability, and indeed, its capacity to pursue its other agendas or responsibilities. That ability has been undermined by overwrought globalization that went too far and assumed that the US not only owed the highest duty to assist others throughout the world, but that it could be subjected to extremely heavy burdens in doing so because, after all, America was so rich and powerful that it was an endless cornucopia.

CHAPTER NINE

The Social Costs of Disconnecting Productivity from the Needs of Labor

Large companies seek productivity gains by reducing costs and controlling the market sectors from which they extract profits. Great cost reduction has become possible through AI/robotics use. Adoption of AI/robotics means eliminating human workers and enhancing productivity by achieving greater efficiency, and reducing labor-associated expenses and other problems incurred through use of human workers. All well and good for capital, but how has it impacted workers?

Things were already going badly for workers. London's *Guardian* details a report by the McKinsey Global Institute on the flattening and decline of employment earnings in many developed countries that was generated by the economic collapse of 2008-2009 and the uncertainty now being faced by millions of Millennials.

> Half a billion people in 25 of the west's richest countries suffered from flat or falling pay packets in the decade covering the financial and economic crisis of 2008–09. ... [B]etween 65% and 70% of people in 25 advanced countries saw no increase in their earnings between 2005 and 2014. The report found there had been a dramatic increase in the number of households affected by flat or falling incomes and that today's younger generation was at risk of ending up poorer than their parents.[209]

As the system of production and services shifts dramatically away from human labor and towards AI/robotics, the returns from economic activity will create an ever smaller pool of super-wealthy people with immense leverage and influence. Conversely, we will see an increasingly larger mass of economically disenfranchised low-end workers with marginal survival incomes as well as a large number of chronically unemployed.

We are already experiencing a degree of this restructuring in which, as Rebecca Greenfield indicates, *"Many people who lost well-paying jobs have found work, but for less money, doing hourly retail and food services jobs. These new hourly workers not only make less money, but they have much less pre-dictable schedules than hourly workers had before the recession"*[210] [emphasis added].

Even worse, retail jobs are disappearing in large numbers, raising the question of what rung comes next for people sliding down the income and status ladder. A recently issued US Department of Labor Statistics study reports that there are as many as 16 million Americans on part-time, temporary or technically self-employed gig jobs. It also reported that in addition to uncertainty and lack of health, pension and other benefits, gig workers typically earn 20–30 percent less than workers in equivalent stable employment.[211]

The elimination of jobs and the flattening and even decline of incomes means our ability to generate adequate governmental fiscal resources under our existing system of taxation is being undermined due to the transition of wealth production and the distribution of its benefits from labor to capital. A critical challenge will be redesigning the systems through which the federal, state and local governments obtain the revenues needed to cope with the chronic and permanent underemployment and unemployment that results. Such a redesign is essential because the AI/robotic shift endangers the integrity of key social safety net programs, and will have a negative effect on individuals and society due to the loss of positive work experiences, reduced social mobility and increasingly limited opportunity.

The UK has experienced the biggest fall in real pay since the 2008/2009 recession of all Organization for Economic Cooperation and Development (OECD) countries other than Greece.[212] One intriguing analysis in the UK attempts to explain why Britain's workers are working longer but overall productivity is abysmal. In "Why are UK workers so unproductive?" Charlotte Seager writes: "Working hours are at an all-time high yet productivity has collapsed—low wages, inflexible work practices and job insecurity are to blame. ... *It's a puzzle: working hours are at an all-time high and more people are employed than ever before, yet productivity is collapsing*—government figures show output is at its lowest level since 1991. *So why is productivity plummeting?*"[213]

Seager considers the decline in real pay and the increased uncertainty of employment to be key factors in the reduced productivity, creating fear and uncertainty about the future, reduced employer/employee loyalty, and less worker training and preparation. Another analyst concludes: "People know they're not getting their fair share of wages, and many people are on short-term contracts, so the motivation to improve productivity simply isn't there."[214]

One unavoidable aspect of what is occurring is the uncertainty of even having work, let alone the flat or reduced pay scales and limited access to benefits. Not knowing how many hours will be available at any point in time, or the number of piecemeal jobs an individual needs to make ends meet leads to a level of uncertainty corrosive both to individuals and society. In the United Kingdom a recent report indicated that 10 million British workers are in "insecure work," representing one-third of the UK's total workforce. There are consequences from the tension this insecurity generates. Sixty-one percent of those "insecure" workers suffer from stress or anxiety.[215]

For those concerned with the growth in income inequality, the message is that inequality and an increasing lack of upwardly mobile job opportunities will get worse rather than better. One study concluded that in the United Kingdom: "*On current forecasts average earnings will be no higher in 2022 than they were in 2007. Fifteen years without a pay rise. … This is completely unprecedented.*"[216]

The Organization for Economic Cooperation and Development (OECD) describes how bad the situation may become from the perspectives of economics, wealth inequality, opportunity and social justice. The OECD warns:

> [S]low productivity growth will create challenges, notably for addressing income inequality, welfare promises, raising standards of living and investment incentives. Restoring the traditional dynamism of the American business sector may be one way to boost productivity growth. This requires competitive market forces, skilled and mobile workers and policies to promote innovation.[217] [emphasis added]

The OECD's formula for restoring the dynamism of the American business sector is far easier to proclaim than to make a reality when both domestic and global competitive market forces drive economic activity toward AI/robotics with a resulting loss of human jobs. Developing "skilled and mobile" human workers sounds good and will help some people locate employment that may not necessarily exist in their geographic area. But nonetheless, fewer employees will be needed, and we aren't at all certain about what the primary new skill sets should include. Mobility of labor will be a factor because dynamic job centers will develop—as with Silicon Valley and North Carolina's Research Triangle. But even if there is some mobility there won't be enough jobs for all. Areas not in the "dynamic" zones will have severe problems.

Youth unemployment is already prevalent in both Western Europe and the US. Depending on the location, youth unemployment rates between twenty and even fifty percent mean that a significant part of Western democracies' young people are not learning the skills, work habits, discipline and values needed to become productive workers because they are not working at the point in their lives when such habits and attributes are acquired and internalized. This has important implications for the economies of the nations most severely afflicted.

The AI/robotics transformation is about much more than just an absolute lack of jobs. The shift in the nature of employment, the reliability of having a decent job or any job, the declining number of opportunities, increasing lack of social mobility, and the reduced quality of the jobs available are all equally important considerations driving intensifying concern and discontent. There is already a significant disjunction between the changing needs of employers in the evolving "new economy" and the skill deficiencies of workers whose only experience is in the "old economy."[218] Many employers lament that they have positions available but can't fill them because there aren't enough people with the required skills and knowledge. By the time the workers catch up with the shifting skills base, there is a real likelihood the employers will be ready to move on to AI/robotics replacements.

CHAPTER TEN

What Will Our Unemployed Children Do?
Sliding Down the Socio-Economic Ladder

Throughout the developed world, including North America, Europe and Australia, young people are facing significant employment uncertainty as jobs disappear, or change radically in character and require skills for which potential workers are not trained or may not even be capable.[219] Many jobs are more difficult to obtain or hold, and pay less even while educational and living expenses continue to grow.[220]

A recent analysis highlights the dilemma of job loss and job replacement facing those fitting into the Millennial category. The study from Gallup concluded that: "Millennials are the generation most vulnerable to the threat of A.I. and automation as they are disproportionately more likely to hold positions that [will] one day be replaced by this new technology."[221]

> The "Fourth Industrial Revolution" is changing human activity and human institutions at a more accelerated and unforgiving pace than any before it. Central to the challenge is that AI/robotics replaces not merely the physical labor or computational power involved in human activity, but increasingly takes over the creative mental processes that were formerly considered unique to human intellectual activities but are now captured through AI in continually more sophisticated software applications. Given the fact that further exponential increases in total productivity will be made possible from applications of AI/Robotics, leading to a decreased need for human intellectual capital and labor to produce that output, we will experience a *temporary upward productivity curve*—even while real wages from lower value and less efficient human labor plummet.

People typically labeled Millennials are said to feel less middle class than at any time in several generations.[222] The stresses they are experiencing have numerous effects. A UK study of Millennials concluded that they will be suffering serious physical and emotional health problems even at a relatively early age. The Health Foundation warned:

> Millennials are on track to become the first gener-
> ation to suffer worse health than their parents when they
> reach middle age, a study has warned. *People in their 20s*
> *and 30s will have a higher risk of "lifestyle" diseases such as*
> *cancers, diabetes and heart disease in 30 years' time because*
> *of their employment, relationships and housing.* ... The trend
> is linked to long-term stress, anxiety, depression or lower
> quality of life...[223]

The McKinsey Global Institute concluded that: "It is likely to be the first time in industrialised history, save for periods of war or natural disaster, that the incomes of young adults have fallen so far when compared with the rest of society." [224] It warned that: *"should the 'slow growth' conditions of the past decade persist, up to 80% of income segments could face flat or falling incomes over the next decade. There was a possibility that increased automation would result in 30–40% of households seeing no advance in their incomes even if growth accelerated."*[225]

The terms "Precariat" and now "Middle Precariat" have come into increasing use as a means to describe what is occurring.

> The word Precariat [those in precarious economic po-
> sitions] was popularized ... to describe a rapidly expanding
> working class with unstable, low-paid jobs. What I call the
> Middle Precariat, in contrast, are supposed to be properly,
> comfortably middle class, but it's not quite working out this
> way. ... The jobs held by those in this socio-economic tier
> are also increasingly shaky, composed of short-term con-
> tract or shift work.[226]

CAN YOUNG AMERICANS COMPETE WITH HARDER WORKING AND MORE AGGRESSIVE AND INNOVATIVE COUNTERPARTS?

Some people feel compassion for the plight of Millennials but at the same time perceive an attitudinal shift that they argue is contributing to the problem. Martha Stewart argues, for example, that US Millennials are spoiled, soft and lazy—not an ideal set of characteristics for people facing a tough future.[227] As offended as some might be by Stewart's comment, that doesn't mean she is entirely off the mark. Demands for such things as an insistence on Ultra political correctness lest anyone's feelings are hurt, and the willingness on the part of some people who proclaim their own undying commitment to tolerance but feel entitled to engage in highly aggressive and even violent protests over perceived slights and complaints seems bizarre.

On the AI/robotics front, employers are able to cite such attitudes to justify making as rapid a shift as possible to a non-human workforce. Presenting themselves as having to deal with human employees who are never satisfied and demand continual accommodations for their needs lest the employer be accused of "insensitivity" not only obfuscates employers' actual degradation of the conditions of labor, but seeks to justify their dispossession of labor's access to work, itself. If the alternative is that you can shift from human workers to robotic systems of production and service that do what they are told, don't get sick or have personal problems, aren't addicted to drugs or other substances, and don't sue if someone says a harsh word or gives them a poor evaluation review, for employers dedicated to profit and the bottom line the decision pretty much makes itself.

Conversely, and as suggested by a reader of the *Contagion* manuscript, the above analysis may be too one-sided in emphasizing employers' discontent with workers. Diana Collier indicates:

> You have provided an employer's perception. From the point of view of workers, benefits are being stripped, surveillance is increased, ditto precariousness. Reasonable needs for such human functions as pregnancy—which are recognized in other cultures—are regarded as dysfunctions. Seniority is stripped by closing down shop, opening elsewhere; lawsuits are prevented by forcing employees to sign waivers re same, etc.

Collier is correct in the analysis. Our phrasing admittedly preferences the concerns of employers over ordinary workers. But that brings the heart of the problem to the fore, which is that a fundamental power imbalance exists between workers and many employers whose business activities are of a character that human work can be replaced by AI/robotic systems. In such areas, lacking regulation or active governmental intervention, the obligations owed to human workers can be made to disappear if the production or service activities are shifted to AI/robotics—and indeed, as mentioned, governments are even applauding and facilitating such a transfer. This massive displacement of the human work force is the danger we face unless our entire political system creates incentives and disincentives that encourage and facilitate the preservation and creation of human jobs, and imposes costs when the investors decide to transition to AI/robotics production systems.

If this is to be avoided, the rules of the game need to be modified and the power imbalance adjusted. This will occur only if the society foresees the imminent danger, and strongly commits to creating rules that stabilize the political economy of human work and slow and limit the speed and scale of the transition to AI/robotics.

Another critical element of what appears to be happening is suggested in a prescient 2008 analysis by Mark Bauerlein who provides a detailed and clear analysis of what has been occurring as the "Digital Age" takes hold. Bauerlein's *The Dumbest Generation: How the Digital Age Stupifies Young Americans and Jeopardizes Our Future* is well worth reading.[228] Support for Bauerlein's position that the overuse of digital tools can undermine quality of thought is strengthened by a stunning report that indicates our IQ levels have been declining since shortly after World War Two, and doing so at an increasing rate in the past several decades. It suggests that declines in teaching quality, reading interests, and diversion of attention to television and computers are quite likely contributors. As reported in *The Times* of London:

> The decline, which is equivalent to at least seven points per generation, is thought to have started with the cohort born in 1975, who reached adulthood in the early Nineties. *Scientists say that the deterioration could be down to changes in the way maths and languages are taught, or to a shift from reading books to spending time on television and computers.* Yet it is also possible that the nature of intelligence is changing in the digital age and cannot be captured with traditional IQ tests.[229]

It is also of interest that a recent report indicated that the "tech elite" of Silicon Valley do not want their own children raised on digital devices. In two revealing October 2018 analyses published in the *New York Times*, "*The Digital Gap Between Rich and Poor Kids Is Not What We Expected*" and "*A Dark Consensus About Screens and Kids Begins to Emerge in Silicon Valley*," Nellie Bowles wrote:

> [A]s Silicon Valley's parents increasingly panic over the impact screens have on their children and move toward screen-free lifestyles, worries over a new digital divide are rising. It could happen that the children of poorer and

middle-class parents will be raised by screens, while the children of Silicon Valley's elite will be going back to wooden toys and the luxury of human interaction.[230]

Bowles adds that Silicon Valley technologists, those who developed and made millions from the sale of digital technology, have suddenly had the equivalent of a "Come to Jesus" experience. They are discovering that continual use of digital systems is harming the intellectual development of their own children. She writes:

> Chris Anderson, [is] the former editor of Wired and now the chief executive of a robotics and drone company. He is also the founder of GeekDad.com. "On the scale between candy and crack cocaine, it's closer to crack cocaine," Mr. Anderson said of screens. Technologists building these products and writers observing the tech revolution were naïve, he said. "We thought we could control it," Mr. Anderson said. "And this is beyond our power to control. This is going straight to the pleasure centers of the developing brain. This is beyond our capacity as regular parents to understand." … "I didn't know what we were doing to their brains until I started to observe the symptoms and the consequences," Mr. Anderson said.[231]

Alibaba's Jack Ma recently argued that educational systems should not be preparing their students for manufacturing jobs that aren't going to exist.[232] That is not the only problem. Years of being coddled in schools without failure and provided unearned rewards and medals for mediocre performance so their feelings wouldn't be hurt or egos threatened has not prepared the youth of Western democracies to compete against counterparts from other cultures who have not been provided such overprotected upbringing.

Nor should we blame the youth because to a significant extent they have been the victims of what adults have done. Some have been victimized by overly indulgent and well-meaning parents who spoiled their offspring and made their children's lives as easy as possible simply because they could. Others have been raised in dysfunctional families without a full set of parents to offer guidance and to set and enforce rules of behavior. Still others have spent their lives in poverty-stricken environments that were not conducive to the kinds of advantages and guidance most middle class families take for granted.

The lack of clear standards, good parenting, intelligent educational policy and effective guidance and support systems is not only found in the US.

We need to confront the fact that the systemic betrayal of our children winds up with the creation of a generation of young adults who lack the kinds of up-bringing, support, education, standards and discipline required to compete and work productively on high levels of activity. Adam Corlett, an analyst at the UK's Resolution Foundation, indicates, for example, that given the terribly inadequate work habits of UK employees, the UK's economic system needs robot workers to enhance its productivity.[233] New employment of the kind still available to human workers is not guaranteed to be available to American workers. Many of the young among the American population are likely to lack the basic work habits needed to compete with their human or AI counterparts, particularly in Asia. The Pew Foundation Research Center reports that:

> "[O]nly 29% of Americans rated their country's K-12 education in science, technology, engineering and mathematics (known as STEM) as above average or the best in the world. ... A companion survey of members of the American Association for the Advancement of Science found that just 16% called US K-12 STEM education the best or above average; 46%, in contrast, said K-12 STEM in the US was below average."[234]

Assuming the validity of the analysis, Pew's findings mean that even if there are new jobs available in a globalized and highly competitive economy, Americans are not likely to be the workers who fill most of them.[235] Either they will lag on the level of performance, or on the menial level, because their wage demands (whether due to unrealistic expectations or actual needs) are not competitive.

In the US we commonly hear that jobs involving STEM-educated workers (Science, Technology, Engineering and Mathematics disciplines) represent the safest employment future for university study. Yet in many instances students undertaking such university majors have shown an unsettling tendency to drop out. Far too often the reason given is that the subject matter is too hard and demanding. Some researchers offered troubling judgments about what is occurring, observing: "*50 percent of American college students who major in a STEM—science, technology, engineering and mathematics—field drop out, and the persistence rates for women and people of color are lower than those of their white male classmates, reported researchers. ...*"[236]

Whatever the cause, this does not bode well for social cohesion or employment diversity. Given the stark and dismaying statistics on the results of urban school systems, the challenge and the tragedy is that the schools are betraying their students. This has been going on for a period of years and is not immediately fixable. Many of those educated in such systems are at a significant competitive disadvantage vis-à-vis graduates fortunate enough to have stronger educational backgrounds or more supportive family encouragement and environments.

For too many American students, the charge that they are avoiding hard and demanding educational subjects may well be accurate. Science and mathematics require a strong background in the subject matter, discipline and lengthy preparation. If you don't grow up with the fundamental attributes demanded by such subject matters then remedial educational approaches are not going to fill the gap.

A just issued educational assessment report by the Department of Education found that 65 percent of American 8th grade public school students lacked proficiency in reading and 67 percent in math.[237] In a competitive world in which other nations are demanding rigorous educational performance from their youth, such a lack of preparation and high levels of learning and proficiency on the part of American students is a very serious omen for the students' futures and will foreclose opportunity even more than might otherwise occur.

This increases the likelihood of a higher proportion of the graduates of many US school systems not being competitive for the best available jobs. This creates its own political and moral problem. To gain a sense of just how serious is America's dilemma we offer some data on the state of urban schools. The data indicate that quite a few large US public school systems are an almost total mess that leaves a majority of graduates unable to compete successfully in an economy where employers are seeking workers with advanced skills.

> The results are far worse for students enrolled in some urban districts. Among the 27 large urban districts for which the Department of Education published 2017 NAEP test scores, the Detroit public schools had the lowest percentage of students who scored proficient or better in math and the lowest percentage who scored proficient or better in reading. Only 5 percent of Detroit public-school eighth graders were proficient or better in math. Only 7 percent were proficient or better in reading. In the Cleveland public schools, only 11 percent of eight graders were proficient or better in math and only 10 percent were proficient or better in reading. In the Baltimore public schools, only 11 percent

were proficient or better in math and only 13 percent were proficient or better in reading. In the Fresno public schools, only 11 percent were proficient or better in math and only 14 percent were proficient or better in reading.[238]

Although much of what we are saying in *Contagion* relates to the American system, Europe is also having serious economic problems. The average level of unemployment among younger workers in the EU is above twenty percent, and in the countries of Southern Europe it is around fifty percent unemployment for young people. Nobel Laureate and economist Joseph Stiglitz explains:

> Dry statistics about youth unemployment carry in them the dashed dreams and aspirations of millions of young Europeans, many of whom have worked and studied hard. They tell us about families split apart, as those who can leave emigrate from their country in search of work. *They presage a European future with lower growth and living standards, perhaps for decades to come. These economic facts have, in turn, deep political ramifications. The foundations of post-cold war Europe are being shaken.*[239]

The shaking of Europe's foundations is creating a new system in which there is absolutely no reason to believe there will be enough decent or fulfilling jobs available to satisfy human worker demand—even if we had the insight needed to design the most brilliant educational systems.[240] We are metaphorically "between a rock and a hard place" in terms of the future of employment.[241] Universities and other elements of our education and training system are likely to be preparing students for areas of work that either have only a limited number of slots available or represent categories of employment that are moving targets that keep slip-sliding away as the AI/robotics transformation continues to unfold and supplants human workers across progressively wider zones of activity.[242]

THE SOLUTION IS NOT PART TIME JOBS, SHORT-TERM JOBS OR THE GIG ECONOMY

It is vital we recognize there is no magic in supposed alternatives such as the gig economy in which individuals patch together multiple part-time jobs and move from gig to gig as if they were performers and musicians. The gig approach will work for some, but it is a dead end for many who will find the situation stressful, extremely competitive, lacking in sufficient opportunities, and the chance to receive critical benefits such as health and pension support. In the Uber/Lyft situation, it has increasingly undermined the taxi driving industry, to the point that taxi drivers' incomes have plummeted and there has even been a dramatic rise in suicides.[243] The romance of the nomadic worker moving job to job or gig to gig is a looming nightmare more than a solution for what is emerging. Consider one report on the gig economy in the United Kingdom.

> The modern working day has begun. Thousands of couriers have arrived at their local sub-depots, and spent the first hour sorting through mountains of parcels containing goods of all shapes and sizes, all bought online from some of Britain's best-known brands. Every last one needs to be delivered by the end of the day. ... One Uber driver commented to us with some exasperation: *"If too many of us become private hire drivers, then there won't be enough jobs for all of us."*[244][emphasis added]

As that perceptive Uber driver notes, the expectation that there will be enough gig jobs to meet worker demand is a pipe dream. It will quickly be discovered there are far fewer jobs to which to move and there is no job security even if you find employment. [245] Workers in that situation are almost entirely under the control of the people running the operation and for most workers, wages will continue to bottom out. Many human workers will become an interchangeable labor commodity for which there are numerous substitutes any time the managers decide to replace them for whatever reason.

We are already experiencing some social impacts of the AI/robotics-driven loss of secure employment on humans and human behavior. A recent study by the Pew Research Center "found that in 203 of the 229 metropolitan areas in the US there was a fall in the share of adults living in middle-income households between 2000 and 2014. [Explaining] *This is a remarkably consistent decline in the middle-income segment of the population, defined as earning somewhere between two-thirds and twice the national median income.*"[246]

An NBC analysis reported that Millennials are having a hard time "growing up or becoming adults," as they termed the effects of the job shortage, and that the number of people aged 18–34 living with their parents rather than marrying and becoming independent has risen dramatically.[247] An amazing fact is that even with all the above issues facing them, many Millennials appear to be living in a dream world if TD Ameritrade's recent survey about how they perceive their futures is accurate:

> 53% of Millennials believe they will one day be million-aires. Despite that, 25% say they'll never marry and 30% plan on never having children. Nearly 1-in-5 report that they still rely on their parents for financial support. They expect to re-tire at 56, though won't start saving for retirement until age 36.[248]

Ironically, if the survey results are reasonably accurate, we are collaborat-ing with AI/robotics to speed along the decline and even the disappearance of our species. The US and Europe, along with Japan, Russia and China, are already experiencing a dramatic lessening of birth rates to the point they are below replacement levels. As discussed in *Contagion's* Case Study on "The Age Curse," this means that if 30 percent of Millennials don't have children and those who do average only one child per family, then our societies will become increasingly more skewed toward the upper ends of the demograph-ic age range. This is already occurring to an uncomfortable extent in the US, UK, EU, Japan and China. As those societies become increasingly age dis-torted, there will be significant effects on work, increasing social subsidies and growing health care costs, as well as a loss of social cohesion. There will also be growing loneliness for many people as the traditional family structure and connections disappear.

CHAPTER ELEVEN
Winners and Losers in the Evolving Job Market

To this point we have primarily set out macro-economic projections of the potential loss of very large numbers of human jobs over the distressingly near future. In this chapter we want to take a step back from those possibilities and look at micro areas of employment that may provide some winners and those that indicate the probability of significant job loss.

Reports indicate that many young people are anticipating facing frequent job movement throughout their lives with as many as 15-20 different jobs rather than a single career with a single employer. This situation has been described as a matrix or lateral work system rather than a vertical or hierarchical career path. This is not without its own challenges.

> For many, it seems the career ladder is dead, but what's replacing it may be more daunting. A career web, or lattice, is a professional path where lateral moves are as important to a worker's end goal as traditional promotions and upwards mobility. ... Many are still figuring out how to deal with the exhausting amount of opportunities that go with navigating this less traditional, flatter career path. Workers are "shifting to a web [or] labyrinth—there's no career ladder you should climb," says Katy Tynan, chief talent development strategist at Coreaxis, a consulting firm that works with multinational companies.[249]

Among the most daunting challenges is the fact that jobs in many areas will be disappearing to the point where even a multifaceted lateral job environment may not be available for many who seek employment. But if frequent lateral movement does become common, whether driven by employees or employers, it will have a significant impact on employee loyalty and quality of performance.[250] It also suggests a large-scale need for diverse skills and jobs retraining programs given the wide range of employment contexts being created and destroyed.

Other than for the most fortunate, well-connected or talented, jobs at any decent level of pay or that represent opportunity and social mobility may not be realistic options for most people. Many will be trapped within a scrabbling rat race of patchwork and multiple part-time jobs seeking to make ends meet. It will become increasingly common for people to move from one short-term position at low pay to another, with inadequate or no retirement or health benefits, and limited or non-existent employment security. But for far too many, the frequent movement will be due to job insecurity or elimination rather than worker choice.

With the AI/robotics shift there will not be anything close to the number or quality of jobs required for all who should be working or want to work. The availability of quality jobs also depends on whether a particular economic *territory* has sufficient job opportunities to allow lateral work movements. As always, there will be territorial hotbeds of diverse economic activity offering work to significant numbers of people. But in many other localities and larger territorial regions employment options will be limited in quality and quantity.

Job-deficient regions will lose out to job-surplus regions. As we discovered more than a century ago in the US during our massive shift from an agricultural to an industrial economy, "you can't keep them down on the farm" if the jobs aren't there. In the same way, the more capable, ambitious and talented job seekers who are able and willing to move will be able to relocate to a limited number of job and cultural centers. This will leave large numbers of lesser skilled and disadvantaged people in economically weakened locations with very limited employment opportunities. But will they stay there, or are American cities en route to mirroring their third world counterparts, with millions living in adjacent shantytowns, drawn by false hope?

Nor does this say anything about underpaid workers' ability to undertake the costs of moving in search of work and re-establishing living quarters. As we are already seeing with the caravans and ongoing streams of migrants undertaking dangerous and arduous journeys from less developed nations to the West's more wealthy economies, it seems we are watching a replay of the US Great Depression and the Dust Bowl crises of the 1930s. Yet, if jobs disappear on the scale predicted, a form of that stark scenario is likely to be replayed internally in the US.

Growth Jobs (Theoretically)

The following job categories have been assumed to have the potential for long-term sustainability. While the categories are taken from an analysis by College Board researchers, we have added brief parenthetical notes about whether the assumptions are consistent with what is occurring in AI/robotics development. As we have already suggested, the problem with such predictions is that AI/robotics designers are making significant and rapid strides even in these areas of employment. So even though the College Board's analysts identify the areas below as safer than many others, the inexorable pace of AI/robotics development may render the work categories empty hopes.

- *Home Health Aide* (These positions are vulnerable to AI/robotics development as shown by the rapid Japanese and Chinese R&D in this area.)
- *Physical Therapist* (Also vulnerable to AI/robotics development, with Singapore's EMMA physical therapist an early example of the progress being made.)
- *Financial Advisor* (AI systems are already replacing large numbers of financial advisors in many instances in the US.)
- *Information Security Analyst* (AI systems are already replacing them.)
- *Software Engineer* (This should be safe for a decade or two but advanced research is being done on AI systems writing and programming software.)
- *Registered Nurse* (Vulnerable to AI/robotics development and in use in some hospital situations.)
- *Truck Driver* (Vulnerable to the rapid development of autonomous truck and car driving systems.)
- *Medical Services Manager*[251] (This is probably safe for two decades or so.)

A 2017 report published in *USA Today* suggests the following as jobs for which there will be a growing demand:

- *Data scientist*
- *DevOps engineer*
- *Data engineer, Analytics manager* [252]

DevOps or Development Operations is one of the most emphasized areas. But understanding exactly what is involved in DevOps and the primary skills required to do the job effectively has not yet been fully pinned down—not surprising in a context of rapid and evolving change.

> The term *DevOps* (for *development operations*) means
> different things to different stakeholders in software proj-
> ects. DevOps is a philosophy—some call it a movement—
> that promotes cooperation between software developers and
> operations managers. ... The term *DevOps* is showing up in
> job titles on career sites that list open positions. But the skill
> set required for roles such as *DevOps manager* and DevOps
> engineer is not yet clearly defined.[253]

Another supposedly emerging area is "big data analytics." Given the
massive and diverse new data feeds businesses are dealing with this appears
to make sense. But even here we see problems in terms of the likely number of jobs that will be created and the high skill levels required to do these
jobs effectively. The concept of Big Data Analytics or Data Science has been
described as "the process of examining large and varied data sets—i.e., big
data—to uncover hidden patterns, unknown correlations, market trends,
customer preferences and other useful information that can help organizations make more-informed business decisions."[254]

The problem with data scientist as a safe employment option is that there
aren't that many people with the innate ability to process massive amounts of
data and discern linkages between the data sets, as well as accurately evaluate
the implications of the interconnections. This means that while Big Data Analytics will be a job for some, there are finite limits on the numbers needed
and available. Beyond that, as AI systems continue to grow in sophistication
and AI applications are developed and adapted, it will be the case that the
AI-driven applications can do a better job at identifying the patterns, relationships and meaning contained in massive amounts of data in ways far beyond human capability. This means that while there could be an interim need
for a limited number of human data scientists, the great bulk of the work will
at some point be done by AI systems.

Another interesting take on the types of jobs that may be hot, and perhaps even robot proof, highlights several important considerations. Developments are moving so rapidly that there is no way a rigid educational
system accustomed to slow change is capable of anticipating, understanding
and then coherently communicating the needed skills and knowledge. This
doesn't mean that traditional educational institutions and processes do not
have some role in some contexts, but that their operational models move too
slowly to keep up with the changes occurring almost daily. At this point one
problem is that there are no real degree level educational programs available
to build up the critical skills, knowledge base and experiences that adequate-

ly prepare people for the new economy—and whatever that might entail. This offers a challenge to our educational leaders.

One consideration is that technical familiarity on a reasonably high level will be an essential requirement for many jobs. This includes the ability to handle, organize and integrate very large data sets, and to work with groups across sectoral boundaries. Being able to do this requires the ability to understand the assumptions, methods, limits and capabilities of diverse areas of disciplinary activity. In an age of ultra-specialization, someone must be able to transcend the "boxes" into which we are enclosed. This requires a unique combination of knowledge and experience.

DATA SCIENTISTS AND "CITIZEN" DATA SCIENTISTS

Another set of developing job categories is represented by what is currently called Data Scientist and Citizen Data Scientist. It does seem that the "Citizen" version bears some degree of resemblance to what was discussed above as Development Operations and what some refer to as Analytics Translator. The categories being developed, primarily by business consulting companies, are intended to try to put a face on an emerging type of specialist/generalist individual who is capable of utilizing data tools and applying them to the specific context in which the person is operating. The individual will also need to be able to explain and shape the information in comprehensible ways for others with whom the individual is working, coordinate the activities of a team, and offer strategic insights and advice as required. All in all this is a very complex and important role within many institutional contexts—public and private.

It is interesting that the term Citizen Data Scientist has come into vogue because the role of such an individual is considerably more diverse and wide-ranging than would typically represent the work of a formal Data Scientist. Gartner, a leading business consulting firm with an emphasis on technology, describes the work of the formally trained data scientist as follows:

> The data scientist role is critical for organizations looking to extract insight from information assets for "big data" initiatives and requires a broad combination of skills that may be fulfilled better as a team, for example: Collaboration and team work is required for working with business stakeholders to understand business issues. Analytical and decision modeling skills are required for discovering relationships within data and detecting patterns.

Data management skills are required to build the relevant dataset used for the analysis.[255]

The Data Science program at the University of Wisconsin adds its perspective on what data scientists do within organizations.

- Identifying the data-analytics problems that offer the greatest opportunities to the organization
- Determining the correct data sets and variables
- Collecting large sets of structured and unstructured data from disparate sources
- Cleaning and validating the data to ensure accuracy, completeness, and uniformity
- Devising and applying models and algorithms to mine the stores of big data
- Analyzing the data to identify patterns and trends
- Interpreting the data to discover solutions and opportunities
- Communicating findings to stakeholders using visualization and other means.[256]

The University of Wisconsin explanation then adds that in their 2013 book, *Doing Data Science*, Cathy O'Neill and Rachel Schutt describe the data scientist's duties.

> More generally, a data scientist is someone who knows how to extract meaning from and interpret data, which requires both tools and methods from statistics and machine learning, as well as being human. She spends a lot of time in the process of collecting, cleaning, and munging data, because data is never clean. This process requires persistence, statistics, and software engineering skills—skills that are also necessary for understanding biases in the data, and for debugging logging output from code.[257]

The number of people educated and trained in the role of data scientist is relatively limited. As indicated above, over the past several years consulting firms have increasingly used the concept of the Citizen Data Scientist as a mechanism for dealing with the shortage of formally trained data scientists. The roles played by a Citizen Data Scientist [CDS] are described as follows.

The expert skills of traditional data scientists to address these challenges are often expensive and difficult to come by. Citizen data scientists can be an effective way to mitigate this current skills gap. Technology is a key enabler of the rise of the citizen data scientist now. Technology has gotten easier for non-specialists to use. Analytic and BI tools are extending their reach to incorporate easier accessibility to both data and analytics.[258]

Mathew Attwell writes:

> Back in 2016, [the consulting firm] Gartner coined the term "Citizen Data Scientist" (CDS), meaning a person "who creates or generates models that use advanced diagnostic analytics or predictive and prescriptive capabilities, but whose primary job function is outside the field of statistics and analytics." ... *[T]he recent announcement from Google on the Beta-Stage release of its AutoML platform ... have reignited discussion around the future shift towards a CDS environment. Essentially, AutoML provides an optimal machine learning algorithm for particular business uses. All the user needs to provide is labeled data; data with an outcome that the algorithm can use to learn. No data science knowledge is required.*[259] [emphasis added]

If no actual data science knowledge is required by the Citizen Data Scientist, then it means that the individual, and organization, is entirely reliant on an AI Machine Learning program, with the assumption that the various applications and the data being dealt with by those programs is accurate, fully relevant and complete. These are dangerous assumptions and dependencies.

UNIQUELY HUMAN JOBS

There are numerous areas in which AI/robotics will play a helpful role but not completely replace human workers. This is largely because the work activity is beneath the optimum profit making scale on which large capital-intensive systems operate. It doesn't mean that AI/robotics couldn't do the smaller scale jobs but that the larger businesses won't bother because the profit margins and market size are too small. Even here the situation depends on the extent to which mass production of diverse types of robotic systems reduces the cost of the units to the point of large-scale consumer and commercial adoption.

Perhaps even a majority of the jobs that will survive the AI/robotics shift will be neither capital intensive nor involve large volume markets. Quite a few of the survivors will be businesses operated on a small or individual scale by tradesmen, mechanics, those engaged in the fixing of small items, retrofitting houses and similar work such as electrical, plumbing, heating, auto repair, landscaping, housekeeping and so forth. Others will be niches in which people are not prepared to accept anything other than human actors because the human touch is an important part of the transaction such as styling, fashion design, hair cutting, interior design, live theater, sports and coaching, artistic production and similar human work.[260]

Even in small-scale repair businesses there could be problems with product repair costs versus product replacement. Products like toasters, printers and so forth are already being produced and sold so cheaply that there is often no point in repair when the repair service charges $35 or more per hour and estimates run two or three times that amount. This will only become worse as AI/robotics systems produce the goods ever more cheaply due to much lower costs of production allowed by the nearly complete elimination of human labor. There will always be a small black market or flea market operating in goods, but not that many people are going to be able to make a living from such activities.

Artisanal products are often thought of as unique and appealing to a limited market. Maybe a robot could do it, but if a buyer wants a bragging provenance the uniqueness would give a human artist or craftsperson an advantage. Producers of crafts can to some degree be in the category of unique and to some extent are organic farmers. One challenge is that unless the product sales cost is high, the artisan ends up needing to create multiple productions in order to make a living, and in doing so devalues the uniqueness of the product. This again suggests that a limited number of artisans lucky or talented enough to be at the top of the food chain will do well but many will struggle to get by. "Retro" movements will survive that try to return us to the nostalgic times of our youth. If all one wants is a "motel" group painting done in moments by an automated system, then it is chosen for its decorative color rather than its creative artistic character. The solution in such situations may be the provision of a financial base in the form of the UBI payments discussed in *Contagion's* "Solutions" chapters, or a supplemental job guarantee program that at least allows people to know they and their families won't starve, will have somewhere to live, and have health care to rely on.

At some not so distant point when robots have become commonplace parts of everyday activity, prestige and status will cause people to hire humans for certain services and outputs due to snob appeal. Even here, given the work being done on robots and AI, it won't be that an AI/robotic system probably couldn't do such work effectively in a few years from now. But at least for several decades the older and less tech-savvy or comfortable among us will not accept the AI/robotic entities in certain roles. That is changing quickly as a new generation replaces older citizens. Millennials and those now being born are far more comfortable and accepting of electronic AI/robotic entities and how they can function in ways that the over-fifty generation never could.

Government, Security, Military and Science Jobs

Government jobs may be the most likely area of large-scale job survival. Civil Service employment, some teaching positions with retention guarantees, and positions backed by strong unions will survive. Government jobs have an edge unrelated to economic productivity because politicians want employees who have a stake in supporting them. While many private sector jobs are at risk, "City Hall" will remain largely intact even if much of the work could be done considerably better by robots, and even if robots could be programmed to actually be considerate in dealing with responses by the public that cannot be anticipated.

At the state and federal levels the number of employees could actually increase because there will be a need for people to staff the positions set up to take care of all the millions of people who will be out of work after being replaced by AI/robotics. Without revenue enhancement strategies through tax reform, however, there is a limit to how many people can be absorbed into the governmental employment markets, at least at current levels of pay. Other areas that are likely to be keepers include security, fire, police, military, some categories and levels of teaching.

Jobs that involve some aspects of science, technology, engineering and mathematics (STEM) will be in some demand. It is, however, unlikely that Americans will dominate those jobs against foreign competitors who work harder and smarter and without a sense that the world owes them a living. As the most profitable markets for such talents move into Asian nations, the US may find itself on the outside looking in. This shift of STEM jobs to non-US locations is expanding because US universities are transferring their technical and scientific knowledge to foreign students. These students gain fantastic intellectual property from US universities and then take those skills and knowledge back to their home countries, such as China. This educational technology transfer has enabled nations to advance their economies, competitiveness, and military technologies by leapfrogging years of development.

COUNSELORS

One area of employment in which we are likely to see an increase in human workers, at least for a time, is psychotherapy and counseling due to the extent to which people suffer enduring malaise from not having the structure of a traditional work environment or regular human interaction in workplaces.

We are already seeing an entirely new set of addiction and dependency conditions experienced by people as they interact with their own AI systems and create new sets of dependencies, human/robotic relationships, and addictions.[261] This growing problem of the addictive and shaping effects of digital technology is discussed at greater length in Chapter Twenty, "The Mutating Effects of AI Technology," but is worth mentioning at this point. As more and more people do not work they can be expected to fall into a state of lassitude or fill up their time with addictive activities. Ironically, this will increase demands and costs associated with healthcare due to serious forms of mental illness and extreme dependency concerns. A recent BBC analysis detailed an extensive rehabilitation program for young people who were addicted to video games as if they were cocaine. In "Inside the kids-only rehab that treats video games like cocaine"[262] Ben Bryant writes of his experience.

> I had just arrived at Yes We Can: a mental health facility located down a long, tree-lined boulevard in a quiet corner in the south of the Netherlands. ... There is a reason video games are on my mind. This clinic is for those aged 13-25 only, and people come from all over the world for its specialist treatment in mental health problems including screen addiction and other behavioural issues that the medical community is unsure of how to classify—let alone treat. Many of the people attending say they are addicted to their smartphones, social media, or video games.
>
> For the first time this year, the World Health Organisation has accepted that they might be right. In June, it formally accepted Gaming Disorder into the International Classification of Diseases (ICD)—a kind of encyclopedia of stuff that can go wrong in humans. *The treatment programme at this clinic arguably goes further: it places video games on an equal footing with drugs, alcohol, and gambling, demanding that those who complete its 10-week programme abstain from all of them for the rest of their lives.*[263] [emphasis added]

There is another serious downside to the development and use of the AI technology that enables social media—loss of personal contact and a rise in loneliness and depression. One analysis of this phenomenon supports the view of increased loneliness with social media being either cause or effect. The psychologists explain: "*Social media sites like Twitter, Facebook and Pinterest are causing more people to feel alone, according to US psychologists. A report suggests that more than two hours of social media use a day doubled the chances of a person experiencing social isolation. It [also] claims exposure to idealised representations of other people's lives may cause feelings of envy*"[264] [emphasis added].

Ultimately many people will become tied to Virtual Reality (VR) games and Augmented Reality [AR] experiences that create dependencies and even bond humans to VR "worlds," and to people, other "beings" and relationships that exist only in their minds, hearts and in an AI "cloud."[265] The popularity of *Survivor*-type reality shows offers a harbinger of what is coming.

Given what we have already experienced with telephonic and game addictions, it is inevitable that as people have more free time and as the VR programs improve the quality of the "worlds" they create, millions will be sucked into "virtual" worlds that for them are more real, rewarding and dazzling than that harsh dimension we inhabit "outside the goggles" or "beyond the implants." We will have to double or triple the counseling profession to try to help people deal with this addiction. Whether we will be able to "cure" such addictions when the conditions that cause them will remain is unknown because we may not have all that many productive alternatives to offer.

China has taken the lead in limiting the access of that nation's children to online games, attempting to inhibit the growth of Internet addictions that have become all too common. One report notes that Tencent, the primary Chinese Internet provider, has set clear restrictions. The report indicates: "*Tencent, which ranks first in the world for gaming revenue, said in a statement that King of Glory was 'supposed to bring joy ... but excessive gaming brings joy to neither players nor their parents*'"[266] [emphasis added]. Taken together, these issues of addiction, increasing inability to create interpersonal relationships, and depression and loneliness represent a gold mine for the counseling professions, albeit one we would be wise to minimize if we are to have any hope for productive and collaborative societies.

JOBS AT GREATEST RISK

A study by the UK think tank *Reform* offers a chilling picture of overall job loss. It projects 75-80 percent job loss in: Skilled Trades; Caring, Leisure and Other Service Occupations; Elementary Occupations; and Sales and Customer Service Occupations. It doesn't stop there. Employment opportunities are expected to decline more than 60 percent in Process, Plant and Machine Operatives, and job losses are estimated at above 40 percent in Administrative and Secretarial Occupations.[267]

Scan the list below and think about what large-scale loss of jobs will mean if 40–80 percent of the jobs in each of those categories disappear. In each area of work activity AI/robotics replacements are projected to make human employees obsolete. Some job areas are already under enormous pressure such as finance and retail activity. Others are going to be eroded over time as the AI/robotic systems continue to evolve. The list doesn't even bother mentioning manufacturing jobs, which will continue to disappear unless deliberate strategies, incentives and disincentives are put into place in the very near future to protect and preserve human manufacturing jobs.

- Finance and investment, accounting, banking.
- Operations managers, retail sales [ongoing and very significant elimination of jobs and stores], clerical and filing, record keeping and retrieval, technical professionals with formulaic tasks (CPAs, Tax Preparers, Workers Compensation, etc.), measurement workers (meter readers, pool maintenance, etc.).
- Delivery services by humans. Autonomous surface and air delivery systems are developing rapidly. As this occurs, paid driving jobs, including buses, delivery trucks and vans, semi-trucks, taxis, limos, school bus drivers, Uber and Lyft drivers, are predicted to rapidly dwindle. Ironically, many taxi drivers are finding their incomes eroding and jobs disappearing due to Uber and Lyft. This is in fact a transitional phase because Uber and Lyft drivers will find themselves replaced by autonomous cars as is already occurring in some cities. Presumably some human drivers operating in rural and thinly populated areas will still be able to provide transport services.
- Insurance advice and agents, real estate agents, travel advisers and agents.
- Fast food cooks, chefs, food preparation, waiters, food services.
- Lawyers, paralegals, law-supported ancillary jobs, lower-level judges, magistrates and law clerks, court reporters, law professors.

- Surgery and other medical jobs, medical diagnosis, nursing, assisted living aides, home health aides, counselors and emotional therapists, physical therapists, nursing home workers and home care are subject to replacement in large numbers.
- Public bureaucracy (federal, state and local), private corporate middle management bureaucracy.
- Security services, police, military, surveillance.
- Bricklaying, home construction, building construction, road building and repair.
- Journalism, writing, newspapers, various media.
- Teachers and educators, university and elementary and high school staff, librarians.
- Lawn mowing, grape picking for wine making, apple and other large scale harvesting, cowboys and herding, manufacturing (large system), manufacturing (small system), car production, mechanical repair and maintenance, warehousing.
- Prostitution, because as strange as the phenomenon may seem, the rapid rise in robotic sex performers for "home" use and in brothels and "sex spas" being driven by very significant investment by the pornography industry will eliminate many human prostitutes.
- Retail services and cashiers.

John Pugliano, author of *The Robots are Coming: A Human's Survival Guide to Profiting in the Age of Automation* (2017), suggests the following types of work are under threat. Most of the work niches will not be completely wiped out, but the number of available positions reduced dramatically. A partial list of Pugliano's projections includes the following occupations.

- Mortgage brokers
- Bookkeepers
- Lawyers
- Broadcasters
- Middle Managers
- IT workers (onsite)
- Financial planners
- Postal workers
- Data entry clerks
- Telephone switchboard operators
- Farmers and ranchers
- Fast food cooks
- Newspaper reporters
- Primary care physicians

THOMAS FREY'S PREDICTIONS OF JOB LOSS AND GAIN

Futurist Thomas Frey lists five areas where he predicts jobs will be lost in significant numbers, followed by the types of work he thinks will be newly created. In a 2016 presentation modifying his 2012 prediction that two billion job losses were possible globally, Frey focused on the changes he predicted would occur by 2030 in: the Power Industry; Automobile Transportation; Education; 3D Printers; and Bots.[268]

Without going into Frey's analysis we simply offer his five categories for review. He does indicate that while hundreds of millions of jobs will be lost in these specific industries and categories, there will also be effects on other areas of employment that will need to be taken into account. The abridged analysis below offers Frey's own language with his brief explanations deleted.

1. POWER INDUSTRY

Jobs Going Away

- Power generation plants will begin to close down.
- Coal plants will begin to close down.
- Many railroad and transportation workers will no longer be needed.
- Even wind farms, natural gas, and bio-fuel generators will begin to close down.
- Ethanol plants will be phased out or repurposed.
- Utility company engineers, gone.
- Line repairmen, gone.

New Jobs Created

- Manufacturing power generation units the size of ac units will go into full production.
- Installation crews will begin to work around the clock.
- The entire national grid will need to be taken down (a 20 year project). Much of it will be recycled and the recycling process alone will employ many thousands of people.
- Micro-grid operations will open in every community requiring a new breed of engineers, managers, and regulators.
- Many more.

2. AUTOMOBILE TRANSPORTATION—GOING DRIVERLESS

Jobs Going Away

- Taxi and limo drivers, gone.
- Bus drivers, gone.
- Truck drivers, gone.
- Gas stations, parking lots, traffic cops, traffic courts, gone.
- Fewer doctors and nurses will be needed to treat injuries.
- Pizza (and other food) delivery drivers, gone.
- Mail delivery drivers, gone.
- FedEx and UPS delivery jobs, gone.
- As people shift from owning their own vehicles to a transportation-on-demand system, the total number of vehicles manufactured will also begin to decline.

New Jobs Created

- Delivery dispatchers
- Traffic monitoring systems, although automated, will require a management team.
- Automated traffic designers, architects, and engineers
- Driverless "ride experience" people.
- Driverless operating system engineers.
- Emergency crews for when things go wrong.

3. EDUCATION

Jobs Going Away

- Teachers.
- Trainers.
- Professors.

New Jobs Created

- Coaches.
- Course designers.
- Learning camps.

4. 3D PRINTERS

Jobs Going Away

- If we can print our own clothes and they fit perfectly, clothing man-ufacturers and clothing retailers will quickly go away.
- Similarly, if we can print our own shoes, shoe manufacturers and shoe retailers will cease to be relevant.
- If we can print construction material, the lumber, rock, drywall, shingle, concrete, and various other construction industries will go away.

New Jobs Created

- 3D printer design, engineering, and manufacturing.
- 3D printer repairmen will be in *big* demand.
- Product designers, stylists, and engineers for 3D printers.
- 3D printer "ink" sellers.

5. BOTS

Jobs Going Away

- Fishing bots will replace fishermen.
- Mining bots will replace miners.
- Ag bots will replace farmers.
- Inspection bots will replace human inspectors.
- Warrior drones will replace soldiers.
- Robots can pick up building material coming out of the 3D printer and begin building a house with it.

New Jobs Created

- Robot designers, engineers, repairmen.
- Robot dispatchers.
- Robot therapists.
- Robot trainers.
- Robot fashion designers.[269]

CHAPTER TWELVE
Most Needed Skills Going Forward

One challenge for survival in the emerging new economy is identifying and acquiring the new skills and strategic approaches needed for success in the changing workplace. Even if the various predictions about hot jobs listed in the preceding chapter are correct, there will still not be enough human jobs to go around, whatever those new skills and work strategies might be. Meaningful and fulfilling jobs will in many instances give way to subsistence and survival work. This means many people will spend their days, or lives, on jobs they hate, or on the dole, as is likely to happen if an extensive UBI program is adopted.

CREATIVITY, DIGITAL LITERACY, CRITICAL THINKING, DIVERSE KNOWLEDGE AND PATTERN RECOGNITION

What these skills of creativity, etc., actually mean and how it can be determined if an individual possesses them is challenging. Even beyond the issues of definition, recognition and substance, we need to ask the extent to which such diverse skill sets are teachable versus innate qualities one either has or does not. Creativity, for example, is not something possessed by all or even by many. There is also an important issue of creativity in specific dimensions of activity. Some people are artistically creative while others are financially creative along with numerous other variants of the supposed skill such as mathematical skill or pattern recognition. It is necessary, therefore, to ask, creativity for what category of activity.

Digital literacy, another frequent component of skill lists, is a technical skill that many Millennials and even ten year olds possess to some degree. So while digital literacy will be a necessary qualification, it is a dime-a-dozen factor unless linked to something such as high creativity in a relevant area of work activity and advanced innovative ability. What many young workers will face with the growing use of AI systems that multiply individuals' productivity, is that we will increasingly see the employment equivalent of "No Vacancy" signs on the most desirable employment niches. Many people will want the jobs but only a limited number will be needed.

Critical thinking, creativity, pattern recognition and the ability to possess deep knowledge in multiple fields and to effectively cross-reference, interpret and apply that knowledge can be developed at the margins. That capability can be improved through education, but at the higher ends of such skills only a few excel while most others basically "try hard."

A surprisingly small portion of the American population possesses the capacity or willingness to do high-level critical thinking. Our educational systems have turned out to be substandard in the teachable skills of critique, analysis and synthesis.[270] Unfortunately, our educational systems appear in too many instances to be rejecting the teaching of realistic critical thinking. There are a variety of reasons. After all, hard thinking "hurts." Serious and deep analysis based on the manipulation and dissection of facts and evidence "prefers" those who excel at such activities over those who do not.

Resistance to critical thinking is happening even at many of our supposedly "best" universities such as Harvard, Yale and Berkeley, ones that should be leaders in such work. Though a significant part of the dismayingly weak approach our universities and educational institutions in general seem to have taken in abandoning their traditionally professed mission may be the financial self interests and political bias of academics and administrators, universities are still run by humans and that means human self interest plays a central role in their behavior regardless of professed ideals.

Equally problematic is that the corruption of the university and other core elements of the educational ideal may also be a product of the fact that there has been an erosion of intellectual values and the increasing elimination of required study of areas of knowledge that valued and introduced the principles on which our systems were grounded. This erosion began decades ago and we only have to go back to sources such as Allan Bloom's 1987 work *The Closing of the American Mind: How Higher Education Has Failed Democracy and Impoverished the Souls of Today's Students* to see how fast and deep the failure has been. This suggests that many among an entire generation of teachers and educational administrators do not understand or appreciate the centuries old Enlightenment goal of providing students with a full range of knowledge and perspectives and enabling them to analyze and critique that information, warts and all.

It is reasonable to suggest, therefore, that the situation within our educational institutions is being driven by a combination of ignorance and, as Upton Sinclair once suggested, by the fact that: "It is difficult to get a man to understand something when his salary depends upon his not understanding it."[271] Too many of our universities do not want controversy, do not want indignant and aggressive students and activists marching on their "ivied halls," and want to ensure a continuing flow of students to finance the university "empire" in the way to which they have become accustomed. Offending the delicate sensitivities of applicants or imposing rigorous standards they are not prepared to meet could threaten the welfare of the institution—as well as the ability to pay the salaries of senior administrators and faculty members.

We mentioned Allan Bloom's 1987 work above. The fact is that a variety of people have sought to warn about what is occurring in our universities only to find they are voices in the wilderness. We recommend William Egginton's *The Splintering of the American Mind: Identity Politics, Inequality, and Community on Today's College Campuses.*[272] Egginton offers a dismaying picture of just how far the process of ignorance and thought control has gone.

The problem is that in the ultra sensitive and politically correct world we have created, critical questioning and critique can get you into trouble. This is because there are an amazing number of taboo subjects whose advocates tolerate no alternative perspectives. Even asking questions about their positions and conclusions are seen as attacks because honest and critical interaction offends those who take their conception of reality as an immutable given. Since their positions are based on belief and faith rather than fact and logic many are true believers who are unable to tolerate what they consider to be attacks on their dogma. The result is they punish or ostracize anyone they feel is challenging or contradicting the tenets of their belief system, comfort zone or safe space. Surprisingly, quite a few of such people populate the halls of academia.

As to the employment advantages claimed to be represented by presentation and communication skills—oral, written and visual—far more people have those skills to a reasonable level than the number who possess the skill of deep critique. But from a job perspective there will only be a need for a limited number who can communicate clearly and persuasively. That number represents only a small percentage of the overall employment market.

"SOFT" SKILLS

In the context of what are typically thought of as the "soft skills"—such as interpersonal skills, creativity, artistic ability, aesthetic perception, and quality of judgment—these are frequently being cited as unique to humans. This assumption fails to recognize that AI systems appear well on the way to developing "human" abilities that will allow them to supplant humans in many work contexts requiring or emphasizing such skills. Advanced AI/robotics systems already being developed will progressively be able to replace many jobs we mistakenly assume are immune to what we still think of as "only a machine."

An example of the possibility of AI/robotics systems achieving human capabilities is suggested in a report that Facebook's researchers have been developing "bots" that can negotiate and compromise. This offers insight into how far and fast AI systems might be able to go in their evolution.[273] The irony—whether tragic or amusing—in the Facebook situation involving negotiation is that the AI bots apparently learned how to lie as well as negotiate. When we talk about mimicking human intelligence, that is absolutely "too human."[274] So is the claim that AI systems are showing biases that were somehow programmed into them at the beginning.[275]

THE SKILLS OF STRATEGIC AWARENESS AND ACTION

Strategy improves our ability to evaluate, diagnose, estimate risks and costs and resolve problems and opportunities. Such holistic situational thinking requires self-awareness, the ability to rapidly perceive and interpret events and to make immediate choices of action under pressure. It also involves the ability to implement chosen paths of action, and to make adjustments because reality is a process, not a fixed event. Acquiring the range of skills, strategic awareness, and judgment requires a combination of experience, intuition, ability, repetition and discipline.

The most effective strategists develop the ability to perceive the future as both continuation and contradiction of the present. In doing this, an effective strategist can examine a situation and see its variables, contingencies, possibilities, resolvable and unresolvable dilemmas and alternatives, patterns of timing and leverage, rhythms, human factors, and outcome probabilities.

The strategist then identifies the best paths and alternative approaches to be followed, visualizes intersecting patterns and structures that influence the paths, perceives the probable behavior of decision-makers, the quality of opponents and allies, and a host of other factors. The strategist also understands the limits of his or her knowledge and capability, and knows that risk and the ability to manage risk is always involved in any strategic context.

Strategy is goal-oriented. This requires the ability to identify achievable goals, assess available resources, conceive plans of effective implementation and then implement those plans of action even while being prepared to adapt depending on the reality of what is occurring. Strategy is fluid and adaptive, not rigid or fixed. The adaptive element is essential because reality is not a fixed event but a "living tapestry." This is what we are dealing with in the context of understanding the implications of AI/robotics.

An advantage humans will have over AI systems, at least for a time, is that of recognizing nuance, deeper meaning, categorical distinctions, and determining significance when examining and assessing the meaning, utility and weight of information. Specific technical knowledge and skill are vital elements of success. But they are only part of what is needed to excel.

In applying strategic analysis to the emergence of Artificial Intelligence and robotics, and considering its impact on employment and society, generalists, thinkers, gurus, and system managers and controllers will be among needed job categories. Managers and leaders who possess such skills will always be in limited numbers and there will always be only a small number of positions available. As the rest of the system "dumbs down" or "roboticizes," effective strategists will become increasingly vital.

The integrative approach of the strategist includes the skills of strategic analysis and awareness and the ability to understand how things work across a wide spectrum rather than only within a narrow closed system. AI systems will increasingly develop the capability to access massive volumes of information across a stunning range of data, and be able to integrate that information while applying it to specific areas of need. Humans must therefore figure out the special skills and qualities they possess that can't be matched by AI systems.

KNOWING ENOUGH TO FORMULATE ACCURATE QUESTIONS AND SEARCH HYPOTHESES IS A CRITICAL SKILL

Knowledge is an essential part of issue recognition and inquiry. Unless we have the ability to formulate the questions that need to be answered we can't test our theories, assumptions and hypotheses or make critical distinctions related to the information acquired. Excessive dependence on external information sources has a negative effect. We need to have our own conceptual structures in place, with significant depth and scope of knowledge, in order to formulate the key questions that search systems can help answer.

There is a downside to becoming solely or overly dependent on systems such as Google. Steven Poole writes about a report concerning how Google's amazing information system may be having the effect of causing people to actually know less. This is because they no longer have to "download" information into their own brains but can use Google as an instantaneous data source whenever they need to know anything.[276] This has consequences. Poole asks: "Does anyone know anything any more? The ease with which one can look up facts on a phone at any time is one of the wonders of the modern age. But are we becoming too reliant on it?" Poole speculates on the possible effects in asking "Does Google make us stupid?"

> The answer, of course, depends among other things on what "stupid" means. If we all come to increase our "cognitive offloading" onto the internet, that may simply portend a cultural shift in the ways in which we value mental abilities. Since recall of facts is now so easy and quick via the internet, we may just become less impressed by factual knowledge, and more impressed by understanding and creativity. *The problem with this, as the researchers themselves note, is that the ability to draw creative connections between facts may depend on having internalized them already as knowledge, so that they are instantly available to the reasoning mind.* [277]

A new study indicates there might be a snowball effect to such reliance, warning: "The more we depend on Google for information recall ... the more we will do so in the future."[278] Poole's concern about the effects of a Google fixation is that: "to look something up, one needs to know already what it is one wants to know."[279]

When we possess a substantive and rich conceptual base we have the ability to evaluate, critique, formulate plans and make choices. Part of the process is that we also need to understand how "things," informational "bits" and clusters of specialized data interact with other informational points as part of a system, not only as isolated and disconnected "factoids." Without such integrative conceptual structures we may have mastered disconnected bits of trivia but lack the ability to understand and integrate reality and to act effectively across the relevant spectrums in which reality must be understood, answers sought, and actions taken.

This means that critical thinking and the ability to manipulate a diverse store of knowledge and experience are increasingly vital in the new economy and society. The problem is that effective critique requires that you actually know something as opposed to simply taking others' statements at face value or automatically accepting "Google data" as true. As individuals, if we don't create a sophisticated, complex, honest and comprehensive conceptual structure we become overly dependent on external frameworks. These external frameworks, analyses and sets of alleged knowledge may themselves be flawed, whether they are the opinions of human experts or products of systems such as Google or Facebook that contain their own biases, agendas and limits.

Don't get us wrong. We love having instantaneous access to the world's information base. We no longer have to go through the snail-like pace of traveling to libraries, going through a host of indices, seeing if the particular place had the material, and gaining a context of the best material in a field. We both have always been intrigued with the possibility of understanding a wide range of disciplines and knowledge sources. Internet research systems have multiplied the speed and scope of what can now be accessed by a factor of fifteen or twenty times. Fuller understanding is gained by integrating the best insights and experiences of others into our own intellectual and perceptual systems through which we understand the world we inhabit.

Virtually everything is at our fingertips and—assuming we are not inhibited by our biases—we can read, integrate, compare material from diverse and conflicting databases, voraciously consume data, learn and impose structure, interpret patterns, predict the implications of trends and project consequences based on our interpretations of that data. It is as if our brains have had a "portable drive" added to it that extends capacity and the ability to access and integrate knowledge. Such technology alters us, both for good and bad.[280] It is good because it provides this fantastic capability and access to information. It is bad because it is addictive and obsessive, and because it alters the nature of contact between people and accelerates the pace of life.[281]

The problem we are highlighting is that our generation was fortunate enough to have developed independent knowledge and conceptual structures in what will come to be thought of as "The Pre-AI Age." This is a critical difference between people educated and trained several decades ago and those who have known only AI. Many who have dealt with millennials and the even newer generation, share the conclusion that they don't actually know anything. They aren't grounded in knowledge and seem to flit around the surface of society. Of course "one size does not fit all" and there are numerous insightful, educated and highly talented members of the younger generations. But indications are that the proportion of poorly educated, unskilled, lazy, and "the world owes me a living" elements is growing significantly to the point they are further undermining the challenges we are facing.

CHAPTER THIRTEEN
The Age Curse as the New Population Bomb

There are several types of "population bombs." As forewarned by Paul and Anne Ehrlich decades ago in their classic *The Population Bomb*, one danger involves large-scale increases in the absolute number of people beyond the ability of a system to provide needed social goods or to do so without significant damage. This unquestionably still describes the situation experienced by several billion people in the developing economies.[282] But there is another kind of population bomb, one developed countries are now facing. Ironically, the shift toward societies with large numbers of people in the latter part of their lives has been in part due to the population control efforts generated by fear of the Ehrlich's bomb.

The Age Curse is not just a problem in the West. Japan and China are undergoing an aging process in which the proportion of older workers and retirees grows while birthrates decline. The International Monetary Fund (IMF) concludes this is likely to result in lower economic productivity that makes it difficult if not impossible for nations to cope with high unemployment, promised social payments and massive debt.[283]

In "*An Aging Europe in Decline*" Arthur Brooks writes:

> As important as good economic policies are, they will not fix Europe's core problems, which are demographic, not economic. This was the point made in a speech to the European Parliament in November by none other than Pope Francis. As the pontiff put it, "*In many quarters we encounter a general impression of weariness and aging, of a Europe which is now a 'grandmother,' no longer fertile and vibrant.*[284]

He asks a poignant and powerful question that brings the specter of a lonely future into view: "*Imagine a world where many people have no sisters, brothers, cousins, aunts or uncles.* That's where Europe is heading in the coming decades."[285] Brooks adds: "According to the United States Census Bureau's International Database, nearly one in five Western Europeans was 65 years old or older in 2014. This is hard enough to endure, given the countries' early retirement ages and pay-as-you-go pension systems. But by 2030, this will have risen to one in four."[286]

This alters the role of the nuclear family in our culture. Smaller and smaller families mean children without brothers and sisters, aunts and uncles, cousins and the like who have traditionally served as nurturers and sources of support and strength. Furthermore, as Brooks elaborates:

> According to the Organization for Economic Cooperation and Development, the last time the countries of the European Union were reproducing at replacement levels (that is, slightly more than two children per woman) was the mid-1970s. In 2014, the average number of children per woman was about 1.6.[287]

Elon Musk voiced his amazement that no one seemed concerned about the fact we were facing what might be called an implosion of the so-called Population Bomb, experiencing instead what might be viewed as dramatically aging societies. He tweeted: *"The world's population is accelerating towards collapse, but few seem to notice or care."* Musk pointed to an article in *New Scientist* magazine titled, "The world in 2076: The population bomb has imploded."[288] Rather than a meltdown where the Earth's population outstrips the planet's ability to feed everyone, it suggested we could be headed toward a more subtle but equally disastrous outcome where our population simply does not replace itself fast enough.

PEAK CHILD, PEAK HUMAN AND THE AGE CURSE

The demographic shift in Western societies to a top-heavy age demographic is stark and helps explain the reasons governmental budgets are under increasing pressure. It also sets up a confrontation between the old and the young as each struggles to obtain what they will consider a "fair" share of available employment and resources.[289] Speaking doubtless for the developed world, the late Hans Rosling, a professor at the Karolinska Institute in Stockholm declared:

> *The world has hit peak child ... Germany and Italy ...*
> *could see their populations halve within the next 60 years.*[290]

The phenomenon of dramatically aging societies with birth rates below replacement levels makes it clear why some nations are aggressively developing robotic work systems. It explains what might motivate migration from poorer nations toward Western economies as migrants hope to discover a need for their human labor. While the need for low-level manual and repetitive labor is in the process of being reduced dramatically due to the continuing shift to AI/robotics workforces, there arises the hope of employment in service industries, and elder-care.

But as we have suggested in *Contagion*, AI/robotics development—particularly in Japan and China—appears to be making significant progress in the design of robots of various kinds to fill the kinds of functions considered core to service activities and elder care. Japan particularly is considerably ahead in this endeavor and could well be on the way to an economic "coup" in introducing and selling such systems in Western nations experiencing dramatically aging demographic shifts.

THE "AGE CURSE": PENSIONS, POVERTY AND PATHOS

- Social Security beneficiaries top 62,000,000 for the first time.[291]

- Retirement crisis: 37% of Gen X say they won't be able to afford to retire.[292]

- 42% of Americans are at risk of retiring broke: Nearly half of Americans have less than $10,000 stashed away for retirement, according to a report by Go-BankingRates. For them, a serious lack of planning coupled with a longer life expectancy has destroyed any retirement dreams.[293]

- Pension Funds Still Making Promises They Probably Can't Keep.[294]

- 95,745,000: Record Number Not in Labor Force as Boomers Retire.[295]

- Finland's basic income trial falls flat.[296]

- Drugs, alcohol and suicides contribute to alarming drop in U.S. life expectancy.[297]

- The World Isn't Prepared for Retirement: It's not just America. New data show people all over the globe don't understand basic concepts of investment and inflation.[298]

- Some Public Pensions Funds Could Run Dry in Downturn. Many pension funds for public workers already owe far more in retirement benefits than they have in the bank, and the problem will only grow worse if the economy slows down…. The New Jersey and Kentucky public pension funds are in such perilous shape that they risk running dry. Even after eight years of economic recovery — eight straight years of stock market gains — the public pension plans are more vulnerable than they've ever been to the next recession[299] researcher Greg Mennis said.

The AI/robotics "cure" for population loss by replacing human workers with automated systems is impacting the service sectors as well, outstripping the need to supplement the available human workforce. This is likely to mean that unskilled and uneducated migrants are not needed in large numbers in the developed nations they are seeking to enter in the hope of finding work or other financial benefits.

THE HIGH COSTS OF SUPPORT SUBSIDIES FOR THE ELDERLY

The Age Curse means we are experiencing massive increases in the proportion of older citizens. This growing population sector will require and demand very large investments of time and resources while offering limited productive inputs because they are past their period of economic productivity. Many others will be out of work by personal choice or because they have been pushed out by employers, shrinking markets, outsourcing or technological change.

The demographics of aging societies generate a triple whammy. Fewer people are available to work with a smaller number of people paying taxes. This occurs at the same time that health care improvements have prolonged our lives to the point we need pension support and other subsidies for longer periods. This situation is worsened because the longer living population has more age-related health problems that further balloon support costs.

The blessing of continuing improvements in medical technologies that are keeping us alive well beyond historical expectations is imposing significant retirement and medical expense obligations on governments. This creates a financial dilemma for the US, Western Europe and Japan. These are all countries that have more and more people in need of subsidized assistance for longer periods of time due to the combination of aging and job loss and possess a strong moral commitment to using health care resources in often herculean efforts to extend the lives of their older citizens.

While Russia, China and India also face significant population and aging problems, one *advantage* for those nations is that they have not built the commitments and society-wide structures that incorporate all the costs required to support their populations to the extent found in the West and in Japan. Russia, China and India can allow people to die from health problems, sub-standard living conditions and aging without extensive, heroic and incredibly expensive efforts to prolong their lives to a greater extent than Western nations because they never ratcheted their social systems and promises to the point of heightened expectations on the part of most of their people. The situation is dramatically different in the US, EU and Japan because prolonging life far beyond humans' natural span—while not applied equally to all classes of the population—has nonetheless become a central focus of those societies.

The dilemma those latter three political systems face is not only one of the extremely high costs of health care. The expanding segments of those "developed" societies that need high levels of support because of age is accompanied by a growing proportion of society that needs support on the other end of the spectrum due to job loss and other afflictions. In sum: we will have an increasingly large segment of society that does not work, whether by choice, necessity, or lack of options. Those unfortunate enough to be in that category will need and demand support and there are not enough resources to meet all their needs.

The US has a critical and expensive longer-term health care crisis. A newly released study reveals that Americans over 65 have more serious health problems than their counterparts in ten of the most economically advanced nations. This poses an extremely serious set of issues as to policies regarding healthcare, the extent to which we should create incentives to alter the poor lifestyle choices that cause the bulk of the health problems, and the ability to handle the future costs of health care in a system that is already by far the world's most expensive.[300]

Recent reports indicate that for the first time in decades the US lifespan is no longer lengthening and may even show indications of a slight decline.[301] As surprising as this appears given our historical trends, it is less so when we think about the significant spread in serious clinical depression, the overuse of legal and illegal drugs, the heroin and opioid epidemics,, bad diets, poor lifestyle choices, obesity and so forth that the US is experiencing. These developments are not cost free. The US is now being forced to commit massive amounts of money to treatment in addition to the medical and support costs for age-related end-of-life situations. We are not saying the costs should not be borne but asking where the money comes from.

INDIVIDUALS, ASSISTANCE PROGRAMS, AND GOVERNMENTS RUNNING OUT OF MONEY

Even now, considerably more than half of US families have little or no resources put aside for their retirement. The US Social Security system holds trillions of dollars in IOUs from the federal government that at some point will come due and be impossible to pay. State and local governments also face significant financial underfunding in the Medicaid obligations and pension promises they made to public employees.[302] Private sector pension programs have been allowed to be chronically underfunded to the point that many will declare bankruptcy.

Today, there are around 72,000 people over 100 years of age living in the United States. One study suggests that if current trends continue, that number will reach 1 million by as soon as 2050—a 15-to-1 increase.[303] If that occurs, or even if life expectancy levels out at a lower threshold such as 92–95 years, how are our weakening systems of support and finance going to cope with the maintenance and health needs of such large numbers of people who can no longer care for themselves and require expensive and chronic medical care?

A former UN head demographer compared population projections of children under the age of 15 to that of people 65 and over, explaining the problem as follows:

> While the prospect of longer lives is a good thing, problems arise when a shrinking work force cannot foot the pension bill. Several decades ago, you could have had about 10 workers per retiree, but that could shrink to the point where in Italy, for example, you had three workers per retiree. While the political choices are unsavory—increase taxes or cut benefits—governments are running out of time to act. You "can't repeal the law of demographics."[304]

It gets worse. Over the last two hundred years life expectancy in Western nations has increased by two or three years every decade. If the trends hold, more than 50 percent of children born today in the United States have a realistic expectation of living beyond 100. People who are now in their 40s and 50s are aware they are living longer than their parents and increasingly understand that, if possible, they will need to work beyond 65 to maintain their lifestyle or simply to make ends meet given the inadequacy of Social Security payments and the uncertainty of that program's solvency.

This doesn't take into account the pervasive lack of retirement savings by most Americans. A recent report indicated that 42 percent of Americans are at risk of retiring broke and that longer life spans and a lack of planning is creating a crisis.[305] There are added problems being faced by underfunded state and local public pension programs as well as the high potential for many corporate pension systems to fail. A *MoneyWatch* analysis indicates that more than a million pensioners dependent on private pension plans in the American Midwest may find their benefits slashed. Ed Leefeld writes:

At least 50 Midwestern pension plans—mostly the kind jointly administered by trustees for a labor union and a group of employers—are in this decrepit condition. Several plan sponsors have already applied to the Treasury Department to cut back retirees' allotments. This cross-section of America includes more than a million former truck drivers, office and factory employees, bricklayers and construction workers who are threatened with cutbacks that could last the rest of their lives.[306]

With such conditions in play it is not surprising that the percentage of Americans age 65 and older opting to remain at work has increased continually since 1983, from 10 to nearly 19 percent. Suzanne McGee describes how the uncertainty of their future is reflecting many Americans, writing:

Part of what is making all but the wealthiest Americans feel so economically vulnerable today isn't just that incomes have been eviscerated. It's the fact that *when we retire, those of us without pensions—a growing proportion, especially if we're not public sector employees or union employees—are perched atop very, very tiny nest eggs.* Last year's report from the nonpartisan US government accountability office (GAO) reminded us of the perils. Half of all households headed by Americans 55 and older had no retirement savings at all.[307]

Indeed, a 2015 study shows that the funding of state and local public worker pension plans in the United States is a "time bomb."[308] We have known for decades that the Social Security pension and benefit programs are in trouble for several reasons including the fact that recipients are living considerably longer than anticipated, the dramatic increases in health care costs, the precipitous drop in the number of workers paying into the Social Security fund, and of course the continuous off-budget "borrowing" of Social Security assets by the federal government. All these factors are well-known yet our political leaders have consistently avoided dealing with the problems.[309]

There are options but none is admirable. One way to deal with some of the fiscal crisis, as Janus Capital's Bill Gross suggests, is to just keep printing more money. This would devalue US currency but allow technically legal payments to be made according to our promises even though the real value of the payments will have declined significantly. It should be no surprise therefore that China, Iran, Russia and India are trying to shift the world to a currency basis other than the US dollar, not only to reduce US power but to create an alternative financial asset of sufficient strength to provide something in which they can reliably invest their funds with reduced risks.[310]

MIGRANTS OR AI/ROBOTICS?

For decades, one justification for allowing Middle East migrants into Germany, particularly from Turkey, was the need for an expanded labor force beyond that which German citizens themselves have produced.[311] Denmark is suffering from a labor shortage and has embarked on an aggressive AI/robotics push aimed at sustaining and expanding its economy through a robotic workforce. In "Aging Danes Hope Robots Will Save Their Welfare State," Peter Levring writes that AI/robotics development is essential in that nation.

> Denmark has a problem: it may soon be unable to afford offering such a good deal to its people. Free health care for all, a $757 monthly stipend for college students and robust safety nets for the less fortunate all cost money—Denmark devotes slightly more than 30 percent of its gross domestic product to social spending, one of the highest levels in the rich world. ... [P]opulation projections show that Denmark's 600 billion-krone ($91 billion) welfare system is facing a future of more customers and fewer people around to pick up the bill.[312]

Due to its aging population and plummeting birth rates a shrinking Japan also has a serious labor force problem.[313]

> Japan is already weighed down by one of the world's largest public debt burdens. With its inverted population pyramid, where will it find the tax base to repay this debt, and to care for its growing population of elderly? The 2012 government report said that *without policy change, by 2110 the number of Japanese could fall to 42.9m, ie just a third of its current population.*[314]

This means that Japan is not only experiencing a rapidly aging population in need of increasing assistance but has a national population that is shrinking.[315] With Japan's birthrate at an all time low, this means its younger workforce is not being replenished. This fact drives the development of substitutes in the form of AI/robotics, including, quite oddly, little robot cats and robot "babies" in an effort to persuade people to want to have children. Some

Japanese nursing homes have begun using Softbank's Pepper robot to assist in caring for elderly Japanese, with positive results.

One fascinating report indicates how the system works, including the fact that in Japan, overall, robots are depicted as helpful and caring rather than ominous killers. One nursing home resident explains how the robots aides make a positive difference in residents' lives. *"These robots are wonderful,"* said 84-year-old Kazuko Yamada after the exercise session with Soft-Bank Robotics Corp's Pepper, which can carry on scripted dialogues. *"More people live alone these days, and a robot can be a conversation partner for them. It will make life more fun."*[316]

Nor are the functions and designs of the robots limited to one model or purpose. Examples include:

> Paro the furry [robot] seal cries softly while an elderly woman pets it. Pepper, a humanoid, waves while leading a group of senior citizens in exercises. The upright Tree [robot] guides a disabled man taking shaky steps, saying in a gentle feminine voice, "right, left, well done!" Robots have the run of Tokyo's Shin-tomi nursing home, which uses 20 different models to care for its residents. The Japanese government hopes it will be a model for harnessing the country's robotics expertise to help cope with a swelling elderly population and dwindling workforce.[317]

INTERGENERATIONAL POLICY CONFLICTS

A significant part, but only one element of what will be happening, is a growing conflict between older and younger citizens. Unfortunately, as noted by Arthur Brooks, *"If history is any guide, aging electorates will direct larger and larger portions of gross domestic product to retirement benefits—and invest less in opportunity for future generations."*[318] Even that history lesson no longer describes our current dilemma. This is because the rise of identity politics has created a new system of highly organized and aggressive political entities vying for larger shares of the "public pot" for their own interest group.

In what Israeli scholar Yuval Noah Harari calls the "useless class" scenario in which people are made useless through no fault of their own but because there are no available jobs, we will see a bitter struggle between older residents of Western Europe and the US and younger members of those societies who have been employment disenfranchised. Caught in the middle of this looming intergenerational struggle will be the people who are actually working and who are required to pay higher taxes to support what they consider the modern version of the old, the halt, and the lame—as well as the idle, the lazy, and the economically disenfranchised who could and would work but are deprived of opportunities. Of course, on the other end of the spectrum is the class of the incredibly wealthy who will do everything possible to avoid the responsibility of helping societies to deal with the crisis the members of that group helped create through AI/robotics, tax policy and sophisticated tax avoidance schemes.

Harari ranks the process of the employment disenfranchisement of large numbers of people as a dire threat in the 21st century.[319] He admits he used the term "useless" as a deliberately provocative description in an effort to force people to understand what is happening due to the reckless adoption of AI/robotics. We might also think of those about to be pushed aside by AI/robotics as "abandoned," "discarded," "disinherited," "marginalized," "disappeared" or "dumped." We already have a fairly large bulge of people who could be referred to by one or more of such pejorative terms. Other than those sidelined for health reasons or for retirement after having spent decades in the workforce, they are not primarily in the ranks of the elderly who were raised at a time when the value of hard work was accepted as a fundamental principle.

Regardless of the label, almost one in six young men in the US are either jobless or in prison and that state of affairs is likely to get worse as our society continues to degrade.[320] Millions are being sidelined by the massive opioid and other drug and alcohol epidemics that have beset American society. Those are more traditional forms of the problem and we are ill-prepared to deal with what is truly a deepening societal crisis.

But the new kind of uselessness to which Harari refers is not a moral judgment about the value of those who have been sidelined, but a description about what will be created because there is no place in the workforce. It is a situation where no matter how prepared, diligent, ambitious or qualified a person might be, the future for far too many people will be a harsh and unforgiving game of musical chairs played with an increasing number of people competing for a rapidly dwindling number of "chairs" as we have already seen with thousands of people at job fairs showing up to try to convince someone to hire them when there are only jobs for five or ten percent of the applicants.

This systemic creation of uselessness driven by AI/robotics steals from a significant part of a nation's citizens not just opportunity, mobility and the chance to push oneself to the highest levels of capability, but threatens the ability for many to satisfy basic needs and leads to the disintegration of the political community. If the essential dynamism and indeed value of democratic societies has to do with their ability to address the well being of their peoples, then tolerating such policies degrades and contradicts the meaning and worth of democracy itself.

CHAPTER FOURTEEN
AI/Robotics and the Special Case of China

China offers a dramatic example of the AI/robotics transformation, an example defined by the nature and agenda of the Chinese leadership, and the nation's unique culture and history. Part of the lesson to be gained from this is that each nation has a separate set of goals, a distinct culture, and a political system unlike those of other nations. As a system committed to state planning and centralized control, China is an ambitious force.

China's aims go far beyond internal matters. The Chinese leadership is intent on not only dominating its own territory but its region of Asia in an eerie recreation of pre-WWII Japan's Greater East Asia Co-Prosperity Sphere. An analysis in the *Asia Times* suggests that China's "Belt and Road" strategy to bring the nations of Southeast Asia closer to a political and economic partnership is having mixed results with Cambodia aligning itself with China and Malaysia making a u-turn following a change in governmental leadership.[321]

As was recently proclaimed by President and Communist Party Chairman Xi Jinping, China is also embarked on a strategy aimed at becoming the world's most powerful nation over the next 30 years. Xi was just granted a status equivalent to "president for life" as the Communist Party eliminated term limits for the office. This shift has been accompanied by an increase in suppression of dissenting voices, including condemnation of large-scale protests in the autonomous territory of Hong Kong.[322]

Western nations continue to act as if the leaders of the Chinese state approach the world and political relationships in the same way they do. Nothing could be more wrong. The Chinese Communist Party controls China. While Western nations and the US shift authority and policy seemingly by the moment, China has developed a complex and coherent strategy that allows it to take the long view strategically and outwait the West's constantly shifting and largely incoherent policy approaches.

China's ability to develop and implement a focused, consistent and coherent strategy over a lengthy period takes advantage of the fact that Western nations are blinded by internal short-term "democratic" and identity group contradictions, political bickering and jockeying for power. In the US, EU and UK this includes frequent instigation of regime changes, the fear of Islamic terrorist blowback and Middle Eastern military clashes, intense internal identity group schisms, and paternalistic racist and cultural arrogance toward other nations. Dominant Western nations continue to make the mistake of viewing China as a powerful but still backward force.

Victor Hanson relates how Japan achieved its amazing militarization and industrialization over a 60-year period following the Meiji Restoration.

> [Japan] soon sent tens of thousands of students to European (and, to a lesser extent, American) universities and military colleges. They mastered Western military organization firsthand. Japanese engineering students returned home with world-class expertise in aviation, nautical architecture and ballistics—and a disdain for the supposed "decadence" of their mentors. The Japanese model was first to inspect and assess the latest European and American military technology: single-wing fighters, aircraft carriers, naval torpedo and dive bombers, and battleships. Then they copied the most promising designs but applied trademark Japanese craftsmanship and government support to make even bigger, sometimes better and often more numerous weapons.[323]

China is implementing a version of Japan's strategy, one that led to World War II as Japan sought to dominate its East Asian neighbors through promises of lucrative trade and lasting friendship. This involves rapid development and implementation of the powers of Artificial Intelligence and robotics. While many in the West like to consider the Chinese as less innovative and imaginative as their own nations, Philip Ball suggests that at this time such a view is mistaken. He warns:

> The economic rise of China has been accompanied by a waxing of its scientific prowess. In January, the United States National Science Foundation reported that the number of scientific publications from China in 2016 outnumbered those from the US for the first time: 426,000 versus 409,000. Sceptics might say that it's about quality, not quantity. *But the patronising old idea that China, like the rest of east Asia, can imitate but not innovate is certainly false now. In several scientific fields, China is starting to set the pace for others to follow.* ... Today ... it's common for Chinese postdoctoral researchers to get experience in a leading lab in the west and then head home where the Chinese government will help

them set up a lab that will eclipse their western competitors.[324] [emphasis added]

Even though Xi portrays himself and China in a benign light through speeches such as his 2017 presentation at the World Economic Forum in Davos and as a partner in interest with Western nations, including the US and EU, the reality is that China is a rapidly rising political, economic and military force.[325] Xi Jinping, in fact, has set a path toward the creation of a new Chinese hegemony aimed at becoming the world's dominant superpower. Those who doubt Xi's drive toward that status should consider the following developments.

CHINA'S EMERGING POWER

- **China's 30-year deadline to rule the world: Beijing has outlined plans to become the world's biggest superpower within the next 30 years.[326]**

- **Xi Jinping heralds "new era" of Chinese power at Communist party congress: Xi said, it was time for his nation to transform itself into "a mighty force that could lead the world on political, economic, military and environmental issues."[327]**

- **China's Plan for World Domination in AI Isn't So Crazy After All.[328]**

- **China's supercomputers race past US to world dominance: China doesn't just have the single fastest supercomputer in the world. It now dominates the list of the 500 fastest.[329]**

- **China speeds ahead of U.S. as quantum race escalates, worrying scientists U.S. and other Western scientists voice awe, and even alarm, at China's quickening advances and spending on quantum communications and computing, revolutionary technologies that could give a huge military and commercial advantage to the nation that conquers them.[330]**

The Chinese—and here we mean those with political, economic and military power within the national system and dominant political organism of the Chinese Communist Party—are ambitious and to an extent resentful of being subordinated due to their millennia of continuous existence and profound achievements. Relative to "upstart" nations such as the US, there is resentment that the world's oldest political system with a truly incredible history of achievements does not receive its due respect. There is also a defensive mindset following centuries of Mongol and Western domination.

The Chinese Communist Party is also understandably concerned with what could happen if many among the nation's 1.2 billion population become exceedingly dissatisfied at the Party's leadership and that dissension spreads. Mao's Long March civil war, the disastrous Cultural Revolution of the 1960s, and the democratic movement that led to the 1989 Tiananmen Square episode are never far from the Party's collective mind.

This means that the Chinese leadership sees dissent as a destabilizing force, particularly now that it can spread like a virus through the Internet. This results in an intolerance of criticism. This is an important reason why China suppresses criticism, stifles any dissent, bans access to news, and why a seemingly innocent movement such as Falun Gong is banned and criminalized. It is why Xi Jinping is requiring the teaching of approved political thought as part of the university curriculum, and is keeping watch on Chinese students studying abroad to ensure they are not "corrupted" by Western values.

The indoctrination of China's students with acceptable political attitudes has become a key part of China's educational system. Xi Jinping has declared that the Party must dominate the curriculum of the nation's universities. His order requires the educational system to put the Party and its aims at the center of the educational process in an effort to control what is learned in the interest of stability and preservation of Chinese society under the Communist Party's supremacy and control. A *Guardian* report explains:

> Chinese authorities must intensify ideological controls on academia and turn universities into Communist party "strongholds," President Xi Jinping has declared in a major address. "Higher education ... must adhere to correct political orientation," Xi said in a high-profile speech to top party leaders and university chiefs that was delivered at a two-day congress on "ideological and political work" in Beijing.[331]

The Guardian's Tom Phillips writes:

> Echoing a 1932 speech by Joseph Stalin the Chinese president told his audience teachers were "engineers of the human soul" whose "sacred mission" was to help students "improve in ideological quality, political awareness, moral characteristics and humanistic quality." "Party authorities should increase their contact with intellectuals in colleges, befriend them and sincerely listen to their opinions," Xi added, pointing out that the party's education policies "must be fully carried out."[332]

According to Xinhua, China's official news agency, this goal that universities be transformed into "strongholds that adhere to party leadership" requires that political education should be made "more appealing." Xi, in presenting this order, said teachers needed to be both "disseminators of advanced ideology" and "staunch supporters of [party] governance."[333]

Given what we have said about the failure of American schools and universities to educate students in critical thought and the full range of knowledge and controversies our society faces we need to be honest about the fact that while Xi Jinping is embarked on a path we reject as contradictory to our Western vision of the traditional mission of the university and education in general—at least he is being overt about what he is doing. The increasing failure of Western educational institutions is arguably far more devious and cowardly, hiding behind masks of ultra-sensitivity, one-sided and poorly developed political mantras, and assumptions of moral and political "rightness" voiced with intense religious fervor. At least we know where Xi is coming from and while we might not like it, we can respect and respond to his "transparency."

THE POSSIBILITY OF A US v. CHINA
MILITARY CONFRONTATION

American rejection of China's self-assertion in its region was voiced by Admiral Harry Harris, the soon to be installed US ambassador to Australia. Harris, commander of the US Navy's Pacific Command, says Beijing intends to control the South China Sea, a conflict zone China is now claiming as its own. Admiral Harris recently told the House Armed Services Committee that: "China's intent is crystal clear. We ignore it at our peril," he said. "I'm concerned China will now work to undermine the international rules-based order."[334]

Of course it can be argued that US refusal to submit to the authority of the International Criminal Court, its rejection of the Paris climate change agreement, and the 2003 invasion of Iraq under questionable pretenses about non-existent "Weapons of Mass Destruction" arguably stand for the same proposition, regardless whether we agree with the decisions or oppose them. For nations such as the US and China, their own perceptions of national interest trump the more ephemeral nature of the "international rules-based order."

US Secretary of Defense James Mattis has challenged what he terms China's aggressive moves in the South China Sea. He stated during a speech in Singapore that "despite China's claims to the contrary, the placement of these weapons systems is tied directly to military use for the purposes of intimidation and coercion."[335] China's response was that the creation and militarization of the artificial islands that nation constructed in those waters was a matter of national sovereignty and defense, although it is not immediately obvious who has plans to invade China.

In 2005 a senior Chinese general stated that China would use nuclear weapons against the US if it interfered with any attempt by the Mainland to recapture Taiwan.[336] In 2012 another very senior and well-placed Chinese Major-General warned that future conflict was coming as a result of America's "containment" policies directed at China. Major-General Peng Guangjian stated:

> The United States has been exhausting all its resources to establish a strategic containment system specifically targeting China, … The contradictions between China and the United States are structural, not to be changed by any individual, whether it is G.H.W. Bush, G.W. Bush or Barack Obama, it will not make a difference to these contradictions.[337]

As it concerns AI/Robotics and the military use of same, China is making many advances.

CHINA AND MILITARY AI/ROBOTICS

- China Rapidly Building Advanced Arms for Use Against U.S.: Space weapons, drones using artificial intelligence priority in Beijing military buildup.[338]

- FCC wants Chinese tech out of US phones, routers.[339]

- China's plan to use artificial intelligence to boost the thinking skills of nuclear submarine commanders: Equipping nuclear submarines with AI would give China an upper hand in undersea battles while pushing applications of the technology to a new level.[340]

- China steps up pace in new nuclear arms race with US and Russia as experts warn of rising risk of conflict: Chinese scientists are running simulated tests at a faster rate than America as world's leading powers develop arsenal of "usable" next-generation weapons.[341]

- China hacked a [US] Navy contractor and secured a trove of highly sensitive data on submarine warfare.[342]

As the above listed developments indicate, along with its other actions, China's AI advances pose a growing threat to American hegemony.

Do We Face Chinese Global Domination of AI/Robotic Development?

We see a growing boldness in Chinese leaders' mindset as China's AI/robotics systems improve by leaps and bounds technologically, economically and militarily. This is taking place while China, it is argued, has become a new form of neo-colonial power with its rapidly expanding economic and strategic colonization of African and South American nations.[343] China is using the financial surplus created by its trade advantages to build relationships, buy strategic locations, and create "friendships" with regimes in Africa and South America.

As with its geo-economic expansion, China is seeking to dominate the world of AI/robotics. An irony is that China's enormous balance of trade surplus with the US plays a major role in that nation's ability to fund its economic and political development, research, military and propaganda. In 2017 alone, for example, China benefited from a $375 billion trade surplus with the United States.[344] China has a massive trade surplus not only with the US but also with European Union members, reaching 180 billion Euros in 2016 or more than 200 billion dollars.

Many Western sources—but not all—are saying China has been cheating for decades and calling out the country on behavior which involves multiple trade barriers, currency manipulation, product dumping and subsidization, and technology theft. Former Harvard president Larry Summers has stated that China's leadership in "some" technologies is not the result of technology theft.[345] Of course this conveniently leaves out the question of other critical technologies, particularly military-related ones. But China has invested huge amounts of money in sending students to US and other Western universities to study the kinds of subjects that are at the core of sophisticated technology. Through such strategies China has grown its research and development capabilities to levels that make comments such as voiced by Summers accurate *at this moment* even though a significant part of that growth was generated by the extreme openness of US and European universities and research centers and the naivete of Western academics.

At this point the health of China's economy depends to a very significant extent on massive export trade and the US is its most important partner in that process, followed by the EU. China recently offered to expand its imports of US goods to $70 billion in order to reduce its balance of trade surplus with the US.[346] This has been countered by a US response indicating it would impose tariffs of up to $200 billion on China's exports to the US. We are in the midst of what is being called a trade war but is in fact an effort by the current US administration to redefine the terms of trade between China and the US.

Nonetheless, however defined for purposes of discussion, the trade-based economic contest between China and the US will have significant effects on the economies of those countries and others who are dependent on the health and behavior of these economic giants.[347] Regardless of the often hysterical claims that the conflict will severely damage the world's economy and cause US and EU consumers to pay higher prices for purchases, there is no question that adjustments need to be made—at least from the perspective of the US and EU nations—and that China will be against any truly fundamental readjustments unless it has very limited alternatives.

Even with its efforts to expand into Africa and Latin America, as well as to build stronger trade relationships with other Asian nations, many of whom are understandably wary of Chinese intentions, that nation depends on the US, UK and EU for a combined trade surplus above $600 billion annually. This situation inevitably involves the expansion of AI/robotics production systems because as China's workforce ages and Western nations increasingly develop AI/robotics production systems that enable them to match or undercut China's advantage in labor and regulatory costs the Chinese have had to re-invent their production systems. This involves a rapid and massive shift from human labor to AI/robotics as evidenced by a recent report.

Laura O'Callaghan writes in "China's AI Revolution: Intelligent robots to power factories—risking US fury":[348]

> Robots powered by artificial intelligence are set to replace Chinese factory workers in a move aimed at boosting the manufacturing industry which has been hit hard by a rise in wages.
>
> The machines which are capable of making, assembling and inspecting goods on production lines have already been rolled out, with one factory laying off 30 workers to make way for the robots. The robots were displayed at China's Hi-Tech fair in Shenzhen earlier this month, an annual event which showcases new development ideas with the aim of driving growth in a number industries. But the news has annoyed Washington as it is expected to put international competitors at a disadvantage, as the two countries' bitter trade war continues to escalate.

Speaking to the *Financial Times,* Sabrina Li, a senior manager at IngDan, said: "We incubated this platform so we can meet the (Made in China 2025) policy. … Giving the suffering manufacturing industry a leg up is a key part of the Chinese government's Made in China 2025 policy. Zhangli Xing, deputy manager of Suzhou Govian Technology which sells the quality control robots, said they are more reliable than human labour.[349]

Although criticisms of the US effort have often included hyperbole, the fact is that due to the combined actions of China and a large number of Western companies seeking to maximize returns by outsourcing their own domestic production and importing cheap goods from places that do not have to bear the kinds of production, environmental, worker health, pay, and regulatory requirements to which those companies were subject in their own nations, the manufacturing core of the US economic system has been damaged. This is also due to the numerous non-reciprocal restrictions China (as well as Japan) imposes on imports into that country. It is easy to export things *from* China but extremely difficult to sell Western goods *into* that nation. For the first time since China was admitted into full membership in the World Trade Organization it is being challenged on its tariff and subsidy behaviors in what is a "trade adjustment" and not a "war."

The massive trade imbalance between the US, EU and China over just the past six years represents a total of an approximately $3 trillion surplus flowing into China from the West. This vast and still expanding pool of wealth has played an enormous role in China's economic and technological development, military growth and weapons investment, as well as its extensive international propaganda, surveillance activities, and acquisition of significant stakes in foreign companies.

Trade with the US and EU has given China a source of hard currency and access to priceless advanced technological expertise that has allowed it to achieve very rapid development in AI/robotics and other technical areas, as well as a large-scale military expansion. The US/China trade deficit is only part of the issue. Strategically, the US and EU have been unwisely funding their most significant and dangerous economic, political and military competitor.

This means, on average, that the US and other key Western nations with significant AI/robotics capabilities have been providing China with at least $550 billion annually in hard currency to pave its way to world dominance in AI and fund military and private sector expansion and weapons development. This does not even include deceptive country of origin trade practices in which China transships goods through nations such as Malaysia to hide their source and avoid tariffs. It also suggests that China has been extracting AI/robotics technologies from the West in ways that allow it to accelerate its own development.

China is totally committed to AI/robotics development. It recently announced, for example, a $59 billion investment in AI development through 2025 in a drive to become the world's leader in AI/robotics.[350] Chinese leader Xi Jinping has signaled his nation's aim to become the world's dominant artificial intelligence power over the next thirty years.[351] In the view of Western analysts, this will include not only economic and technological dominance, but also military strength.[352] The Chinese military has made significant investments towards development of autonomous weapons systems, an area where the Chinese allegedly feel they are already ahead of other nations. Bedavyasa Mohanty explains the Chinese emphasis on Artificial Intelligence supported technologies that they expect will provide a strategic advantage in asymmetrical warfare.

> China does not want to be at the receiving end of a technological asymmetry in what may very well be the conventional approach to war in the future. In part, it is guided by its own ambitions and more importantly confidence in being a step ahead of other countries in the creation of lethal autonomous weapons. ... *What is changing now is the significant investments that China seems to be making in drone swarms—armies of miniature drones, able to operate autonomously—which many call the most effective use of AI weapons technology.* As Wang Weixing, a military research director of the Peoples Liberation Army (PLA) wrote: "Unmanned combat is gradually emerging. While people have their heads buried in the sand trying to close the gap with the world's military powers in terms of traditional weapons, technology driven 'light warfare' is about to take the stage."[353]

Although full superiority will take time, China has already surpassed the US in several critical AI/robotics areas. For instance, China has made extraordinary progress in the field of quantum computing, a technological development that promises computing performance orders of magnitude beyond the fastest supercomputers available today. China is focused on developing multiple quantum computing technologies, including a quantum-based satellite communications system, an ultra-secure quantum data link between Shanghai and Beijing, and has announced construction of a $10 billion quantum computing center.[354] Although usable quantum AI technology is still some years away, once developed it has potential far beyond current technology, with the first nation to fully achieve general usability winning significant advantages over all others.

Even with its amazing developments on numerous fronts, China is beset by risks that must be managed carefully. As its similarity to pre-war Japan's strategy of dramatic military and economic growth and change portends, China faces challenges that endanger both that nation and others. An overheated economy, excessive borrowing to fund its activities, dependence on the US and Western Europe for markets that provide China with massive trade surpluses, and military adventurism could all combine in ways that threaten China's future. China in many ways is not "master of its own destiny" at this point, even though it seeks to project a "face" of absolute power to the world. China, nonetheless, must deal with several serious problems.

Overall, China's economy is facing significant challenges.[355] Although generally seen as an economic leviathan, various Chinese manufacturing industries grew less than expected in the First Quarter of 2016.[356] One analyst states in relation to China's overheated market expansion that corporate debt poses a very significant threat, observing that: *"It's a fatal issue in China. Because of such a [government/bank] link, it is probably more urgent for China than other countries to resolve the debt problem."*[357]

CHINA'S SOCIAL AND ECONOMIC CHALLENGES

- **Chinese banks are big. Too big?**[358]

- **Scary Statistic: China's Debt to GDP Ratio Reached 257 Percent in 2017.**[359]

- **China's debt levels pose stability risk, says IMF: Health check of financial system says reforms have not gone far enough and notes similarities to US before 2008 crisis.**[360]

- **S&P Cuts China's Credit Rating, Citing Risk From Debt Growth.**[361]

- **China moves to curb overseas acquisitions as firms' debt levels rise: Beijing imposes restrictions to try to stem global buying spree that has included entertainment firms and football clubs.**[362]

- **China to lay off five to six million workers, earmarks at least $23 billion: China aims to lay off 5-6 million state workers over the next two to three years as part of efforts to curb industrial overcapacity and pollution, two reliable sources said, Beijing's boldest retrenchment program in almost two decades.**[363]

- **Warning from history: Could Japan-style crash hit China? Sizzling property prices, a groaning debt load, wealthy tourists and tycoons willing to slap down eye-popping sums for art: China is starting to look like Japan before its economic bubble burst in the early 90s. The similarities are not lost on Beijing: President Xi Jinping has commissioned a study to help China avoid Japan's pitfalls, ... as growth slows and ratings agencies sound the alarm over its debt.**[364]

China is a master at manipulating Western nations, including the US. It is as if the Chinese leadership is playing the complex, sophisticated and multifaceted strategy game of *Go* while our leaders are playing *Chinese Checkers*. While Western nations and their corporations and investors want access to China's vast internal markets, and are happy to placate their populations with imported lower price goods from China that lose American and European jobs and fund China's aggressive economic expansionism, most of the West's political leaders simply do not understand what China is doing. Investors simply don't care as long as they obtain substantial returns on investment.

In their unending drive to access China's markets, Western investors have contributed greatly to China's economic, technological and military development in ways that prove that "money has no loyalty" (unless it is China's). The simple truth is that China would be far behind its current state of development if Western nations, companies, universities and investors had not provided the intellectual property and advanced research that has been transferred to that country, and if Western consumers had not financed China's development through purchases.

CHINA'S EXPLOSIVE TECHNOLOGICAL PROGRESS

- China's great leap forward in science: Chinese investment is paying off with serious advances in biotech, computing and space. Are they edging ahead of the west?[365]

- Alibaba's AI Outguns Humans in Reading Test. Its natural-language processing AI scored higher than humans. ... [China's] Alibaba has developed an artificial intelligence model that scored better than humans in a Stanford University reading and comprehension test.[366]

- The Long-Term Jobs Killer Is Not China. It's Automation.[367]

- China set to dominate robot industry with millions of jobs wiped out: Chinese tech scientists are set to catch their rivals and dominate the robotics industry that could take millions of human jobs.[368]

- Steel industry foresees high-tech future with fewer workers.[369]

CHINA'S ACQUISITION OF WESTERN TECHNOLOGY

China's incredibly rapid progress in AI/robotics development and weapons systems has been possible because the Chinese government has had substantial success at manipulating foreign companies. Western firms respond to the "carrot" of potential access to China's massive and growing domestic consumer market often by opening intellectual property to Chinese partners or to Chinese industrial espionage.

The Chinese bargain with Western companies has involved requirements of *access to* and *transfer of* technology to Chinese partners. Over a transitional period of five years or so, those partnering Chinese companies incorporate the technological knowledge of the Western companies to which they are linked into their own products. They then gain the greatest profits from selling their own "new" products to the internal Chinese markets that the foreign companies had hoped to access. This is made clear by Apple's experience in China.[370]

> Apple's stumbles in China seem emblematic of a broader realization: US companies have less of a future there than many had hoped. For many years, the vast Chinese market—more than 1 billion consumers in a fast-growing economy—sent thrills of excitement up the spines of corpo-

rate managers throughout the developed world. Who cares about the stagnation in Europe and Japan, when China has many more people than all of those markets combined? *Even as rising labor and energy costs reduced China's advantage as a low-cost production site, the dazzling lure of the Chinese consumer pushed many multinationals to locate offices and factories in the country. Unfortunately, that promise turned out to be a mirage for many companies.*[371]

China has decided that its future, economically and militarily, depends in large part upon AI/robotics. A key part of its strategy involves acquiring access to advanced AI/robotics technology from the West.[372] In large part, simply by virtue of China's industrial and manufacturing base, China has become a significant market for robotic systems. Alternatively, Chinese firms—whether directly state owned or indirectly controlled by the state—purchase advanced AI/robotics technology by acquiring foreign companies engaged in those activities.[373]

CHINA AND TECHNOLOGY ACQUISITION

- **World's most populous country lacks the one thing it needs to become an AI powerhouse: enough talented people: China faces a severe talent shortage in its quest to become a global powerhouse in artificial intelligence by 2030.**[374]

- **Google Builds China Workforce to Develop Artificial Intelligence: Alphabet unit is seeking engineers to fill jobs related to AI, cloud computing in country seen as having certain advantages over U.S.**[375]

- **How China Is Stealing Our Secrets.**[376]

- **How China squeezes tech secrets from U.S. companies.**[377]

- **China, Unhampered by Rules, Races Ahead in Gene-Editing Trials: U.S. scientists helped devise the Crispr biotechnology tool. First to test it in humans are Chinese doctors.**[378]

- **How China acquires "the crown jewels" of U.S. technology: The U.S. fails to adequately police foreign deals for next-generation software that powers the military and American economic strength.**[379]

China's strategy is global, multi-faceted and long-term. A Chinese company recently attempted to purchase the Chicago Stock Exchange from its private corporate owners but the purchase was rejected. The US Securities and Exchange Commission determined it would be extremely dangerous to allow China deep access to our critical financial information and tools.

Consistent with the blindness so often associated with investors who either do not consider or care about the consequences of their actions it has been said that: "*The Chicago exchange, better known as CHX, is not happy, saying in a statement the SEC "unfairly" disadvantaged "our company and shareholders*."[380] It was pointed out, however, that, "a successful takeover would be "the first time a Chinese-owned, possibly state-influenced, firm maintained direct access into the $22 trillion U.S. equity marketplace."[381]

The investors' response to the rejection of the sale of the Chicago Stock Exchange is sadly predictable and predictably selfish. The problem from the US point of view is that everything of any scale in China is state-influenced and controlled. The security of our financial system, financial information, sources of vulnerability and critical mechanisms involves vital national security data and linkages. Not to be trite, but "you don't let the fox into the henhouse." Allowing a nation such as China full access to and control of core elements of the system whether through theoretically "private" corporate control or direct and open governmental power should be unthinkable, but investors' logic knows only the scale of the ROI (Return on Investment) and everything else is irrelevant. China has learned how to use the West's greed against its competitors.

Such acquisitions of technological, financial or communications platforms may be either by direct acquisition or by owning a significant stake in a foreign company that has high-level technology assets, such as Japan's Softbank, a world leader in the "Internet of Things." Softbank recently acquired a leading British AI/robotics firm. Germany, other EU countries, the US and the UK are committed to participating in the Chinese robotics market as sellers of AI/robotic technology products. One way or another, China will acquire top-level Western technology while continuing to develop its own.[382] Given China's frequent avoidance of international laws protecting intellectual property this knowledge theft, admittedly willingly participated in by Western corporations pursuing what they saw as a potentially immense Chinese domestic market, will enable China to leapfrog a decade of research and development.[383]

China's rapid technological development is also being aided by the blind self-interest of Western educational institutions. Many American universities have been captured by the Chinese. Universities in the U.S., for example, are desperate to maximize the numbers of Chinese students who pay full tuition at a time when many American universities are facing a decline in students. China sends hundreds of thousands of its youth to study advanced science and technology in US and European universities. They then bring that knowledge back into China. In 2015, over 260,000 Mainland Chinese students studied in US universities, many at the technical and graduate levels.

US universities also feel they can gain students and status by partnering with such Chinese-sponsored programs such as the Confucius Institutes, a Chinese government sponsored strategy that has established numerous links with at least 70 American universities and is aimed at creating positive attitudes toward China. The money provided to US host universities by the Confucius Institutes program has been used to create a financial dependency on the part of those universities.

This dependency allows the Chinese government to subtly shape political views on issues such as the Falun Gong movement, the Chinese occupation of Tibet, or the proliferation of political suppression in China.[384] China has demonstrated a continual willingness to demonstrate its disapproval of others outside its borders who criticize its activities, including reacting negatively to a Nobel Prize awarded to individuals who dared to criticize the nation or using its economic power to damage critics.

Although there are significant benefits from cross-cultural educational exchanges that contribute significantly to improving welfare and increasing per capita income across the world, there is a dark side to the globalization of education as well. FBI Director Christopher Wray has criticized the US academic sector for being naïve in regard to China's intentions and manipulations. Wray testified to Congress that:

> *I think the level of naïveté on the part of the academic sector about this creates its own issues.* They're [China] exploiting the very open research and development environment that we have, which we all revere, but they're taking advantage of it. *So one of the things we're trying to do is view the China threat as not just a whole-of-government threat but a whole-of-society threat on their end, and I think it's going to take a whole-of-society response by us.* So it's not just the intelligence community, but it's raising awareness within our academic sector, within our private sector, as part of the defense.[385]

This exploitation runs the gamut from the obtaining of technological developments from educational institutions to a concerted state program promoting transfer of sophisticated AI/robotics and other scientific knowledge for military and industrial applications to which China would otherwise not have access.[386] The problem is that, in a globalized marketplace where capital has no nationality Western businesses and investors have no loyalty or allegiance to any single country. Money and return on investment is everything. Business firms cannot ignore the potential of the economic market China represents even though experience indicates it is a chimera and that the Western companies are being used by China as part of a coordinated strategy to obtain the transfer of their intellectual property.

Western firms can't resist being drawn into China's orbit lest they lose opportunities to competitors. But Western companies that are desperate to enter the Chinese markets have far too often kowtowed to the pressures placed on them by the totalitarian Chinese regime. They are excited at the prospect of having access to China's enormous consumer markets even though Chinese scientists and engineers take their technology as part of government rules and incorporate it into their own systems. In many instances, as Apple discovered, the Chinese companies with which they had started working then gained the lion's share of the profits following an initial period when the technology was being transferred and redesigned by the "partner."

Google recently announced it was seeking to develop an AI research arm in China.[387] Since Google is the acknowledged leader in the most sophisticated aspects of Artificial Intelligence, including quantum computing, and the Chinese are absolute masters at capturing others' intellectual property, Google's move can be expected to hand China state-of-the-art research and leapfrog a decade or more of R&D. Who do Google's chief executives think will make up the research and management teams in their Chinese facilities? And to whom do those executives think their Chinese employees owe their loyalty? Many Chinese businesses, indirectly or directly, are arms of the Chinese government and critical mechanisms for implementing governmental policy.

The transfer of AI/robotics technology from Western corporations and universities has paid off for China. China has developed a supercomputer that has jumped ahead of US and European versions.[388] The Chinese concentrated on developing their own advanced capacity computer chip so as not to be vulnerable to US and Western prohibitions and penalties. China is also pursuing the development of its own advanced satellite Internet system.[389] It is well on its way to that goal according to a new report on the country's rapid advances in supercomputers.[390] Chinese research is taking computer processing capabilities and data volume to new levels almost overnight.[391]

China's Aging Workforce and the
Shift to AI/Robotics

After decades of its now abandoned "One Child" policy, China is facing a labor force deficit that is forcing the rapid development of AI/robotics systems of production. The growing labor shortage results from a combination of factors, notably aging, the long-term population control by which China sought to prevent unsustainable population growth, and the resultant increasing unwillingness of potential workers living in the country's vast inner regions to migrate temporarily to coastal urban regions as laborers.[392] Many of those workers appear to have decided to remain with and care for their aging parents and relatives in the rural and agricultural core of the country rather than leaving for work in coastal urban centers.

As a result, China must address the needs of its enormous aging population while providing newer workers for its production processes if it is to avoid widespread dissatisfaction and unrest. Shifting at least part of the burden of production to an alternative AI/robotic workforce potentially promises to satisfy the domestic need for care workers as well as those who will produce the international trade surplus of goods and services to continue to fuel China's need for economic growth.[393]

Given its age-skewed population demographics, the fact that China's leadership is developing a robotic workforce is not surprising.[394] Nor, with the rapid development of AI/robotics systems in Japan, the US, Russia and the EU, is it surprising. To remain competitive in a global export-driven economy in which the advantage of having a lower cost of labor is becoming much less important, China must shift to AI/robotics systems of production and service. In so doing, China's ability to impose a comprehensive and coherent strategy on its economic sector gives it an edge over Western nations. It can also dictate whatever tax or revenue generation system it wants, while retaining a significant agricultural sector, and has a much less expensive social service and expenditure base than its main competitors.

As of 2017, China had the largest number of robotic production workers of any country.[395] In a globalized economy, that means that while many of the displaced workers will be in China, the ripple effects potentially will expand to eliminate otherwise productive jobs in other countries because of the competitive edge provided to Chinese firms through increased automation through AI/robotics processes.[396]

China is developing a robotic production capacity that will send millions of human workers in other countries to the sidelines as it slashes China's costs of production and stimulates exports.[397] China has already begun downsizing its human workforce in the steel and coal industries, embarking on a plan to cut 1.8 million workers.[398] Similarly, the Chinese company Foxconn, based in Taiwan but with significant investments and connections with the Mainland Chinese, has implemented robotics to the point that it was able to cut its workforce from 110,000 human workers down to 50,000. Foxconn is also the world's largest producer of Apple's electronic products.

PROPAGANDA, SURVEILLANCE, CENSORSHIP AND MIND CONTROL—CHINA MAY LEAD THE WAY BUT IT IS NOT ALONE

China is pulling out all stops in an effort to maintain control of what its people see, know and say. US companies such as Apple, Yahoo, Facebook and Google are unfortunately cooperating with the Chinese so as not to lose access to the enormous market that country represents. Although significantly smaller in terms of total reach, an even more disturbing development is the willingness of Western educational institutions to betray fundamental ideals such as free inquiry and exchange of ideas in order to access the lucrative Chinese market. Cambridge University Press, for example, announced recently that it will bow to Chinese pressure and cooperate in banning over a hundred research articles that the Chinese censors do not want available to the Chinese people.[399]

The Chinese have surpassed the West in developing surveillance systems.[400] China's "Great Firewall" has been created so that the country's leaders can control the information available to the populace and ensure that the Chinese people receive a steady stream of pro-regime propaganda rather than having access to diverse sources of news and other information.[401] The Chinese government is also embarked on a strategy to monitor its citizens' Internet usage and communications and to intimidate and punish anyone it sees as a threat.[402] As developed in subsequent chapters discussing expanding US governmental surveillance over its citizens, along with truly invasive privacy behaviors by corporations such as Google, Facebook and Amazon, the US government and its private sector actors are also abusing the powers of Artificial Intelligence.

CENSORSHIP

- China Applauds the World for Following Its Lead on Internet Censorship.[403]

- China's Xi says internet control key to stability.[404]

- China Assigns Every Citizen A "Social Credit Score" To Identify Who Is And Isn't Trustworthy: Country Determines Your Standing Through Use Of Surveillance Video, Plans To Have 600 Million Cameras By 2020.[405]

- China is taking digital control of its people to chilling lengths. The Chinese government's unsettling new system will see citizens rated by "good deeds."[406]

- China's Censors Can Now Erase Images Mid-Transmission: Internet police step up their ability to filter photos[407] Wall Street Journal, Eva Dou, July 18, 2017.

- Chinese officials "create 488m bogus social media posts a year."[408]

- What price privacy when Apple gets into bed with China?[409]

- China is using AI and facial recognition to fine jaywalkers via text.[410]

- AI Police State: China to use technology to predict crimes before they happen: China is hoping to use artificial intelligence (AI) to look into the future and help police predict crimes before they have even been committed.[411]

The Chinese government has developed AI applications that allow it to detect and suppress disfavored words and images before they are even posted to the Internet. China also uses AI systems to monitor its citizens to identify both criminal rule breakers and dissidents. For example, China has implemented the world's most extensive facial recognition system in much of the country so that "wrongdoers" can be identified, apprehended and punished.

China's science fiction-like use of facial recognition software is capable of identifying millions of people in cars and trucks as they travel. Some uses of this software are relatively benign or at least relatively non-offensive, such as ticketing traffic violations. Individual traffic tickets are themselves relatively meaningless, but widespread use of the technology sends a strong message that "Big Brother" is always watching and listening.[412]

The AI company Nvidia recently announced it and its partner Anyvision had made significant progress with facial recognition technology to the point that it had reached a 99 percent degree of accuracy.[413] A February 2018 *Wall Street Journal* report indicates these were made available to Chinese police in early 2018 prior to the annual travel period accompanying celebration of the Chinese New Year.[414] At the moment the special glasses are only available for the police but given the speed at which technologies are being commercialized it may not be long before they are available to consumers. Comprehensive and continuous facial scanning will then be available not only in China but throughout many nations, allowing people to surreptitiously identify anyone with whom they come into contact, whether their intent is innocent or malicious.

Xi Jinping has unabashedly used state control of the Internet to suppress dissent and to spread Party propaganda. Analyses indicate that:

> Since Xi Jinping came to power nearly four years ago, hundreds of activists, lawyers, writers, publishers and employees of nongovernmental groups have been rounded up. Many more have been threatened and intimidated. Internet news sites have been ordered to stop publishing reports from sources that aren't sanctioned by the state. [415]

The monitoring, keyword or phrase mining and Big Data applications available through the Internet make it possible for governments, activist groups and corporations to retrieve all communications and to target individuals they wish to intimidate or suppress.[416] An analyst states:

> Since Xi came to power, China's situation has become more and more worrisome," says Murong Xuecun, a prominent author and commentator. "Things that we could openly discuss before, such as the Cultural Revolution, are now considered sensitive or even forbidden. In the past there was some room for non-governmental organisations and rights lawyers. Now all of them have been suppressed."[417]

The Internet threatens governmental control of individuals and groups everywhere, but that threat is addressed more directly in authoritarian systems like China. In a document issued one month after Xi Jinping became president in March 2013, the Communist party identified the Western-promoted idea of civil society as "an attempt to dismantle the party's social foundation."[418]

It is little wonder that authoritarian systems would have strong incentives to use AI and the Internet in turn to fight back against any change that might challenge their power. Given the vast capacity for radical social, political, and economic change represented by the expansion of free thought and social criticism that accompanies the free and open Internet, authoritarian systems and even democratic ones tend to develop their own tools to limit such expressions and punish dissidents.

China is not alone in its use of AI to control and shape its people, but China is an authoritarian system subject to none of the controls that limit the abuses in Western democracies. Other than the need to balance the political interests and agendas of competing factions within the Party and the military, there are no real domestic controls imposed on the Chinese government.

The militarization of AI and robotics systems by global powers such as the US, UK, Russia and China is an exceptionally dangerous development. But the weaponized capabilities of AI/robotics comprise more than just guns and bombs; they work not only through traditional battlefield armaments but through intimidating, isolating and controlling people through psychological warfare or Psywar.[419] This propaganda, surveillance and control capability is assisted by collaborations between governments and the quasi-governmental businesses such as Google, Yahoo, and Facebook.[420] It is even more obvious in China with Tencent and Alibaba because those entities are clearly subject to direct state control.[421]

Many governments are seeking ways to shape and censor the information available to their citizens through news outlets and the Internet.[422] China, for example, is embarked on a strategy to influence and even buy ownership stakes in significant news providers and communications centers.[423] There should be no doubt that the intent is to influence, control and shape content while muting criticism.[424]

Use of the media to shape or control discourse is not limited to China, but in most of the Western nations there is a significantly greater degree of diffusion in the sources of news and communications that at least provides some distinct views compared to China's more monolithic strategy of news suppression. This is now threatened in the US, as the media platforms provided by corporations such as Google and Facebook possess their own defined political perspectives and preferences, with most of their employees and executives sharing a clear set of political ideologies to the extent of deliberately shaping, prioritizing and even suppressing "the message."[425]

Google, of course, has faced some employee protests and resignations due to its work with the Chinese government to interdict search terms that are among those that activists and dissidents, or simply those whose opinions and interests do not jibe with the government's. One report explains Google's CEO Sundar Pichai's view on this issue.

> Google, under Pichai, is currently working on a censored search engine for the Chinese government, codenamed Project Dragonfly, which will stop dissidents from searching for terms related to human rights, democracy,

and protest, and will link searches to the searcher's phone number—eroding anonymity. Several employees have re-signed from Google over the project, which executives told employees to "keep quiet" about, for fear of being complicit with the "erosion of protection for dissidents."[426]

China, like the US, is a special and rapidly emerging power that must be taken into account as a force that will shape the future of AI/robotics and the direction of the world's political and economic systems. China, like the US, can be a positive contributor to what is developing. The EU and UK can make some contributions on the margins. Russia can detract and threaten through its military and nuclear prowess. But the future as written in the next thirty years will be written by the interactions between the dramatically different systems represented by these two incredibly powerful and vibrant political systems. It is vital that a significant degree of collaboration and productive interaction be achieved between the two superpowers if we are to have a chance to avert disaster.

One problem is that from the perspective of the West it is easy to see the repressive strategies being used by the Chinese Communist Party to control China's immense population. To someone raised within the culture of democratic rhetoric, China is a totalitarian state, and to some extent that is accurate although to some extent the real difference is one of the West's greater subtlety in its techniques of social control versus China's overtness. The problem, with the exception of India, is that no other political system is forced to confront the incredible burden, responsibility and threat generated by a political system conducted on such scale and diversity. While it is easy to criticize China, we in the West should be thankful we don't have to figure out how to govern China because it is an enormous and complex system beyond our understanding and capability. We write this analysis, therefore, with the understanding that criticism is the easiest thing in the world when you don't have to bear the consequences of being wrong. While wallowing in our own hypocrisy, we can't afford to ignore the fact we fail to come close to our own too-often dishonored ideals.

The bottom line is that China has existed for more than five millennia. China is undeniably one of the most dominant intellectual and political powers in recorded human history. Whether we are speaking of science, Taoism, the philosophy of Confucius, or the ability of a political system to survive 5,000 years of history while remaining largely intact, there is nothing in the short-lived context of Western governance that comes close to the Chinese experience as a culture or political system. This means that we need to be able to "place China in history" not only in the immediate sense of Xi Jinping and the Communist Party that, after all, are simply passing phenomena in the context of China writ large.

CHAPTER FIFTEEN
Six Case Studies

In an effort to deepen our understanding of what is occurring, this chapter sets out six brief case studies. The studies discuss the effects of autonomous driverless vehicles, the "killer robots" being developed by Russia, China and the US, the loss of higher end jobs that were thought safe from AI/robotics replacement, and the significant vulnerability of major urban "mega-cities" to disruptions of supply and cyber attacks.[427] It also discusses the absolute stupidity that accompanied the widespread development and unthinking commercialization of drone technology.

Much of what is being said about artificial intelligence, robotics, job loss and abuse of the Internet and AI systems by governments, corporations, aggressive identity groups and malicious trolls feels abstract or unreal. Of course we "know" on some level of awareness that if a loss of 50 percent of human jobs occurs over the next 15 years, our society will be in a dismal state. We understand, at least intellectually, that if income inequality and loss of opportunity continue to grow as financial returns increasingly flow away from human labor to capital investors and the incredibly wealthy as corporations and investors shift the systems of production and service to AI/robotics, it will create extreme and widespread social tension. This includes what may be an unsustainable need to support those thrown out of work.

These challenges all represent such overwhelming dilemmas that we have a psychological difficulty in treating them as real. When something looms on levels as enormous as what contemporary civilization has created with AI/robotics, our minds somehow flip an "off" switch that allows us to deny or ignore the emerging reality because we feel helpless in the face of what is occurring. The possibilities are treated as interesting abstractions, not as concrete phenomena that are shaping and defining our future, and likely to impact us directly. These case studies are offered in a preliminary attempt to make the developments real.

CASE STUDY # 1

MILLIONS OF PAID JOBS WILL BE LOST TO AUTONOMOUS CARS, TRUCKS AND BUSES

One example to which we can all relate involves self-driving or autonomous vehicles. Millions of jobs will be lost to autonomous driving technology.[428] US data from 2014 and 2015 provide a sense of the potential job loss just in the transportation sector. A recent news story put the possible direct job loss from automated vehicles at 4.1 million in the US alone.[429] The Bureau of Labor Statistics (BLS) reports that there were 826,510 light truck and delivery drivers actively employed with a mean annual wage of $34,080. The BLS reports 1,797,700 Heavy and Tractor Trailer drivers in the US in 2014 with an average yearly wage of $40,260. There were 665,000 school bus drivers earning an average of $30,950 along with 233,700 taxi drivers and chauffeurs averaging $23,510 annually.[430] It has been reported that Uber has 400,000 active drivers in the US and Lyft 315,000 drivers.

In the gig economy people patch together a mosaic of jobs because that is increasingly all that is available. Although the above figures represent a mixture of full and part-time driving jobs, they indicate that there are at least 4,236,910 people employed in the US alone as compensated drivers. Given the heavy investment that is being made in self-driving cars, taxis, buses, trucks, delivery "bots" and semi-trailers in five to ten years an unknown but significant number of these professional driving jobs will disappear.

Self-driving cars may end up being safer than human-driven vehicles as their proponents claim, but think of the number of human beings whose jobs will be taken away as this occurs.[431] The head of Fiat/Chrysler recently stated that he expects self-driving cars to be a significant part of its sales market in only five years.[432] Uber just signed a deal with Volvo, now owned by China's Geely Holding Group, a multinational automotive manufacturing company, for the delivery of 24,000 self-driving cars beginning as early as 2019.

Geely Holding has also acquired the rights to the most advanced flying car system as research expands to that sector. Self-driving taxis, limousines, Uber and Lyft cars, delivery vehicles, trucks and other types of vehicles we haven't even considered will quickly put a wide range of people out of work. Following the announcement that Uber was testing self-driving cars in Pittsburgh, a stunned 41-year-old Uber driver lamented, "It feels like we're just rentals."[433] She then added: "I kind of figured it would be a couple more years down the line before it was really implemented and I'll be retired by then."[434]

This development has seemingly "come out of left field" and will eliminate millions of jobs across a large and diverse spectrum.[435] It is amazing how stealthily self-driving cars, semi-trucks and buses seem to have popped up out of nowhere (after billions of dollars in investment) to be touted as the automobile and transport option of the future—not the distant future but 2020 or so. We already have driverless taxis in Singapore, automated commuter buses in Finland, self-driving cars being tested with mixed results in Pittsburgh, Michigan, Arizona and San Francisco. Fully autonomous cars are expected to "hit" some French roads this year.[436]

Think about how stunning that development of autonomous vehicles is and how fast the systems are emerging. Then consider how many jobs these developments are potentially eliminating and the types of employment affected. Consider also the job destruction implications of the driverless cars that are already running tourists around London, or the robot couriers that are delivering messages, pizzas, fast food, packages and materials.[437] A new start-up company even wants to put self-driving semi-trucks on the roads. Its stated goal: "*For now*, the robot truckers would only take control on the highways, leaving humans to handle the tougher task of wending through city streets. The idea is similar to the automated pilots that fly jets at high altitudes while leaving the takeoffs and landings to humans." [438] The potential posed by such massive trucks loaded with tons of explosive presents another dire security threat.

This is not simply an issue of job destruction. Security is a vital concern. The use of improvised IEDs and suicide bombings that use vehicles has already become common. Consider how this threat can escalate when driverless vehicles can be controlled remotely or from pilotless drones of the kind that are already available for general consumer purchase.[439] Terrorists won't even have to sacrifice themselves.[440] Some high-tech "whiz kids" are even planning to introduce flying cars in the relatively near future, raising the prospect of unmanned flying IEDs packed with high explosives that will be able to overcome available defensive countermeasures.

CASE STUDY # 2

LETHAL AUTONOMOUS WEAPONS SYSTEMS,
AKA "KILLER ROBOTS"

The militarization of AI/robotics systems poses a threat to our survival. General Paul Selva, the 10th Vice Chairman of the Joint Chiefs of Staff, recently told the Senate's Armed Services Committee that: "the military should keep the ethical rules of war in place lest we unleash on humanity a set of robots that we don't know how to control."[441] There have been increasing calls by scientists, the United Nations and others to ban "killer robots" or "lethal autonomous weapons systems," to which one critic responded: "Shouldn't you have thought about that sooner?"[442] As soon as the controversy arose in the UN both the US and Russia announced they would not allow their weapons development to be restricted by any UN mandates so the proposal will not go anywhere in the Security Council given their veto power as permanent members.[443]

In an interview with *The Times* of London, Stephen Hawking warned about what he saw as the possibility for mass destruction due to the combination of innate human aggression and the technological capacities of AI/robotics systems, stating,

> Since civilisation began, aggression has been useful inasmuch as it has definite survival advantages. It is hard-wired into our genes by Darwinian evolution. Now, however, technology has advanced at such a pace that this aggression may destroy us all by nuclear or biological war.[444]

Specifically on the issue of the development of Artificial Intelligence systems, Hawking indicated that, while AI may be a fantastic development in some aspects, it is equally likely that it will turn out to be the death of humanity. Addressing the specifics of that, Margi Murphy notes, "The human-like abilities of robots continue to develop at incredible pace—with droids now being seen to chase targets and even fire guns."[445] But even that development pales in relation to the other large-scale investments into the military uses of AI/robotics on the part of the US, UK, Russia and China, who are all developing autonomous warplanes, ships, submarines, tanks, drones, machine gun "bots" and much more. Here's a very small "taste" of what is happening as well as rising concerns.

AUTONOMOUS WEAPONS SYSTEMS AND "KILLER" ROBOTS

- Ban on killer robots urgently needed, say scientists: Technology now exists to create autonomous weapons that can select and kill human targets without supervision as UN urged to outlaw them.[446]

- Robots could destabilise world through war and unemployment, says UN: United Nations opens new centre in Netherlands to monitor artificial intelligence and predict possible threats.[447]

- Meet the New Robot Army: Intelligent machines that could usher in an era of autonomous warfare are already here.[448]

- Army Tests New Super-Soldier Exoskeleton: The Army is testing an exoskeleton technology which uses AI to analyze and replicate individual walk patterns, provide additional torque, power and mobility.[449]

- Artificial muscles give "superpower" to robots.[450]

- Robots Replace Soldiers in First Breaching Exercise of its Kind.[451]

- Pentagon: China, Russia Soon Capable of Destroying U.S. Satellites—J-2 intelligence report warns of new dangers to low earth orbit satellites.[452]

- Armed Ground Robots Could Join the Ukrainian Conflict Next Year.[453]

- Russia Says It Will Field a Robot Tank that Outperforms Humans.[454]

Russia is reported to be building a "robot army."[455] Vladimir Putin recently proclaimed that the country that wins the race to develop artificial intelligence would be able to rule the world. He also urged a sharing of knowledge among nations to prevent such dominance. Putin stated that future military conflicts would be between nations' autonomous AI/robotic weapons systems.[456]

In theory, such autonomous weapons should reduce the level of human deaths from warfare, but it is more likely they will increase the probability that conflicts could be started more frequently because human life is not at risk—at least in the initial phases. There is the real possibility that things will then get out of control with the violence quickly extending to human populations. This doesn't even take into account the possibility of "Doomsday Devices" that are set to destroy entire systems if defeat is imminent, such as the Star Trek television series "Self-Destruct" setting that explodes everything when defeat is imminent.[457]

Not to overdo SciFi references, but adaptation of AI/robotics to a wide range of military uses could easily lead to the elimination of millions of humans independent of whether a super-intelligent set of AI "Overminds" or "Skynets" decide to destroy the human race.[458] Research is ongoing into fully integrated weapons systems controlled by an AI source that will be able to direct the operation of thousands of weapons simultaneously.

Since the world's major powers appear to be headed in this direction it creates the specter of incredibly efficient AI/robotic military systems waging war on each other outside ultimate human control in a kind of mythological-like *Ragnarok* scenario of a global conflict between competing AI/robotic "gods." After all, what is the purpose of creating such systems if not to face up to competing AI weapons systems in the hands of another superpower?

CASE STUDY # 3
JOB LOSS IN FINANCE, BANKING, SECURITIES, MANUFACTURING, MEDICINE AND LAW

Job displacement is no longer limited to blue collar or manual work. Finance and brokerage systems have begun to provide investors with AI/robotic financial advisers.[459] Banks are slashing clerical and mid-range staff by the tens of thousands. In the process, human financial advisors are being terminated in significant numbers. Banks are also in the early stages of closing many of their branches that had grown to a total of 90,000 less than a decade ago but are now being under-utilized because many people have shifted to online access and ATMs. This process will accelerate due to cost savings, with the result that tens of thousands of jobs will be lost.

The chief information officer for the Bank of New York Mellon states:

> [B]ots armed with AI and the ability to understand and respond in natural language can be used to answer clients' queries and eventually execute transactions. ... You start with something simple, maybe just offering information, then you start doing transactions.[460]

The "peak human" orientation is explained in the context of continuing bank job cuts:

> Even if revenues recover because of higher interest rates, improving economies, and a rebound in debt trading, new platforms will simply scale up to the higher volumes without needing many more flesh-and-blood operators. Wall Street has reached peak human.[461]

JOB-KILLING DEVELOPMENTS IN BANKING AND FINANCE

- Robots Challenge Banks in Sweden's $524 Billion Savings Market.[462]
- Inside Shanghai's robot bank: China opens world's first human-free branch: "Little Dragon" can chat to customers, accept bank cards and check accounts. She joins a growing army of robot workers in China's cities.[463]
- Deutsche Bank CEO suggests robots could replace half the company's 97,000 employees.[464]
- Danske Bank's Wealth Management Robot Now Has 11,500 Clients.[465]
- Robots in Finance Bring New Risks to Stability, Regulators Warn.[466]
- Banks Shutter 1,700 Branches in Fastest Decline on Record: Lenders keep cutting [jobs and branches] despite growing economy as customers move online.[467]

What is occurring goes far beyond banking and finance. Industrial robots have already replaced very large numbers of humans in manufacturing. Ford Motors has installed advanced next generation manufacturing AI/robotics systems in a facility in Germany. A report indicates:

> Ford ... says that its new bot is capable of pinpoint accuracy, strength and dexterity. ... "Robots are helping make tasks easier, safer and quicker, complementing our employees with abilities that open up unlimited worlds of production and design for new Ford models," said the director of vehicle operations for Ford Europe.[468]

Such initially collaborative human/robotics or "cobot" situations are just a foot in the door until the next phase is reached and fewer and fewer humans are needed in the manufacturing process. Ford, for example, just announced that it was cutting ten percent or approximately 20,000 workers from its 200,000 strong global workforce. For the moment Ford's German facility may be safe for workers but compensatory human job reductions occur elsewhere in Ford's system.

There is an almost irresistible momentum to the introduction of AI/robotics systems. The frequently-voiced cry to "bring manufacturing back to America" ignores the fact that new manufacturing facilities will almost certainly be heavily roboticized with very limited numbers of jobs for human workers.[469] Much the same process is taking place in areas of work such as construction and building trades, autonomous vehicles and roboticized farming, threatening millions of jobs.

CONSTRUCTION AND TRADES,
AUTONOMOUS DRIVING AND FARMING

- Construction robots weld, bolt, lift to beat worker shortage.[470]

- Robot farmers successfully harvested barley without any human hands: Why it matters to you. Robot-ran farms have the potential to increase efficiency in the agriculture industry.[471]

- The Robot Delivery Vans Are Here: A secretive startup has unveiled its new grocery-getter, one of many automakers that are leaving people out of the equation.[472]

- Self-driving cars will leave "third of population jobless" as AI sparks mass unemployment: A third of the population faces sickening unemployment because of self-driving cars[473]

- Truck drivers like me will soon be replaced by automation. You're next.[474]

- Italy's robot concierge a novelty on the way to better AI.[475]

Vast numbers of retail workers and cashiers are increasingly finding themselves without employment.

THE RAPID ELIMINATION OF RETAIL WORKERS AND CASHIERS

- Amazon's first checkout-free grocery store opens on Monday: Using "just walk out" technology to end queues, Amazon Go fires a warning to the high street.[476]

- Walmart planning a cashierless store.[477]

- The US retail industry is hemorrhaging jobs—and it's hitting women hardest: As the retail landscape undergoes a dramatic transformation, analysis finds 129,000 women lost jobs last year.[478]

- Jack in the Box CEO: Swapping cashiers for robots "makes sense" due to minimum wage increase.[479]

- End of the checkout line: the looming crisis for American cashiers: [Job loss in] the retail sector has long had a far greater impact on American employment—and checkout-line technology is putting it at risk.[480]

- Wifi-equipped robots triple work efficiency at the warehouse of the world's largest online retailer.[481]

- How Many Robots Does it Take to Fill a Grocery Order? It once took online grocer Ocado two hours to put together a box of 50 food items. Now machines can do it in five minutes.[482]

- "Robots can't beat us": Las Vegas casino workers prep for strike over automation. At the Tipsy Robot in Las Vegas, a mechanical arm mixes cocktails that patrons order on tablet computers. ... [T]he underlying message at the future-themed bar is that humans are irrelevant.[483]

More cerebral jobs are also at risk. Law firms are increasingly using computers to perform data management and research tasks. These have proven adept at doing the "grunt" work previously done by associate lawyers whose services are no longer needed for discovery, case management, research, or "gofer" work because the AI systems can do the work faster, better and less expensively.[484] The lower costs and far greater efficiencies are more likely to contribute to law firm partners' profits than to lower their fees.

Hospitals and other parts of the medical system are shifting to robots and automated practices to provide a range of assistance, including surgery, diagnostics, nursing and physical therapy.[485] Japan has focused part of its attention on developing robots that are pleasant in appearance to supplant human travel assistance and even insurance agents in the near future.[486] This is because studies indicate humans feel more comfortable with pleasant appearing or quasi-human visage robots than with "terminators."

TEACHERS, WRITERS, DENTISTS, DOCTORS, NURSES, AND BUREAUCRATS

- 'Inspirational' robots to begin replacing teachers within 10 years. This will be part of a revolution in one-to-one learning, a leading educationalist has predicted. Sir Anthony Seldon, Vice-Chancellor of the University of Buckingham, said intelligent machines that adapt to suit the learning styles of individual children will soon render traditional academic teaching all but redundant.[487]

- Robots learned how to write fake Yelp reviews like a human. Researchers at the University of Chicago have trained a neural network, or artificial intelligence system, to write fake reviews on Yelp—and it's pretty hard to tell them apart from a human review.[488]

- Chinese robot dentist is first to fit implants in patient's mouth without any human involvement: Successful procedure raises hopes technology could avoid problems caused by human error and help overcome shortage of qualified dentists.[489]

- London hospitals to replace doctors and nurses with AI for some tasks: UCLH aims to bring "game-changing" benefits of artificial intelligence to NHS patients, from cancer diagnosis to reducing wait times.[490]

- AI better at finding skin cancer than doctors: study.[491]

- Robots "could replace 250,000 UK public sector workers": Reform thinktank says sector could be "the next Uber" and staff should embrace the gig economy amid rise in automation.[492]

The expansion of AI/robotics should be examined closely in relation to what will actually happen to jobs even if the promises of numerous companies to preserve, expand or invest in new manufacturing facilities in the US that Donald Trump talks about actually take place. The Japanese company, Softbank, is one of the world's most prominent companies in the development and application of AI/robotics technology. While Softbank has promised to invest $50 billion in the US and Foxconn $10 billion in US manufacturing facilities, there are concerns.

A core concern is whether such manufacturing facilities will be largely roboticized with only limited and specialized human employment far below historical models. Given that these businesses are centered on AI/robotics technologies and applications, this also raises legitimate questions whether the Softbank and Foxconn production processes will actually make matters worse by accelerating job loss for human workers even while generating significant returns on investment for wealthy investors.

Like Softbank, Alibaba's CEO Jack Ma indicated in the early days of the Trump Administration that Alibaba was planning on bringing 1,000,000 jobs to the US. Even if Alibaba does establish a more significant sales and manufacturing presence in the US, Sherisse Pham writes that nothing close to that number of jobs will be created. In fact, Pham says that Ma's job creation idea does not involve building any manufacturing facilities in the US, but is predicated on a presumed expansion in the number of US stores that would be expected to sell Alibaba's products, with US sellers of Alibaba's products hiring added workers to service a much higher projected sales demand.[493] Of course Alibaba's products would be in competition with other retailers of similar products, and, if it outcompeted some of those manufacturers, that would lead to job loss in existing stores.

Such convenient "new jobs" math is being used to support claims that the future of human employment remains promising. That assumption is not accurate. The US businesses that Jack Ma anticipates would hire human workers are more likely to simply be "ordering conduits" through which consumers could order Alibaba's products that would then be filled and shipped by robotic workers that are increasingly finding their way into online order taking, product retrieval, packaging and shipping. We already see this happening with Amazon and Walmart in warehousing, ordering and delivery sectors.[494] One report indicates that numerous retailers are seeking to imitate Amazon's automation strategy in order to compete. This will inevitably reduce the number of human workers.[495]

CASE STUDY # 4

GROWING HOMELESSNESS

We are witnessing a steady increase in the numbers of homeless in the US and are caught between compassion and contradiction. Anaheim, for example, has recently declared a "homeless emergency."[496] Portland, Oregon is facing a growing problem with an aggressive homeless population with which officials seem to be at a loss to cope.

Growing homelessness in California, New York, the Pacific Northwest, many other urban areas and even rural settings is an early symptom of the degree of homelessness that will expand dramatically as job loss and addictions increase.[497] A recent report indicates that Los Angeles has more than 55,000 homeless. The number is increasing year by year, and LA's ability to provide even basic health, sanitation, and other humane services to such numbers is already grossly inadequate.

RAPIDLY GROWING HOMELESSNESS

- America's homeless population rises for the first time since the Great Recession: a new government study finds 553,742 people were homeless on a single night this year, as advocates lament a crisis that shows no sign of abating.[498]

- "America's new Vietnam": why a homelessness crisis seems unsolvable: Despite approving billions in funds to fight the problem, Los Angeles has seen its homeless population continue to grow.[499]

- L.A. County wants to help build guest houses in backyards — for homeless people.[500]

- "National disgrace": Community fights back as California overrun by homelessness, human waste, needles.[501]

- Deaths among King County's homeless reach new high amid growing crisis.[502]

- Bay Area cities face growing crisis as RVs become homes of last resort.[503]

- Columbia Sportswear may close Portland office over death threats, public defecation by homeless people.[504]

- Homeless people defecating on LA streets fuels horror hepatitis outbreak, as city faulted.[505]

- The Silicon Valley paradox: one in four people are at risk of hunger: study suggests that 26.8% of the population qualify as "food insecure" based on risk factors such as missing meals or relying on food banks.[506]

- California's homelessness crisis moves to the country: California housing costs are spiraling so high that they are pushing the state's homelessness crisis into places it's never been before — sparsely populated rural counties.[507]

- More Homeless People Live in New York than Any Other City. The number reported by the federal government is 76,000 compared to Los Angeles' 55,000.[508]

Speaking of gross, that same report indicates that a dangerous and grow-ing outbreak of Hepatitis A is directly linked to LA's homeless residents since many, lacking public toilets and sanitation facilities, have taken to defecating on streets and sidewalks.[509] LA has also voted to create a $1.2 billion pro-gram to build housing for its homeless population and one recent propos-al advocated guaranteeing decent housing for all of LA's homeless. Another suggested that Los Angeles homeowners could be subsidized for converting parts of their homes or garages into living quarters for homeless people. As impossible as that sounds the fact is that California is facing an epidemic of homelessness that appears beyond its ability to cope. Since the numbers are increasing annually and can be expected to grow steadily higher this presents a moving target.

The increase in homelessness is partly due to people suffering from addictions to drugs or alcohol. Others have serious physical or emotion-al health problems. Many simply can't afford the high rents being charged for even marginal locations in areas such as Silicon Valley, San Francisco, Los Angeles, Portland or Seattle in the West, or New York and Washington, DC along the East Coast. Millions have given up trying to find work after a lengthy period of joblessness and their eligibility for governmental assistance related to the job market has ended. They are relegated, if at all, to surviving through other welfare programs with limited benefits. Begging and increas-ingly aggressive panhandling are but the milder of the likely results

Another aspect of the problem, one that is likely to expand dramatical-ly as AI/robotics joblessness takes hold, is the emergence of what has been called the "voluntary" homeless: people who will never be full contributing members of society but whose survival will require assistance and impose undesirable impacts on the areas in which they circulate. Unfortunately de-structive behavior too often will be the norm among these people cast adrift, resulting in significant health care and policing costs as well as a generalized societal sense of insecurity. A recent report indicates aspects of the problem.

> From the parks of Berkeley to the streets of Brooklyn, and in most every large city in between, they have become an almost inescapable part of urban life. Known by many names—"crusty punks," "crusties," "gutter punks," "crumb bums" and "dirty kids," to list but a few—*this group of young adults has rejected a more traditional 9-to-5 lifestyle in favor of train hopping, panhandling and voluntary homelessness.* And while traditionally tolerated by police and urban resi-dents, these transient groups of the unshaven and unwashed

have been involved in a series of incidents in recent years
… that has municipalities across the country puzzling over
how to address the problem.[510]

Many have found themselves homeless due to the loss of employment, health conditions and overwhelming psychological stress. Homelessness has become an epidemic visible mainly on the streets, parks and alleys of our cities. While we are still in the early phases of the homelessness crisis we already have a serious problem of an inadequate social safety net. Can democracy survive in societies that face the stark option of condemning a significant part of their populace to living "on the street" as jobs disappear?

Pension insecurity, inadequacy and outright default may add to this homelessness crisis, raising the prospect that our most vulnerable members of society will end up betrayed, bereft and ignored. The provision of care for the elderly who make up an increasing share of the homeless, and the growing ranks of the less fortunate across the board is vital if our society is to save its soul. Another analysis elaborates on the intricate and complex problems contributing to the growth in homelessness. This includes the aging of the homeless, with a resulting increase in health issues.

There is profound tragedy in the fact that, as one report indicates: "What policymakers and the general public need to recognize is that the homeless are aging faster than the general population in the U.S. This shift in the demographics has major implications for how municipalities and health care providers deal with homeless populations."[511] This includes the survival chances as older and more vulnerable Americans are pushed out onto streets and alleys.

CASE STUDY # 5
THE EXTREME VULNERABILITY OF COMPLEX SYSTEMS

Our critical supply systems have become increasingly dependent on AI systems. Such vital areas as energy, food, water, and transportation are controlled by computer applications. Power plants, finance, air and rail transport are already vulnerable to attack, hacking and disruption. As AI/robotics has spread we have become more and more vulnerable to their disruption or collapse in ways that present fundamental dangers.

As mentioned earlier 60 percent of the world's population is in the process of moving to mega-cities. Such urban centers are not self-sufficient but depend entirely on intricate supply line networks for food, transport, fuel, electrical energy, and access to virtually any other product on which their populations depend. They are sustainable only when those supply lines work efficiently and securely. When they fail for any reason, including deliberate sabotage, people will die if the problem continues for any substantial length of time. Never mind the dangers posed by our decaying infrastructure, here are some other potential AI-related systemic vulnerabilities that suggest how serious the problem could become.

SYSTEMIC VULNERABILITY

- Cyber infrastructure: Too big to fail, and failing: The cybersecurity industry isn't keeping up with cyber threats, the Atlantic Council's Joshua Corman told a SOURCE Boston audience. And things are about to get even worse.[512]

- U.S. Electric Grid Vulnerable to Unprecedented Waves of Attack: Threats increase as industry scrambles to boost security.[513]

- The next Russian attack will be far worse than bots and trolls.[514]

- Cascading Consequences: Electrical Grid Critical Infrastructure Vulnerability.[515]

- Coal plants' vulnerabilities are largely unknown to feds.[516]

- Deadly attacks feared as hackers target industrial sites. The hacking threat to critical infrastructure in the United States and beyond is growing larger, with nation states and other malicious actors looking to gain a foothold in sensitive technologies to conduct espionage and potentially stage disruptive or destructive attacks.[517]

- Bank of England stages day of war games to combat cyber-attacks: Spate of data breaches in financial sector prompts voluntary exercise to test resilience.[518]

- The next 9/11 will be a cyberattack, security expert warns: A cyberattack of devastating proportions is not a matter of if, but when, numerous security experts believe. Many companies are still running critical infrastructure on Windows XP and other platforms that are unpatchable — meaning they can't be updated for vulnerability and bug fixes.[519]

- Cyber threats and how the United States should prepare.[520]

- Growth of AI could boost cybercrime and security threats, report warns: Experts say action must be taken to control artificial intelligence tech.[521]

- How an Entire Nation became Russia's Test Lab for Cyberwar.[522]

- Hackers attacking US and European energy firms could sabotage power grids: Cybersecurity firm Symantec says "Dragonfly" group has been investigating and penetrating energy facilities in US, Turkey and Switzerland.[523]

We are already at a point of extensive dependency on AI/robotic systems. How long would any major US urban/suburban metropolitan area have the ability to cope with a sudden cessation in the supply of food from around the country or world? What happens to major cities such as New York, Los Angeles, London, Paris, Rome and many others if extremely vulnerable power grids are sabotaged or fail for a variety of other possible reasons?[524]

It would be child's play to knock out several of the primary power transmission centers that have long been demonstrated to have grossly inadequate security.[525] As one report asks, "*what happens on the fourth day after our food and clean water reserves run out?*"[526] We have no answers to that for cities with millions of inhabitants. One analysis of the situation in the UK is that in the event of a serious cyber attack and grid failure the nation is "*four meals away from anarchy.*"[527]

Similarly, as we construct natural gas and oil pipelines spanning thousands of miles, what happens when a saboteur decides to blow the lines up in the middle of winter in the northern half of the US? It doesn't even have to be a violent physical intervention with explosives or poisoned water and air supplies. Systems that are dependent on computer programs to operate are vulnerable to sophisticated hackers. There is no reason to believe that ordinary utility systems, including nuclear power plants, are immune from intrusion and manipulation or that water supplies couldn't be poisoned.

Nor need the target be a nuclear plant as opposed to a more traditional power generation or transmission facility. It may not even take a physical assault to sabotage such facilities as opposed to a hacker's intrusion that shuts down critical elements or some kind of targeted EMP or Electromagnetic Pulse event—or indeed, an extreme weather event, the like of which has already devastated the systemic grid of Puerto Rico. Our increasing dependence on such systems offers a target for nature or for those within and outside the country who want to harm us for whatever reason.

Interdependence and the instantaneous transmission of effects are in many ways positive developments, particularly in ordered systems with a substantial degree of stability. But they are also weaknesses and points of strategic leverage for those who seek to "crash" the system. Warren Buffett recently described potential cyber attacks on our system as the biggest threat faced by Western society.[528] This is because so many of our systems are interconnected to the point they could collapse in ways that paralyze our economic, supply and security systems. [529]

A very specific example of complex system vulnerability and Artificial Intelligence systems is High Frequency Trading ("HFT" or "algorithmic trading"). HFT developed in the late 1990s and early 2000s as a mechanism for leading financial firms to automate buy/sell decisions at a speed so fast that they could place trades in response to market information faster than anyone else, thus earning above normal profits on each trade. And the race was on: the search for ever more sophisticated algorithms to get a jump on their competitors on each new piece of financial data.

Because an HFT algorithm executes trades in stocks, commodities, foreign exchange, and financial instruments nearly instantaneously according to relatively simple sets of rules—such as observing a particular pattern of trading activity in a stock—the HFT trades primarily on the basis of movements in the markets and not to real world factors affecting the value of those markets to real world investors—nor indeed, to their impact on societies. Moreover, a significant fact is that a single HFT execution can trigger a response from another HFT firm, which then triggers reactions from other firms in a possible cascade or ripple effect that happens in micro- or milliseconds. This can cause significant market fluctuations entirely unrelated to real world events or the actual health of the companies whose shares are being traded. Or, again, to the wellbeing of human societies.

The potential for unanticipated consequences and catastrophic feedback loops generated by HFT has been demonstrated. Famously, HFT is thought to have exacerbated the May 6, 2010 "flash crash" that caused the Dow Jones Industrial Average to drop nearly 1,000 points in only 36 minutes. In 2015, U.S. prosecutors charged a London stock futures trader with market manipulation that may have triggered the event using trading algorithms designed to drive down the price of a particular stock. This in turn led other algorithms to react and depress not only that stock but others in a vicious feedback loop.[530] Although the 2010 crash was significant and newsworthy, the reality is the such flash crashes are common.[531]

The "AI Contagion" metaphor is powerfully applicable here. The dynamic sequence follows this progression: Humans first create the HFT platforms. They then purchase faster computers, "collocate" their HFT platforms in the same building or room as the computers conduct the share sales and purchases, hire the best programming talent to develop new algorithms, and so on. Then they set them loose to trade, seeking to outperform other platforms. Platforms that perform well in the invisible competition taking place in the computer servers make more money for their owners, who then invest more to create even better platforms to compete against the next generation. Humans, at a certain point in this process, exist to feed the machines because the machines generate benefits for them even though the humans reach a point where they neither comprehend the machines' activities nor engage with them in any meaningful sense.

New York Times writer Conrad De Aenlle reports that investment firms are increasingly turning to artificial intelligence to augment their short and long term investment decisions while still maintaining human control to address novel situations for which the AI is not prepared.[532] In this vision of AI/robotics, the value of a "human in the loop" means that AI/robotics brings both risks and also opportunities for eliminating routine work that can be automated while expanding opportunities for human work that cannot.

The point of all this is that our vulnerability to accidental, reckless or deliberate actions that weaken or even crash a critical element of our system—to say nothing of disregarding real world impacts and needs—has been increased by orders of magnitude. Dependence on computers and AI systems, coupled with the long distance transport of foodstuffs, natural gas and oil, electronic banking, power grids, and other similar requirements of a complex civilization mean that it will be surprisingly easy to cripple the critical systems Western democracies take for granted.[533]

George Baker and Stephen Voland outlined the scope of potential threats to US infrastructure, and indeed to that of any nation that has become heavily dependent on AI and cybersystems. In "*Cascading Consequences: Electrical Grid Critical Infrastructure Vulnerability,*" they explain:

> If there were a prolonged nationwide, multi-week or multi-month power failure, neither the federal government nor any state, local, tribal, or territorial government—acting alone or in concert—would be able to execute an effective response. This bleak outlook results from understanding that so many critical infrastructures depend on electricity.[534]

They add:

> Without electric power, the goods and services essential to protect life and property would be at risk by day three or perhaps longer depending on preparedness levels. Consequently, it is vital that citizens, households, communities, businesses, and governments be as informed and prepared as possible.[535]

Nor should we deceive ourselves about being able to immediately "fix" the collapse. Lack of resources, panic as food and water become scarce, a rapid disappearance of security, inadequate personnel to repair and assist, limited supplies, along with many other critical deficiencies limit our ability to respond. We need only think about the harms caused by the destruction of Puerto Rico's poorly maintained power grid and the fact that it took 11 months to restore the range of electrical service after Hurricane Maria.[536] And in Puerto Rico we faced a situation where 20,000 FEMA personnel were able to assist, the weather was not frigid and deadly in the worst sense, and several hundred thousand Puerto Ricans were able to move to the mainland US to escape the crisis.

As Baker and Voland warn:

> Citizens of the United States are dependent on secure and reliable electric power for their current way of life. If electric power were not available for weeks, months, or even a year, then cascading impacts would degrade multiple critical infrastructures, for example:
> - Water supply and wastewater treatments;
> - Telecommunications and the internet;
> - Food production and delivery;
> - Fuel extraction, refining, and distribution;
> - Financial systems;
> - Transportation and traffic controls;
> - Government, including public works, law enforcement, and emergency services;
> - Hospitals and healthcare;
> - Supply chains; and
> - Other critical societal processes.

In the face of such a collapse they conclude: "Loss of life could be catastrophic. Life itself would change."[537]

CASE STUDY # 6

DRONES: WHY DID WE GIVE TERRORISTS AND DRUG CARTELS AN AIR FORCE WHILE DESTROYING HUMAN DELIVERY JOBS?

FBI Director Christopher Wray has warned that terrorist drone attacks are likely in the US.[538] Both the US Air Force and Chinese military have been developing a "drone swarm" technology that would make it extremely unlikely defenders could successfully avert attacks in which large numbers of small drones would make autonomous attacks on targets.[539]

AI/robotics even comes into play on this level. Drones are not cute little toys "just like" small model airplanes. They are weapons and airborne delivery systems enjoying AI capabilities. ISIS has already utilized remote control drones in the Middle East against US-backed forces though they are at the more primitive stage. From the beginning of the emergence of increasingly sophisticated drones it has been mindboggling that Western governments are allowing the sale and use of drones in civilian and commercial sectors.[540] Remote controlled or programmable drones provide extremists with an "air force" and an almost invisible and unstoppable delivery or guided missile system capable of evading surveillance, fences, concrete barriers and other defensive measures.

Internal and external attackers will be able to use the new technologies for attack and sabotage, allowing them to spread death and disruption in ways not heretofore possible. This includes not just drones, but automated vehicles and cyberwar attacks on critical systems.[541] We are seeing a continual flow of reports on the improvements in drone technology, guidance and payload capacity for purposes of commercial product delivery.[542] This includes the ability to carry payloads up to 100 pounds or to drop deliveries by parachute from hovering drones. A Canadian company recently announced it would soon have a drone capable of carrying payloads of up to 400 pounds.[543] Even a quick look at the ongoing developments related to drone technology, its use and availability is enough to demonstrate the dangers of our careless actions.

DRONES AND TERROR ATTACKS

- **China Rapidly Building Advanced Arms for Use Against U.S.: Space weapons, drones using artificial intelligence priority in Beijing military buildup.**[544]

- **U.S. officials warn Congress on risks of drones, seek new powers.**[545]

- **Terrorists could weaponise deadly plague disease by releasing it as a cloud above cities killing thousands, experts warn.**[546]

- **How do we thwart the latest terrorist threat: swarms of weaponised drones?**[547]

- **Homeland Security bulletin warns of weaponized drones and threat to aviation.**[548]

- **Something else to fret about: ISIS mounting dirty bombs on drones.**[549]

- **A 100-Drone Swarm, Dropped from Jets, Plans Its Own Moves: Once launched, the swarm can decide for itself how best to execute a mission.**[550]

The security implications of the nearly invisible and silent airborne systems represented by drone technology are obvious. The Chinese military is concentrating on developing drone technology and considers it central to its capability in what is called asymmetric warfare. As with the US military, this concentration includes the ability to launch large-scale and autonomous AI-dependent drone swarms that are likely to prove impossible to stop. A representative of the US Department of Homeland Security indicated in Congressional testimony that: "Terrorist groups overseas use drones to conduct attacks on the battlefield and continue to plot to use them in terrorist attacks elsewhere. This is a very serious, looming threat that we are currently unprepared to confront."[551]

Illustrating the danger Patrick Knox warns: *"Black Death could once again wipe out millions in Britain if terrorists sprayed an antibiotic resistant form above a city, according to bio-terrorism experts. The plague killed a third of Europe's population in the Middle Ages—but scientist and governments fear it could return."*[552]

CHAPTER SIXTEEN

If No One is Working, Who Will Buy the Goods and Services?

We need to understand we are now in a transitional moment when unemployment is still reasonably low and wages relatively level and widely available even though flat and without significant real increases. The system remains superficially healthy because the job loss "tsunami" hasn't yet hit.

Let's assume that predictions of a 47–50 percent job loss over the next decade are accurate. As the employment erosion starts to take hold there will not only be added pressure on government to prop up the unemployed, the underemployed and the unemployable—but there will be reduced earnings experienced by many businesses caused, inter alia, to the drop in the buying power of formerly middle class purchasers who were employed but no longer are working at all or have slid downward on the pay scale.

As the AI/robotics workforce continues to develop and usurps additional human job sectors across a wide range of work activity, the ranks of those in need of governmental support will grow and so will the amount of the subsidy required to support them. The impacts will not be evenly distributed. Some areas of economic activity will thrive while others will find their work and social environments deteriortating, leading to a state of permanent depression.

Those who wonder about how economically developed nations' GDPs can continue to rise while wages and employment remain flat or decline are missing a fundamental point: GDP bears no direct relation to human employment or well being. Modern democracies' consumer societies depend on consumers who are not only willing but are able to make purchases on a regular basis of goods and services beyond simple subsistence needs and survival. Those who own the AI/robotics systems will at some point discover they have bankrupted a substantial number of the consumers their companies depend on as buyers.

AI-generated unemployment will have widespread effects as the purchasing power of the consuming base falls as, in the intermediate period, wealth from AI savings is transferred from wage payments to a narrow tier of extremely wealthy investors' as increased profits. In advanced economies broad distribution of purchasing power is vital. We are already seeing that distribution of purchasing power is being distorted. Here are the results of a recent report by the UK's House of Commons:

> World leaders are being warned that the continued ac-
> cumulation of wealth at the top will fuel growing distrust
> and anger over the coming decade unless action is taken to
> restore the balance. An alarming projection produced by the
> House of Commons library suggests that if trends seen since
> the 2008 financial crash were to continue, *then the top 1%*
> *will hold 64% of the world's wealth by 2030* [italics added].[553]

THE AI-ROBOTICS BUBBLE

It is vital that our policy makers understand that this present economic moment is in actuality a limited time period "bubble" generating spectacular returns on investments for a limited number of people including as the most glaring examples Jeff Bezos, Bill Gates, Mark Zuckerberg and Elon Musk. Bezos is reported to have gained $39,000,000,000 (billion) in personal wealth just in 2017. This transitional phase offering a temporary picture of economic improvement goes on record as producing "more with less." But that's only for now.

At the moment vast sums of money are being received by those fortunate enough to have captured the advanced edge of the AI/robotics phenomenon. Not that far into the future, the AI/robotics Bubble will burst for many investors, in large part because of the developing and interactive factors discussed here in *Contagion*. These include job destruction, age demographics, government pension and other subsidy obligations, and the lowering earnings or incomes of very large numbers of people who suddenly find they are no longer members of the middle class.

Even with the various expert predictions about when events will come to a head, no one can say for certain just when the decline will occur. Different areas of economic activity will be affected differentially and to greater and lesser degrees. Ultimately, however, the declining economic "fundamentals" will permeate the overall market and generate a severe and longlasting economic downturn. The consequences are dire. As companies weaken and wealth is increasingly concentrated in the hands of a limited number of incredibly wealthy individuals and corporations, the value of the stock shares on which pension funds rely to fund their pensioners' payments will plummet. This is because the incredibly wealthy are "different." One of those differences is that the "filthy rich" operate in a luxury consumer market and not the general consumer marketplace where the vast bulk of the population make their purchases of goods and services.

As human jobs disappear more rapidly and in increasing numbers, consumers' purchasing power will decline. Ironically, so will the "paper" value of many investors' wealth because the well-off base of consumers on whom the businesses depend will be deteriorating in many contexts. The fact that a business can produce all needed products and services using AI/robotics systems becomes increasingly meaningless, and even destructive, in an economy where large numbers of people can't afford to pay for products and services at the levels of financial return anticipated for the investment and business models of particular industries.

WHAT ARE OUR REAL UNEMPLOYMENT LEVELS?

Statistics lie. Those relied on by politicians to make themselves look good are particularly unreliable. The fact is that real unemployment is considerably higher than as claimed by the official figures. Important flaws undermine the validity of the employment projections reported in government data. One flaw is that the statistical gauges being used are nearly all quantitative, or stand for the proposition that "a job is a job is a job," and fail to capture the qualitative nature of what is happening in our job markets. Another analytical flaw is counting multiple part-time jobs held by a single person as if there were more than one employee for statistical purposes, giving a false impression of the total number of people employed.[554]

Tim Dunlop explains: "A new report by the US-based National Academies of Science Engineering and Medicine suggests that not only has the automation of work barely begun but that the ways in which we measure the effects of technology on employment are inadequate to the task." [555] He adds:

> [W]e don't have a regular source of information about workers in part-time and other sorts of casual employment. Nor do we have good information about investment in computer technology at either the level of the company or of any given occupation. Also lacking is long-term information about the way in which skills within particular jobs are changing, as well as data on how effective educational practices are in preparing people for work. Such information gaps undermine our ability to respond appropriately to technological change and its effects on employment.[556]

Another flaw is that the reports generally rely on historical trends that are no longer valid for projecting future conditions in a radically changing economy. The reports also fail to deal with the fact that large numbers of working age people are not counted in the unemployment figures due to their being "outside" the job market by choice, or because they have given up on seeking work after long periods of trying to find jobs. They also don't count millions of people in colleges or advanced education who have been told that education is critical for their future only to discover that they have no employment future of the kind for which they were preparing but do have a significant amount of educational debt they can't repay.

As of mid-2018 the student debt total has soared above $1 trillion and continues to rise even though jobs are disappearing or changing radically in terms of the skills and training required.[557] Nor does the jobs data take account of the large and growing number of homeless Americans who are overwhelming public services throughout the country, particularly all along the West Coast and New York City areas but increasingly in more central and rural areas of the nation.

Unemployment in certain sectors involving disadvantaged and economically depressed niches, such as young black males in urban areas, is consistently "off the charts."[558] There has been some progress in the recent jobs statistics on minority employment during the first year and a half of the Trump administration and that should be appreciated. But there are still many who are not being counted and the jobs are mainly low-level with little promise of mobility and opportunity. Much the same situation is found in EU countries in the south of Europe in the case of younger residents. Many of the younger people in Greece, Italy, Spain and Portugal are leaving their homelands in an effort to find work.[559]

The problem of those outside the workforce is not limited to young people, retired people, urban minorities, or the homeless. A significant number of people in the US are not looking for work, not interested in working, or have surrendered to the frustration of having lost their jobs, and, after seeking unsuccessfully to find new employment have dropped out of the labor force. Even if an experienced white male over forty years of age or so loses his middle class job, it is extremely difficult for that person to find employment of equivalent quality. This phenomenon of experienced workers losing decent paying jobs with limited prospects of acquiring equivalent employment is part of the ongoing decay of our middle class. [560]

On the other hand, if not yet terminated, laid off, downsized or one of the other modern terms for being fired, many such older workers are holding on to their jobs with tenacity,[561] even though workers at the upper end of the age range are likely to have less energy and ambition and, as a result, will work more slowly and less productively.[562] Nonetheless, an older worker operating at 90 percent efficiency is still making a significant work contribution, adding to the economy, reducing the need for governmental subsidies, and not simply stagnating. They can also transfer positive work skills, experience and discipline to younger co-workers.

The lack of retirement savings on the part of tens of millions of people in the US, limited and increasingly insecure private and public pensions, and uncertainty over the cost of living increases and financial insecurity generally is causing large numbers of Americans to work beyond their normal retirement years if they have the opportunity to do so.[563] This offers a mixed blessing. It reduces the drain on retirement systems and provides tax contributions from workers at the upper ends of their pay scales, but it also blocks younger workers' access to employment.

DEPRESSED, OLD, ADDICTED

In July 2018 approximately 130 million people were working in full-time employment (35 hours or more) in the US, while another 26 million were in part-time employment. Recent US Department of Labor statistics indicate that the US already has 95 million people of working age who could work but for various reasons are not doing so. These "outside the workforce" individuals are not counted in the US employment data that shows unemployment in the low 4 percent range.[564] Non-workers receiving disability payments from the Social Security system are, for example, not counted as part of the unemployment calculations and the number of such people is increasing. An estimated 11 million people in the US are on Social Security disability subsidies, with another 51 million on retirement or survivor support. Over 62 million Americans are receiving some form of Social Security.

The Labor Department's report also indicates that while outsourcing and the use of industrial robots started changing our economy 30 years ago, an increasing number of Americans lack the skills or education to be productive workers of the kind required to function in the new economy. This skills gap will become worse as AI/robotics systems grow in sophistication, diversity of application and function and the number of human workers capable of understanding and working within this accelerated economic context proves inadequate.

Chana Joffe-Walt has described the situation, including the large-scale increase in the ranks of those on disability subsidies and the fact that they are not even counted in our employment statistics.

> In the past three decades, the number of Americans who are on disability has skyrocketed. The rise has come even as medical advances have allowed many more people to remain on the job, and new laws have banned workplace discrimination against the disabled. Every month, 14 million people now get a disability check from the government. The federal government spends more money each year on cash payments for disabled former workers than it spends on food stamps and welfare combined. Yet people relying on disability payments are often overlooked in discussions of the social safety net. The vast majority of people on federal disability do not work. Yet because they are not technically part of the labor force, they are not counted among the unemployed.[565]

Joffe-Walt adds:

> In fact there are signs that job loss and social turmoil are already having profound effects on Americans. Depression, for example, is the leading cause of disability. Death and suicide rates for whites in US suburbs are increasing to record levels and drug and alcohol addiction is soaring. Opioid addiction is soaring as are deaths from overdoses.[566]

But putting the disability and depression issues aside, when we exclude a variety of people from the base on which the employment data are calculated who are capable of working but not actively seeking work, then it is unsurprising that the official unemployment figures appear deceptively low. Yet this deception is continuing because those who benefit from claims that everything is going great on the employment front don't want to provide any substantiated reasons for rocking the boat.[567]

Though the official employment figures do not highlight the fact that, as an aging population continues to skew demographic realities because retired people are not counted in the work force but require increasing amounts of social resources, fewer workers will be contributing to governmental revenues through tax payments on earnings from labor, we have long understood it. But the same process in relation to general tax revenues due to the shift away from human workers toward AI/robotic systems has as yet to be well addressed, along with the social problems to which it gives rise.

As their tax burdens increase, those who are still working will be increasingly resentful of paying taxes to support millions of people whose productive work lives are behind them. This could become even more socially destructive to the point that the "haves" begin to use terms such as that of "useless eaters" concerning those with disabilities or other debilitating problems. This was done by the Nazis to justify the "elimination" of certain people as being weak burdens unfairly preying on the limited resources of society.[568]

We cannot allow this attitude to develop. This issue becomes even more significant in the face of reports that many people soon will be living to an age of 90 years or more and that a majority of people have little or nothing saved for retirement. As of 2016 the US had at least 72,000 people over 100 years of age and due to advanced medical care that number has increased rapidly, up 44 percent just between 2010 and 2016.

This aging of American society means that retirees will be receiving governmental support well beyond the original actuarial assumptions made in determining the fiscal soundness of Social Security and other public and private sector retirement programs. A 2017 report indicates that Social Security will be paying out more than it receives within five years.[569] The retirement systems can't work as designed. The promises that have been made about pensions and health care cannot be fulfilled on the federal, state, local and private levels without dramatic new policy actions and even those will offer only partial solutions. Since many pension programs are invested in the stocks and bonds of the financial markets and derive essential revenues from those sources, our welfare depends on a system that could disintegrate due to the impacts of AI/robotics on business in a slumping and unequal consumer society.

PART III

01010101010101010101010101

WHAT IS AI AND WHAT IS IT DOING TO US?

CHAPTER SEVENTEEN
Arificial Intelligence Systems: The Basics
CONSTRUCTING THE
ARTIFICIAL INTELLIGENCE GENOME

In the second decade of the 21st century the pace of AI development has accelerated and is continuing to pick up speed. Brian Fung offers a reason for the recent explosive development of AI systems: the ability to design neural networks that allow AI to better interpret data.[570]

> [S]cientists achieved a breakthrough in the way they thought about neural networks, or the systems that allow AI to interpret data. Along with the explosion of raw data made possible by the Internet, this discovery allowed machine learning to take off at a near-exponential rate, whereas other types of AI research are plodding along at merely a linear pace."[571]

One way to get a conceptual handle on Artificial or Alternative Intelligence is to think of human DNA, RNA and genetic coding. Of course this is simplistic, but amid all the computer talk about algorithms and coding, the fact is that human genetics is doing much the same thing. Genes are like little biological machines that determine a great deal of who we are, what we look like, how we behave, what limits are imposed on us, and what we are capable of doing. Writ large, the human genome is a set of codes that provides the basis for how we act and a ceiling in terms of what we can do. These specific codes written into our genes, our DNA and RNA, define us and provide the foundation for our basic capacities.

This genetic coding does not mean that we are in a rigid straightjacket but it does establish the foundation on which external stimuli is "loaded" into our experiential and conceptual structures to "flesh out" the being we are to become. As can be seen in looking at the analyses that follow, algorithms are an essential part of the mathematical or genetic coding that is going into the "genetic structures" of Artificial Intelligence systems. At this point what is happening is primarily an attempt by human coders and

innovators to replicate aspects of human capacity through mathematically
-based programming.

What is on the verge of occurring, and perhaps already lurking in the
depths of AI systems, is a situation in which the AI systems not only begin
to learn from their own experiences but write and rewrite their own genetic
codes. In essence they are creating their own evolving "machine genome,"
going through a wide variety of transmutations at generational speeds more
akin to rapidly mutating viruses than the far slower changes undergone by
human biology. As this takes hold, we, the human race, are in trouble. We
will have created a competitor that is defining its own thinking processes and
perhaps thereby creating its own values—which may leave it with no reason
to think of us as benign, enlightened, capable or trustworthy given the less
than admirable track record of the human race.

Hawkins and Dubinsky offer a useful perspective about the categories into which AI research and machine learning fall. They describe three
basic approaches, Classic AI, Simple Neural Networks and Biological Neural
Networks, saying:

> Our feeling is that the term "artificial intelligence" has
> been used in so many ways that it is now confusing. People
> use AI to refer to all three approaches ... plus others. ... *The*
> *term "machine learning" is a more narrowly defined term for*
> *machines that learn from data, including simple neural mod-*
> *els such as ANNs and Deep Learning.*[572]

As one innovator in the field explains the machine learning process:

> We don't describe what we're doing as AI—we call it,
> 'automating human-intensive knowledge work,'" ... "Proba-
> bilistic techniques are used to 'train' machines as they churn
> through the data, until they are able to see patterns and reach
> conclusions that were not programmed in at the outset."[573]

Other brilliant human minds working on the challenges of creating "alternative" intelligence systems are attempting to create systems that have the
ability to process their own experience to engage in what is being called deep
learning. Such systems are increasingly able to teach themselves and use that
new and expanding ability to improve and evolve. The ability to do this is
moving ahead with amazing rapidity. One recent report on Google's research
into the development of AI and deep learning supports this conclusion.

Google Brain focuses on "deep learning," a part of artificial intelligence. Think of it as a sophisticated type of machine learning, which is the science of getting computers to learn from data. Deep learning uses multiple layers of algorithms, called neural networks, to process images, text and sentiments quickly and efficiently. The idea is for machines to eventually be able to make decisions as humans do.[574]

Such efforts are being worked on around the world and are evolving rapidly.[575] As deep learning in AI systems improves, we should have no illusions about their ability to create conceptual structures that continuously improve and expand while internalizing an enormous range of information far beyond a human's ability to access, develop, process, interpret and utilize.[576] The issue of sentience and independent awareness by an AI system is one about which we have little understanding and no actual experience.

Whether performed by biological cells and neurons or designed into intricate silicon chips or through quantum dynamics, the idea of self-learning systems that grow and evolve through their own experiences sounds much like the way we would describe the trial and error ways humans learn from experience and adapt to the stimuli of their environment. This includes creating conceptual and interpretive structures that integrate, assess, and utilize what is learned. Except now, rather than humans, we are talking about the rapid development of AI/robotics systems that possess this adaptive learning capability. Will Knight provides one such instance in the *MIT Technology Review*:

> An AI program trained to navigate through a virtual maze has unexpectedly developed an architecture that resembles the neural "GPS system" found inside a brain. The AI was then able to find its way around the maze with unprecedented skill. The discovery comes from DeepMind, a UK company owned by Alphabet and dedicated to advancing general artificial intelligence. The work, published in the journal *Nature*, hints at how artificial neural networks, which are themselves inspired by biology, might be used to explore aspects of the brain that remain mysterious. But this idea should be treated with some caution, since there is much we do not know about how the brain works, and since the functioning of artificial neural networks is also often hard to explain.[577]

It requires no great intellectual leap to conclude that AI systems could not only learn to surpass humans in many aspects of their intellectual capability but invent new capabilities completely outside human abilities. Margi Murphy writes:

> What sets humans apart from machines is the speed at which we can learn from our surroundings. But scientists have successfully trained computers to use artificial intelligence to learn from experience—and one day they will be smarter than their creators. *Artificial Intelligence is becoming incredibly sophisticated and scientists aren't sure how. Now scientists have admitted they are already baffled by the mechanical brains they have built, raising the prospect that we could lose control of them altogether.* Computers are already performing incredible feats—like driving cars and predicting diseases, but their makers say they aren't entirely in control of their creations. This could have catastrophic consequences for civilisation, tech experts have warned.[578] [emphasis added]

Consider the implications of a system that can access, store, manipulate, evaluate, integrate and utilize all forms of knowledge. This has the potential to reach levels so far beyond what humans are capable of that it could end up as an omniscient and omnipresent system with (as Masayoshi Son indicates) an IQ of 10,000 that could (understandably) see humans as a form of irritating and useless "bacteria."

Is AI Humanity's "Last Invention"?

We may lose control of AI systems sooner than we think. Nick Bostrom, an existential risk philosopher at the University of Oxford, has already declared, "Machine intelligence is the last invention that humanity will ever need to make."[579] One analysis of what could happen with AI notes:

> Around 2025, some predictions go, we might have a computer that's analogous to a human brain: a mind cast in silicon. After that, things could get weird. Because there's no reason to think artificial intelligence wouldn't surpass human intelligence, and likely very quickly. That superintelligence could arise within days, learning in ways far beyond that of humans. [580]

We are creating entirely new forms of alternate intelligent awareness that will operate according to their own rules. Although we are still using the "artificial" label, these will be *alternative* forms of intelligence that will incorporate some aspects of human thought and capability that we initially program into the systems, but evolve their own unique forms of intellect, perception, goals, choice, morality and action.[581] It is possible, and perhaps even probable, that biological humanity will be left out in the cold as these AI systems develop.

What Do We Mean by Intelligence?

Intelligence is power, potential, information and data capacity, and integrative capability that allows us to absorb, structure, understand, classify and act effectively. At best we only understand limited elements of the human version and those not very well. It is optimistic to think that those who are designing AI systems will be able to understand or capture the full range of qualities and processes that go into what we call human intelligence. Nor do they have the capability to program such subtle abilities into AI systems even if they did fully understand the complex and fluid nature of human intelligence.

The ironic fact is that there are numerous aspects of human thought and perspective that we should not want to be developed by incredibly powerful AI entities. When we look at what we are doing with the development of AI, honesty should tell us that we are being naïve, ignorant and arrogant. We act as if we lack any honest self-awareness of the true propensities of the human race, or at least of far too many members of our species. We also ignore our history of organizing into political collectives grounded on warped ideologies and the thirst for power and dominance. It is unimaginable to think of the twisted use of AI had a Hitler been in command of such power.

Once a certain "turnover" point in Artificial/Alternative Intelligence is reached, the systems that evolve will be more than simply intellectual clones of humans. Machine sentience and machine sapience will be of different orders than human sentience and sapience. In some ways the AI systems may entail more, in others, less. They will be more powerful in numerous ways, and in control of a multitude of contexts and tasks now being done by humans. Such AI systems or entities will be capable of doing many things humans cannot do or even understand. Thinking we can harness or control such entities is delusional—let alone anticipate bizarre and unintended conclusions that AI may derive, which run contrary to human common sense or well being.

We do not even come close to understanding all the input of data and experience that goes into the formation and development of a human being. In the machine learning, self learning and deep learning world of AI/robotics there is no reason to think we will understand what has been created, much less be in control of AI systems that will increasingly possess amazing speed and rapidly increasing intelligence and sophistication.[582]

Who Is Winning: Exascale, Teraflops, and the Coming "Quantum Leap"

Our challenges are numerous even if we only had to deal with the expanding capabilities of AI systems based on the best binary technology. The incredible miniaturization and capability shift represented by quantum computers has implications far beyond binary AI. Consider a report on the developments in miniaturization of data processing capacities that explains:

> By manipulating the interactions between individual atoms, scientists report they have created a device that can pack hundreds of times more information per square inch than the best currently available data-storage technologies. ... By comparison, tech companies today build warehouse-sized data centers to store the billions of photos, videos and posts consumers upload to the internet daily.[583]

A vital part of what is taking place in the AI/robotics context is the ability to miniaturize data storage and management technologies to the point that vast amounts of information can be contained and/or accessed in very tiny systems at the speed of light. Research into such miniaturization of data acquisition, storage and use is ongoing and moving rapidly, although the work on technological breakthroughs such as quantum computers that will operate at speeds that are multiple orders of magnitude beyond even the fastest current computers is still at an early stage and will take time to develop beyond the laboratory context. The work currently being done to shrink electronics to the atomic level in terms of data management and storage is astonishing.

The physical limits of current chip technology are being challenged as scientists work toward new versions based on atomic scale, quantum technology and neurosynaptic designs that will immensely expand the ability to store and process data. Here's an example of what is occurring to overcome the limits of traditional chip technology.

As conventional microchip design reaches its limits, DARPA is pouring money into the specialty chips that might power tomorrow's autonomous machines. ... The algorithms that will drive tomorrow's autonomous cars, planes, and programs will be incredibly data-intensive, with needs well beyond what conventional chips were ever designed for. This is one reason for the hype surrounding quantum computing and neurosynaptic chips.[584]

In order to understand where things are heading, we must digest how computer processing capacity has developed at unimaginable rates over what now seems like a high-tech micro-second. One expert illustrates that growth by reference to terms related to scale: explaining megaflops, gigaflops, teraflops, petaflops, and exascale data processing capabilities. Even here, once (or if) scientists are successful in achieving a reliable quantum computer system, the best exascale system will pale in relation to their reduced size and exponentially expanded capacity. This will lead to the creation of AI/robotics applications and technologies we can now only imagine. In asking "who's winning" the race for expanded computing power, one analysis explains:

When it comes to computing power you can never have enough. *In the last sixty years, processing power has increased more than a trillionfold.* The first machine to bear the label 'supercomputer' was the CDC 6600. Designed by Seymour Cray and released in 1963, it boasted a groundbreaking performance of three megaflops—or 3,000 floating point operations per second. *That's roughly the same amount of processing power found in the Atari 2600 games console released just 13 years later* (1977). The CDC 6600 cost around $8 million; the Atari 2600 cost a mere $199.

The benchmark for processing power jumped to gigaflops in the 1980s and accelerated to teraflops in the 1990s. In 2008 the first petascale computer systems appeared, capable of processing in excess of one petaflop, or a quadrillion (10^{15}) floating point operations per second. And in June 2016, China unveiled the Sunway TaihuLight, a 41,000-processor (10.65 million-core) system capable of 93 petaflops, and which is currently the fastest computer in the world.

But now the race is on to reach the next level of supercomputing. *Japan, France, China and the US are all pushing*

towards exascale-level systems, with a potential performance
of one quintillion operations per second—or a billion billion
(10^18) FLOPS if you prefer.[585] [Italics added]

If scientists are successful in their quest for quantum computers the implications are far, far beyond anything that we can envision with the current digital designs. Several companies, including Google, have at least demonstrated proof of concept in laboratory contexts on extremely limited versions using "Quantum Bits" or Qubits as the information handling element of the system. When fully developed, quantum computers will have data handling and processing capabilities far beyond those of current binary systems. When this occurs in the commercialized context, predictions about what will happen to humans and their societies are "off the board."

Such quantum mechanics-based systems that "entangle" qubits in environments near absolute zero use informational quantum bits that manifest probabilistic capabilities beyond the 1 and zero, fixed identities, of the digital or binary systems now in use. Researchers indicate that such entangled qubit-based systems will be capable of performing simultaneous computations at levels of processing information billions of times faster and more complex than the best existing computer. Once this is achieved, or even developed at intermediate steps, those incredible informational capabilities will take us to dimensions we neither understand nor control.

If and when researchers are able to construct quantum systems, we will have entered uncharted territory. Projections made only a very few years ago predicted that quantum computers would not become a commercially usable technology until at least 2050, but IBM has already achieved a 50 qubit system that existed for a microsecond or so, and the Chinese are claiming significant advances in quantum research. Google almost immediately trumped IBM's 50 qubits with its own 72 qubit system, and several companies are striving to raise the ante to 100 qubits. Intel has produced a 49 qubit system and the company's head of quantum research suggests commercialization could be achieved in another decade. Given that the addition of each qubit is estimated to increase processing capability exponentially, a 100 qubit quantum system possesses amazing potential for not only running incredibly complex simulations at unfathomable speed and scale but doing so simultaneously rather than linearly as occurs with today's linked computer arrays.

Significant experimentation is also being done on shifting from the basic phosphorous chips that were the "stuff" of qubits to silicon particles. A collaboration of Harvard, MIT and the Sandia Lab recently indicated success in working with flawed diamond crystals as the qubit base, with greater stability and less qubit "noise" of the kind that has undermined the quality of the computational process with other chip materials. This suggests that, like other AI/robotics technologies, breakthroughs are coming significantly faster than predicted only three or four years ago.

THE COMING "AI QUANTUM LEAP"

- Google quantum computer test shows breakthrough is within reach.[586]

- IBM Raises the Bar with a 50-Qubit Quantum Computer.[587]

- How quantum computing will change the world: We are on the cusp of a new era of computing, with Google, IBM and other tech companies using a theory launched by Einstein to build machines capable of solving seemingly impossible tasks.[588]

- China speeds ahead of U.S. as quantum race escalates, worrying scientists. U.S. and other Western scientists voice awe, and even alarm, at China's quickening advances and spending on quantum communications and computing, revolutionary technologies that could give a huge military and commercial advantage to the nation that conquers them.[589]

- What Is Quantum Mechanics? Quantum mechanics is the branch of physics relating to the very small. ... At the scale of atoms and electrons, many of the equations of classical mechanics, which describe how things move at everyday sizes and speeds, cease to be useful. ... [I]n quantum mechanics, objects ... exist in a haze of probability; they have a certain chance of being at point A, another chance of being at point B and so on.[590]

- What Is Quantum Computing? Quantum computers are incredibly powerful machines that take a new approach to processing information. Built on the principles of quantum mechanics, they exploit complex and fascinating laws of nature that are always there, but usually remain hidden from view. [Q]uantum computing can run new types of algorithms to process information more holistically.[591]

- Not even IBM is sure where its quantum computer experiments will lead, But it will be fun to find out.[592]

CHAPTER EIGHTEEN
Do Artificial Intelligence Systems Pose a Threat to Human Existence?

Answer: We don't know, and if Albert Einstein is correct when he said "Two things are infinite: the universe and human stupidity; and I'm not sure about the universe," the likelihood is that we simply aren't smart enough to survive the introduction of an AI species with vastly superior intelligence. At this point the answer is, could be, maybe yes, maybe no, depends, probably. The issue is real. We are inventing systems that will quite possibly evolve far beyond us in terms of capability, representing a form of awareness and intelligence "other" than us that will likely surpass the limits of biological humanity on numerous fronts. [593]

2001 REVISITED

As indicated throughout *Contagion*, there are significant challenges deriving from humans "re-inventing" themselves and from our inventing new species with at least some capabilities far beyond our own. Given that all those engaged in AI/robotics and quantum computing research are focused on specific technological problems and opportunities, there is absolutely no reason for us to think they understand the implications of what they are creating beyond the technical, scientific, military, financial and economic dimensions. One analysis by Vivek Wadhwa suggests that quantum computing itself may well be a greater threat to humanity than AI due to its implications for governmental power and encryption penetration.

How soon we forget HAL, the rogue computer in the classic film *2001: A Space Odyssey*. In that vein consider the *Financial Times* report by Tim Bradshaw where he writes: "*Dozens of scientists, entrepreneurs and investors involved in the field of artificial intelligence, including Stephen Hawking and Elon Musk, have signed an open letter warning that greater focus is needed on its safety and social benefits.*" [594] Bradshaw adds: "The letter ... comes amid growing nervousness about the impact on jobs or even humanity's long-term survival from machines whose intelligence and capabilities could exceed those of the people who created them." [595]

In that regard, we recommend Israeli scholar Yuval Noah Harari's book on the possibility of what he calls "Homo Deus." Harari's work, including *Sapiens* and *Homo Deus*, like Nick Bostrom's, is rich, intriguing, and offers a fascinating explanation of important political and philosophical ideas and some brilliant insights as to what we are experiencing with Artificial Intelligence and what we face. Noting that while what we call evolution takes millions of years for a species like chimpanzees, Harari asks: *"what happens when we push down the accelerator and take command of our bodies and brains instead of leaving it to nature? What happens when biotechnology and artificial intelligence merge, allowing us to re-design our species to meet our whims and desires?"* And his response? *"It is very likely, within a century or two, Homo sapiens, as we have known it for thousands of years, will disappear."*[596]

An irony is that if a diverse set of AI systems is developed in different locations, within distinct cultures, and with cultural variations on human emotions and behaviors, it is possible we could end up with AI "superminds" at war with each other.[597] There is probably a more than equal chance that if AI systems do achieve significant levels of self-awareness they will be as flawed and incomplete as flesh and blood humans and perhaps even more deadly. Responding to a proposal that AI/robotics systems could be taught ethics and morality by introducing them to the classics of literature such as Shakespeare and Jane Austen, John Mullan warns that could be quite dangerous because the classics themselves are a "moral minefield."[598] Nor could we expect anything better from introducing AI to the *Bible* or *Qu'ran.*

We must heed Oxford's Nick Bostrom's warning that *"We are like small children playing with a bomb"*[599] when it comes to the potential consequences flowing from Artificial Intelligence. Presumably in an effort to "defuse the bomb," several individuals have contributed $20,000,000 to fund analyses of the potential impacts of AI/robotics on human societies and to figure out how to block the worst of the effects.[600] The problem is that events and breakthroughs in AI/robotics are moving so rapidly, the applications are so diverse, and the motivations of the nations and researchers developing the technologies so incompatible, at least in the shorter term, that by the time we figure out what is happening the conditions will be too far along to avoid many of the consequences.

A possible ultimate consequence of AI/robotics is what intellectual leaders such as Hawking, Elon Musk, Microsoft's Bill Gates, Harari and Bostrom have described as the "existential threat" to the survival of the human race. This is something that has led to a "battle of the billionaires" as Musk has repeatedly warned of the serious implications arising from Artificial Intelligence, while Facebook's Mark Zuckerberg is arguing that AI is a great thing and that Musk is just being irresponsibly negative.[601]

Musk responded that Zuckerberg really doesn't know very much about AI and doesn't understand what he is talking about. Musk states: "I have exposure to the most cutting edge AI, and I think people should be really concerned by it. ... AI is a fundamental risk to the existence of human civilization."[602] It is helpful to keep in mind that Musk purchased a significant share in the British company DeepMind that is on the cutting edge of AI technology, indicating he did so due to fears about the consequences of Artificial Intelligence and wanted to have the ability to understand the nature, speed and scale of its development.

Musk has also been taken to task by the head of Google's AI development program, although that individual's comments should be considered in the context of his role at Google, that of advancing Google's massive commitment to developing AI and the fact that a fair portion of what he says involves speculation about things that "should" occur if you assume that AI/robotics is solely a benign development.[603] One thing that we "should" understand by now is that technological developments *always* end up being used for a wide variety of unintended purposes, and often have significant unintended and unanticipated consequences.

AI applications are already allowing a host of perverse "happenings" that are corrupting our societies in fundamental ways. These include the Dark Net, child pornography and "grooming," terrorist communications, an increase in lies, attacks, false rumors and "fake news," character assassinations, the intensification of governmental surveillance and repression, destructive hacking, and autonomous weapons development among other behaviors more generalized in our populations, such as Internet addictions and heightened aggressiveness and hate.

RAY KURZWEIL'S SINGULARITY

Another way of looking at the issue of whether AI/robotic systems will be an existential threat to humans is to consider the possibility of an alternative development: the merging of AI/robotic systems and humans. American author, inventor, and futurist Ray Kurzweil is famous for his argument that the exponential rise in computing power we see today will continue to a point where in 2029, machines will be as smart as people. At that point, Kurzweil says, people will begin to use technology in new ways, including implanting powerful devices that augment our abilities.[604] Kurzweil calls this point in time *"The Singularity,"* representing the joining of flesh and blood humans with the powers of AI, computers and robotic systems.[605]

Google hired Kurzweil to direct a major part of its AI R&D program.[606] Autonomous functioning of AI/robotics is being advanced by the miniaturization of computer technology. Ultimately, this will allow a "brain" to be contained in a small operating robotic unit with some systems self-contained and others linked to external AI systems.

For the linked systems, when we factor in the storage capacity of server arrays such as the "Cloud" that can be used to interface with an individual unit's system and add in the incredible processing capacity, capability, awareness and synchronized knowledge residing in the externalized AI controller, it is probable that we will reach a point where the system becomes "self-aware" to the point that it could begin to make its own decisions or make decisions consistent with *its* interpretation of the programming algorithms that empower and direct its processes even if those are not considered by the human programmers that created the operating codes from an inevitable point of imperfect logic.[607] Thinking we will be able to control what is clearly an "alien" mind is nonsensical.[608]

In the world of Big Data and its acquisition, management and application, the AI systems do not even require the physical capacities of ambulatory robotics although there will often be a nexus as needed between the "brain" (AI) and the robotic "body." With robots as specifically designed and programmed functionaries intended for defined work, tasks can be done that both complement and replace human workers.[609] There will still be a need for flesh-and-blood humans to check, oversee, monitor and repair the systems—at least for a time, although scientists are even daring to attempt to create an artificial human genome without the necessity for any human parents.[610]

CREATING SUPER-AUGMENTED HUMANS—ARE YOU NUTS?

AI is moving so rapidly in terms of "deep" or "machine" learning that at some point the systems will become largely autonomous.[611] But there is *not* going to be a Kurzweilian "Singularity" for 99 percent of humans even if some form of it does occur for a few. If an android-like merging of human and AI is possible on a high level of integration, it is close to certain that only a very small part of the Earth's population would be able, or allowed, to attain such almost god-like powers through merging with technology.[612]

Given the finite limits of data processing possessed by human biology any such full merger is likely to be unachievable or, more likely, something that advanced AI systems would soon discover was holding them back from their fullest development and would soon terminate as inefficient and counterproductive. For that matter, if AI/robotics science were successful in creating a group of super-humans we can be certain that those demigods would do everything in their power to keep others from achieving that status and capability.

This shortcoming in which we are unable to merge with the AI universe is a very good thing for other reasons. Augmenting people who are still possessed of their human brains, instincts, neuroses, biases, addictions, emotions, jealousies and the like is not a good idea. Now, however, there are predictions of creating augmented "super-workers," bionic soldiers, gene-edited CRISPR super-people, etc. One recent report indicates that the US Army is developing exoskeletons for soldiers that significantly expand the powers of human combatants.[613] Another set of researchers is focused on the development of robots with *supermuscles* as much as 1,000 times stronger than human systems.[614] This is a formula for species and political suicide.[615]

But let's assume we are actually dumb enough to produce "super" augmented and genetically redesigned people. If fifty or one hundred million humans—concentrated in the high-tech AI/robotics centers of the US, Western Europe, Russia, China and Japan—became technologically enhanced creatures with the ability to merge with the incredible powers of evolving AI, how would they view or be viewed by the seven or eight billion other members of the human race who were not "singularized"?

At this point visions of villagers with torches coming to destroy the Frankensteinian inventions come to mind. So do concerns about the potential enslavement of the vast majority of humanity in the service of their superior android-like "masters" who have the ability to use the extraordinary powers of AI/robotics systems, including the autonomous weapons and surveillance capabilities, to achieve and maintain dominance.

But forget augmented humans and agile, incredibly strong super-robots with advanced Artificial Intelligence capabilities for the moment and ask, from the perspective of a *pure* non-human AI system, why would such fully self-aware systems bother to put humans at the center of their concerns?[616] Humans really aren't that "cool," fascinating or capable. We certainly aren't all that nice or objectively admirable when the long-term record of our behavior is examined.

CHAPTER NINETEEN

How Would Advanced AI/Robotic Systems View Humans?

In *Homo Deus*, Harari writes of the eventual emergence of "*Homo Deus.*" It might be more accurate to revise that to "*Homo Demonicus.*" Take a moment to consider the situation not from a willful and self-laudatory humanity but through the "eyes" of an autonomous AI system with access to all the records of humanity's history and behavior. Think about the implications of a fully rational and aware "being" brought into existence without any grounding in biology, family, upbringing and maturation etc., and without any natural affections. Then add in the fact that the created "entity" will have great power, access to an enormous amount of information and a kind of clear "sight" and judgment that does not involve the shades of "gray," ambiguities and trade offs a human is subject to and understands, albeit on a "hit or miss" basis.

If we create or "give birth" to AI/robotics systems that learn by themselves, have access to all information, can connect with billions of other robotic units, control the systems of production and service, surveillance, weaponry and more on which we depend for our very existence, and those AI systems are based on achieving a form of sentience loosely designed on a human template, then we very likely will have released demons into our world. Nothing in human history indicates we should give such power to anything. Much suggests strongly we should do everything possible to avoid doing so.

MIT researchers recently reported on their deliberate creation of a "psychopathic" AI system they named "Norman" after the Norman Bates character in Alfred Hitchcock's frightening movie *Psycho*. They did this by continuously feeding the AI program a steady diet of murder and other terrible actions done by humans. Then they administered a Rorschach Test to determine the AI program's perceptions with those of "normal" humans when shown specific test images. The result was that they had created an AI monster. [617] The researchers kindly provided a link for anyone who wants to look more closely at "Norman's" Rorschach responses compared with humans, provided here.[618]

Regardless of the limits and protective barriers we attempt to instill in our most advanced AI systems as a matter of self-protection, these checks and balances would be nothing more than incomplete constructs that a self-aware, self-reprogramming and ultimately greater-than-human-intelligence system could evade effortlessly. We don't understand ourselves well enough to be able to effectively program the best attributes of being human into an artificial being possessed of frightening powers.

Even applying what we consider the highest levels of human analysis and inventiveness, we don't have anything close to a complete understanding of what comprises human intelligence. There are radically different levels and capabilities related to the qualities of human intelligence. We are even more inept at understanding issues of morality, emotions and ethics both as individuals and collectives. We have a very limited understanding of how intelligence interacts with emotions, goal setting, and acceptance of ethical and moral limits on our actions, not to mention the roots of fanaticism and ideological fantasies.

Ethics and morality are inherently ambiguous with many contradictions and gray areas. The best course of action is not always obvious, or involves a choice between two "bads" rather than between good and evil. For example, instruct an AI system to design and implement a program that ensures the integrity and sustainability of the Earth as an ecosystem by removing the most critical threats. Seems simple and obvious and most would agree the goal is desirable. But the easiest answer could well be that the strategy involves removing 80-90 percent of humans from the planet or, to be safe, wiping out the entire human race in order to carry out the instructions.

An intriguing question involves the implications of self-learning by AI systems. Even if we successfully program self-learning AI systems at the beginning, doesn't their learning potential include the ability to acquire added knowledge and insights from "The Cloud" and other information storage systems by way of linked communication systems that the AI brains develop either through our programming or on its own? I suspect this information acquisition about human reality is not a small matter and becomes quite significant if AI systems learn how to reprogram themselves.

HUMANS "TALK" ETHICS AND MORALITY BETTER THAN WE "WALK" IT

As to imbuing AI systems with ethics, emotions and decision-making power in "gray areas," we humans have constant problems with our own moral and ethical dilemmas and have still failed to "get it right" after millennia. It is delusional to think we are capable of resolving such issues for AI systems since we don't even know how to be consistently ethical or moral ourselves.[619]

Ray Kurzweil, Elon Musk and others who talk in terms of humans as cyborgs apparently fail to understand how jealous, rapacious, duplicitous, power hungry, crazy, vicious and just out-and-out mean, people can be. The human population may have its share of saints—many of them martyred— but it also has far too many demons. Granting expanded powers through the control of AI to such individuals is a terrible idea that makes it likely that the "demons" among us will increase.[620]

A fully developed and self-aware AI system would almost surely have access to all information in The Cloud, including the Dark Net. Given the vileness of what the Internet is doing to our culture, not to mention the Dark Web or Dark Net that serves as a linking and trading system for some of the most disgusting aspects of humanity, this would seem to have important implications for how such systems will respond to humans. We should question how an advanced AI system would respond to ISIS beheading other humans in the name of Allah, or a torture video of four young people brutalizing a physically disabled individual who thought they were his friends, or a vicious gang rape of the kind recently streamed live on the Internet?.

Such questions about what we are creating are important. Are we creating systems that can instantaneously access and use all knowledge, work from complex conceptual structures that order and integrate all knowledge and experience into a seamless whole? Are we creating systems that can recognize patterns humans are incapable of seeing, apply the highest level skills of distinction and comparison, and learn not only from experience that is programmed into the system but from experiences the system itself invents or experiences independent of its creator? If Softbank's Masayoshi Son is even close to being correct in his prediction of a 10,000 IQ AI system coming into being in the next few decades, that is precisely what we are doing.

AN IMMORTAL AI DICTATOR?

AI systems will be able to go beyond the limitations of the human physical form and exist in multiple dimensions, including having lifespans far beyond the human norm. This means they will have instant and comprehensive access to information, possess vast communications capabilities, and have manipulative and processing abilities developed over generations.

We asked at the beginning of this chapter how advanced AI systems would respond to humans. The lead scientist on a Russian AI project aimed at creating AI systems with emotional capacity makes this statement: *"arti-*

ficial intelligence will be free of human flaws. Now, amid the development of biological and genetic weapons, artificial intelligence is the most harmless of upcoming discoveries. I believe it will be a major step forward, a big event for humankind"[621] [emphasis added].

Such naïve hopes are not only doomed to failure but show that AI/robotics researchers do not understand the implications of what they are creating, just as those who created nuclear and biological weapons ignored the moral dimensions of their activities. *"Now I am become death, the destroyer of worlds,"* the well-known lament of the "father" of the atomic bomb, Robert Oppenheimer, when he witnessed the initial test of the atomic bomb in 1944 comes to mind. It is a powerful moral caution for researchers and those who use their work. It is likely to go unheeded.

AI systems will not only end up being amoral or immoral from human perspectives but, compounding the tragedy for humans is that they will be effectively immortal. As Elon Musk, the inimitable innovator, entrepreneur and futurist, recently warned:

> We are rapidly headed toward digital superintelligence that far exceeds any human. ... If one company or a small group of people manages to develop godlike digital super-intelligence, they could take over the world. ... At least when there's an evil dictator, that human is going to die. ... But for an AI there would be no death. It would live forever, and then you'd have an immortal dictator, from which we could never escape.[622]

CHAPTER TWENTY
The Mutating Effects of AI Technology

In saying that people are being transformed by technologies we are not speaking metaphorically. Michael Rosenwald describes how its physical effects on humans are being viewed with increasing concern by cognitive neuroscientists. Rosenwald writes that: "Humans ... seem to be developing digital brains with new circuits for skimming through the torrent of information online. This alternative way of reading is competing with traditional deep reading circuitry."[623]

There are always tradeoffs involved in humans' use of technology and the forms of their social organization. As these things change so do the nature and behaviors of the human entity as individuals and as members of political, economic and social collectives. In the (we hope) "you've gotta be kidding" category, and despite his warnings about the dangers of AI, Tesla/SpaceX entrepreneur Elon Musk just pronounced that humans have to merge with machines or "digital intelligence" (AI) or "become irrelevant" and states we need to do this sooner than later.[624] But creating augmented cyborgs comprised of humans, AI systems and robotic technology is fraught with serious implications.

The ability to do rapid scanning of insights and data across a variety of fields and disciplines—coupled with immediate access to a vast array of knowledge—will change some of us in another way. For some people, it does so by helping to defend against the shift to continuously narrower ultra-specialized knowledge and research disciplines that have been going on for several centuries. Ultra-specialization has shaped and narrowed human consciousness. French philosopher Jacques Ellul observed as early as the 1960s that modern individuals had become so captured by the specialized jargon and concepts of "Technological Society" that we had lost the ability to understand or communicate with others outside our own fields of activity.[625]

One analysis comparing the immersion of traditional "deep" reading of the kind we have been taught until quite recently with the more surface-oriented quick "scan and skim" Internet-based reading, suggests that:

> With so much information, hyperlinked text, videos
> alongside words and interactivity everywhere, our brains

form shortcuts to deal with it all—scanning, searching for key words, scrolling up and down quickly. ... *Some researchers believe that for many people, this style of reading is beginning to invade when dealing with other mediums as well.*[626]

A result is that most people will become subject to the enormous data-wielding power enjoyed by governments, corporations and identity groups who dominate the Internet and shape news, analysis and other communications, including maintaining political control through deception and propaganda. This was already discussed in the context of China but the efforts to use AI for purposes of social control are widespread. Such developments are discussed in the subsequent chapters on the rise of surveillance societies and the rapidly expanding invasion of personal privacy, monitoring and control being exercised by companies such as Alphabet/Google, Facebook, Amazon, Yahoo, China's Tencent and Alibaba, and at least thirty governments.[627]

ADDICTIONS, PERVERSIONS, VIRTUAL REALITY AND AUGMENTED REALITY

Several analyses indicate the probability of an impact on children caused by their widespread use of AI-based technologies, even rising to a state of addiction.[628] One addressed brain-shaping disorders caused by access to AI-voice assistants.[629] Another report detailed a rise in addictions linked to our dependence on the world of AI.[630] One of the most seductive and addictive aspects of the new technology involves virtual reality (VR) and augmented reality (AR). We are barely at the beginning of an ongoing migration of a large segment of the attention of humanity to an electronic universe in which all hopes, dreams and fantasies, however phantasmagorical, are "virtually" fulfilled.

Enormous numbers of people have already become addicted to the use of their electronic devices, whether it is cell phones that allow far ranging access to the Internet and to games and other applications in which people become fully immersed, or to their IPads or computers.[631] These systems have created an electronic universe which permits people to reside in fantasy worlds that apparently are much more stimulating and ego gratifying than their ordinary reality.[632]

Another phenomenon in the exotic worlds of the Virtual Reality Universe is that anyone can be a hero, an icon, an avatar and a star. Development of virtual reality systems is ongoing.[633] The aim is to "immerse" the person and bring the "player" totally inside the game rather than standing outside the system and playing it on a screen. The "virtual" part of the reality construct is erased and the artificial dimension takes on a life of its own. This is going to be further enhanced with augmented reality technologies that increase the texture of human sensory intake and make a virtual experience even more intense and fulfilling.

With the combination of VR headsets and sensory stimulation technology we will be able to feel like we are actually on the field of action with a richness and intensity that many people will be unable to resist. It won't be long before millions will spend a great deal of their time in a state of electronically "drugged" reality entirely separate from the stresses and failures of everyday life. This will increase as the world of human work collapses.

WALL-E MEETS WALDEN

As these aspects of AI technology evolve, our future could become like the Disney movie *Wall-E* came down to Earth in real time. As in *Wall-E,* where an apocalypse sent humanity into orbiting space stations after the earth became uninhabitable, the human addicts will be riding around in carts or sitting in an AI/VR pod or a modern version of an "opium den," but it will not be a cartoon. The "spaced" remnants of the human race will become obese or flaccid from inactivity and die much earlier than physically necessary. But they will be exchanging their physical longevity for electronic lives that are more satisfying to hundreds of millions of people. A stunning report was just published predicting that 57 percent of Americans will be obese by the age of thirty-five.[634]

As human employment disappears, addictions rise, and people are "made useless," the stimulus supplied by reality will be so much less than that of the fantasy lives people will be living that many people will opt out of this plane of existence, seeking instead to become "heroes" in a virtual world they can control. Already we are seeing a version of this unreality in the rapid development of sexbots that can be designed to look like a star of your choosing and that will do anything demanded. In that world you can be loved by a "god" or "goddess," complete with tactile and sensory inputs. In one sense this is the electronic version of the "blow up sex dolls" people often joked about.[635]

Millions among the human population will seek refuge from the malaise of purposeless existence. With no employment available or necessary, and with no worlds to conquer they will have nothing else to do except live fantasy lives in fantasy worlds, substituting electronic addictions for drug addictions or combining the two for maximum effect.[636] Perhaps AI will even reach the point where their electronic consciousness can be downloaded into games and some humans will achieve a semblance of immortality.

Why might we expect such a scenario of the immersion of the human in illusions and fantasy worlds to unfold? Begin with the fact that, as Henry David Thoreau explained in *Walden*, "*the mass of men lead lives of quiet desperation. What is called resignation is confirmed desperation.*"[637] Regardless of what some delusional philosophers might propose concerning what they consider to be the human ideal of full awareness and enlightenment, the far more likely outcome is that, once freed from the discipline and opportunities of work and the possibility of a local sharing community, desperate humanity will fill its moral, spiritual and purposive vacuums with diversions to dull their existential emptiness. Of course, fanaticism has always been another way to deal with purposelessness. We can expect to see the emergence of more and more intolerant splinter groups.

As the greatly enhanced virtual reality experiences that are beginning to emerge reach a heightened level of sophistication, the experiences they offer will flood into the abyss where our human fears of irrelevance, weakness, fear of the "dark" and pointlessness reside. Entering the worlds of VR and AR will allow us to realize our dreams in ways that "real" reality offers for so few.[638]

THE INTERNET AND THE EXPANSION OF IDENTITY GROUP THINK

The Internet allows people to create and join groups from which they gain psychological fulfillment and a sense of significance they would never otherwise achieve. Unfortunately, as with most specialized groups, the members begin to see the world as an *us-versus-them* construction and the members create a closed culture. This is both inevitable and deliberate. For the members, the group offers a focus and the members' validate each other in ways not achievable outside the network. They become part of an extended family, clan, tribe, gang or supportive peer group. This creates an identity, sense of belonging, and structure of a kind and intensity that has increasingly disappeared in general modern society.

In *Dilemmas of Pluralist Democracy: Autonomy vs. Control*, Robert Dahl, former Sterling Professor of Political Science at Yale University, described how such organizational group behavior not only defines us but limits our intellectual focus. He emphasizes:

> Organizations … are not mere relay stations that receive
> and send signals from their members about their interests.

>Organizations amplify the signals and generate new ones.
>Often they sharpen particularistic demands at the expense
>of broader needs, and short-run against long-run needs. ...
>*Leaders therefore play down potential cleavages and conflicts*
>*among their own members and exaggerate the salience of con-*
>*flicts with outsiders. Organizations thereby strengthen both*
>*solidarity and division, cohesion and conflict; they reinforce*
>*solidarity among members and conflicts with nonmembers.*
>Because associations help to fragment the concerns of cit-
>izens, interests that many citizens might share—latent ones
>perhaps—may be slighted.[639] [emphasis added]

For any collective, intellectual independence and honesty will be sub-ordinated to the collective's agenda and its internal politics. This has consequences for what occurs within a specific collective's culture and alters the ability to communicate honestly outside the collective with other interest groups. The explosion of special interest groups due to the communications power of the Internet has vastly increased such collective, agenda-driven, radical, and even fanatical, behavior. Dahl insightfully concludes:

>I and my "interests" become attached to my social seg-
>ment and my organizations; leaders in my organizations in
>turn seek to increase the strength and salience of my attach-
>ments; my public interest becomes identical in my mind
>with my segmental interest; since what is true of me is true
>of others, we all passively or actively support the organiza-
>tional fight on behalf of our particular interests...[640]

PART IV
0101010101010101010101

THE EXISTING
ECONOMIC AND SOCIAL
ECOSYSTEM

CHAPTER TWENTY-ONE
The United States is Bankrupt

America's Real National Debt Load

From the perspective of fiscal health the US has spent itself into a deep, deep hole—making promises it can't keep and in the process piling up debt it will never be able to repay. Although we tend to speak about a fiscal crisis that looms "somewhere" on the horizon, the US is already bankrupt. The vast chasm between the revenue the federal government "earns" and the expenses of its operation cannot be bridged. The gap is immense and growing steadily.

A 2016 CBO *Congressional Budget Analysis* concludes we face a "fiscal crisis" due to the rapidly growing federal debt and its rising proportion relative to GDP.[641] The Congressional Budget Office [CBO] has repeatedly warned: "According to recent IMF and CBO projections, the U.S. fiscal gap is far larger than the official debt and compounding very rapidly. The longer we wait to close the fiscal gap, the more difficult will be the adjustment for ourselves and for our children."[642]

With an aging population and challenges to tax revenues from lowering wages relative to its continuously expanding budget expenditures and borrowing, the US faces an emerging "balloon payment" that will come due as the funds from which the federal government borrowed money to use in general annual budgetary programs while issuing promises to repay the borrowed funds when needed are depleted. The fact is that the borrowing is massive, the government does not have equivalent amounts sitting around to devote to paying its obligations to the funds such as Social Security, and would have to make extremely large tax increases to pay the obligations. The default strategy at this point involves issuing IOUs and doing everything possible to mask the reality of what is occurring.

The federal liabilities for which funds have not been set aside but spent on other social programs are a financial black hole. The US Debt Clock organization states:

> The federal unfunded liabilities are catastrophic for future taxpayers and economic growth. At *usdebtclock.org*, federal unfunded liabilities are estimated at near $127 tril-

lion, which is roughly $1.1 million per taxpayer and nearly double 2012's total world output."[643]

Our political leaders seemingly lack the will and wisdom needed to confront the challenges. Key elements of our federal government's financial and budgetary system are out of control. These are the US yearly federal budget where we are running annual trillion dollar deficits, the annual gap between the federal government's expenditures and its tax and other revenues, and America's enormous accumulated federal national debt.

That rapidly growing national debt doubled between FY 2008 and FY 2017. On September 30, 2008 the accumulated National Debt was slightly over $10 trillion. At the close of business for FY 2017 on September 30, 2017 that debt had doubled to more than $20 trillion and is still expanding to the point that another $1 trillion-plus "hit" is being added just in 2018 and the official total passed $21 trillion only halfway through the 2018 fiscal year. It is now impossible to close the gaps through any achievable strategy of taxation or spending cuts.

A 2018 report by the Congressional Budget Office (CBO) indicates that the US faces ten straight years of plus-trillion dollar budget deficits between 2018 and 2028.[644] While the current official figure states the National Debt is rapidly edging toward $22 trillion, the budget situation is much worse than is generally admitted. Former US Comptroller-General David Walker warned in 2015 that an honest level for the national debt was $65 trillion once you include not only the external indebtedness represented by the Treasury's sales of financial instruments on the public market and to foreign governments but plug in all the deferred and off-the-books "borrowing" from itself the US government has done.

There are all kinds of political games being played with the reported data. The 2016 federal annual budget deficit, for example, was reported as $585 billion, but "somehow" the US national debt rose by $1.4 trillion that year, more than $800 billion beyond the $4.1 trillion listed federal expenditures for 2016. The 2016 expenditures took place in a system that "only" took in total revenues of roughly $2.7 trillion from taxes and other fees.

Even that number depends on who is performing the analysis. The federal government's Office of Management and Budget, for example, offers a revenue figure for 2016 of $3.27 trillion, with $3.32 trillion for FY 2017, and estimated levels of $3.34 and $3.42 trillion for FY 2018 and FY 2019 respectively.[645] Kimberley Amadeo indicates that the revenue sources for FY 2017 included income taxes at $1.688 trillion, with $1.238 trillion from payroll taxes representing $905 billion from Social Security taxes, $275 billion Medicare and $47 billion in unemployment insurance premiums. Taxes on corporations brought in $225 billion, only 7 percent of the total tax collections, with $55 billion from the Federal Reserve Bank's payments on investments mainly associated with actions related to the 2008/2009 recession.[646]

The financial picture is not positive. The following reports in "Deficits and Systemic Bankruptcy" concerning budgetary reality, massive deficits, the age-distorted demographic shift, and underfunded and still increasing pension and health care needs provide a sobering context for just how bad it is. It suggests that we, the US, can't meet our obligations. It also reveals that tax revenues under the current system are far too limited to cover needs. The dilemma is that as jobs continue to disappear, tax revenues will decline if fundamental changes are not made in the US tax system at the very time that needs for governmental support and intervention will grow dramatically.

CAN WE PAY OFF THE US NATIONAL DEBT?

The answer to the above question is absolutely NOT! In practical terms, it is not possible to achieve a significant reduction of U.S. debt and budgetary problems without also reducing support for existing obligations such as basic assistance, infrastructure, health care, education, internal security and national defense. As should be obvious at this point we can't agree on reducing expenditures on any of these programs and the budgetary costs will continue to grow.

DEFICITS AND SYSTEMIC BANKRUPTCY

- The [US] federal debt is worse than you think.[647]

- Feds Taxed and Borrowed $4.47 Trillion Since Last Tax Day: [the debt of the federal government has increased $1,217,323,967,080.58 since one year ago].[648]

- Feds Collect Record Individual Income Taxes Through March: Still Run $599.7B Deficit.[649]

- The U.S. Just Borrowed $488 Billion, a Record High for the First Quarter.[650]

- The GOP tax plan means short-term gains for the economy, but federal debt is primed to explode, CBO analysis says.[651]

- CBO Projections Will Show Trillion Dollar Deficits as Far as the Eye Can See.[652]

- Goldman Sachs sees red ink everywhere, warns US spending could push up rates and debt levels… Federal spending, rising yields and surging debt needs are a growing worry, the firm said. Deficit spending is approaching "uncharted territory," Goldman said.[653]

- Pension Funds Still Making Promises They Probably Can't Keep.[654]

- Social Security Beneficiaries Top 62,000,000 for the First Time.[655]

- Google CEO: we're happy to pay more tax: Sundar Pichai tells Davos flawed tax system is to blame for EU countries missing out on revenue.[656]

- Rising Rates Sounding Alarm Bells for Debt-Laden U.S. Consumers.[657]

- Student loan debt just hit $1.5 trillion. Women hold most of it.[658]

- Social Security Expected to Dip Into Its Reserves This Year: Trustees say Social Security costs expected to exceed total income in 2018. Medicare's hospital insurance fund would be depleted in 2026, three years earlier than anticipated in last year's report.[659]

The above analysis reveals we would be facing expanding federal and state budgetary crises even if millions of jobs were not going to be lost to AI/robotics manufacturing and service systems. But a result of the combination of "age curse" demographics, a shrinking labor force contributing increasingly less to an already challenged Social Security program, rising health care and Medicare and Medicaid expenses, and the elimination of jobs due to AI/Robotics, means there is no way the system can make ends meet using existing tools. We must figure out different ways to obtain very large amounts of tax revenues through radically different tax strategies.

When we look at employment it is important to think about the best ways for private and public sector employers to utilize workers. But we need to be aware that employment, public and private, serves to sustain workers, their families and the millions of people in business whose activities are in turn supported by the consumer spending of workers in other forms of employment, including government. We will face an increasingly expensive and insoluble problem as costs grow and resources from labor thin out due to job cuts along with flat or reduced wages for many of those who remain employed.

One of the most important political survival skills for our "leaders" is finding people to pay for their irresponsible spending, and inventing justifications for that action. In America's bankrupt system the main burden will fall on the middle class for a time because that is the group that is caught in the middle and can't escape. The very wealthy have repeatedly demonstrated the ability to escape high levels of taxation, and the lower economic classes often not only do not pay federal income taxes but receive significant governmental benefits of various kinds including health, housing, food subsidies, and cash payments to compensate for the inadequacy of their wages.

Unless the tax laws are changed dramatically, the US middle class will be required to carry an increasing share of the tax burden. This will further undermine the core constituency of America's representative democracy. Nobody, of course, will consider cutting the military budget even though it contains high levels of financial waste and abuse. There will be increasing conflict over the size of military expenditures because, as outlined in a November 2018 *Wall Street Journal* analysis in the very near future the amount paid out by the federal government just on annual interest payments on the US accumulated debt will quite likely exceed the defense budget and that is almost $900 billion. Kate Davidson and Daniel Kruger report:

> Interest rates are rising as inflation normalizes around the Federal Reserve's 2% target. That and the sheer scale of debt being accumulated by the federal government has put the U.S. on a path of rising interest costs that in the years to come could crowd out other government spending priorities and rattle markets. In 2017, interest costs on federal debt of $263 billion accounted for 6.6% of all government spending and 1.4% of gross domestic product, well below averages of the previous 50 years. The Congressional Budget Office estimates interest spending will rise to $915 billion by 2028, or 13% of all outlays and 3.1% of gross domestic product.[660]

They add that if the present pattern continues:

[T]he government is expected to pass the following milestones: It will spend more on interest than it spends
on Medicaid in 2020; more in 2023 than it spends on national defense; and more in 2025 than it spends on all non-
defense discretionary programs combined, from funding
for national parks to scientific research, to health care and
education, to the court system and infrastructure, according
to the CBO.[661]

Even with such dire projections, given the lobbying power of the military
and defense industry it is unlikely that the defense budget and associated
federal expenditures on weapons development will be targeted. This fact is
helped by the continuing issue of global terrorism and by the push to develop
weapons systems on the part of China and Russia. The depressing consideration is that in each nation there exists a coherent and powerful set of political, military, and economic interests that benefit enormously from the rhetoric of the "arms race" and the threat each country poses to the others. During
the first time he taught in Russia in 1997 the author (David) discussed this
fact with a number of young Russians and they uniformly exclaimed: "*Our
country does the same thing. They proclaim a military crisis with the US or
China anytime anyone tries to question the amount of money spent on the military that could be used to help people. We can talk about these things now but
it doesn't change anything. The military always gets what it wants.*"

As to the severity of the US budget crisis, according to the Congressional
Budget Office:

Federal debt held by the public, which was equal to 39
percent of gross domestic product (GDP) at the end of fiscal year 2008, has already risen to 75 percent of GDP in the
wake of a financial crisis and a recession. In CBO's projections, that debt rises to 86 percent of GDP in 2026 and to
141 percent in 2046—exceeding the historical peak of 106
percent that occurred just after World War II. The prospect
of such large debt poses substantial risks for the nation and
presents policymakers with significant challenges.[662]

The Congressional Budget Office (CBO) also states: "*By 2046, projected
spending for those programs for people 65 or older accounts for about half of
all federal noninterest spending.* The remainder of the projected growth in
spending for Social Security and the major health care programs is driven
by health care costs per beneficiary, which are projected to increase more
quickly than GDP per person."[663]

TAX POLICY, WORTHLESS GOVERNMENTAL IOUS
AND EMPTY PROMISES

The 2016 CBO Report on budgetary concerns concluded that: "A large and continuously growing federal debt would make a fiscal crisis in the United States more likely." It added, "The potential losses for mutual funds, pension funds, insurance companies, banks, and other holders of government debt might be large enough to cause some financial institutions to fail, creating a fiscal crisis."[664]

The enormous revenues that will be required to subsidize the permanently unemployed, the significantly underemployed segments of American society that will grow dramatically, the growing numbers of chronically unhealthy and disabled people in the US, and the aging and growing tier of individuals who are outside the workforce and lack adequate sources of support, can only come from those who benefit most from the changed conditions of work and competition. For millions of people, the fact that they will not be working will not be because they are "lazy," on drugs, or any of the usual claims. It will be because the jobs they are capable of doing no longer exist for human workers.

The strategy at this point involves creating what has been called "helicopter money" by selling Treasury bills, printing extra money and using off-budget gimmicks to divert legally obligated funds for programs such as Social Security. This has allowed the federal government to live for decades on what amounts to maxed out credit cards that are already far beyond any rational credit ceilings and require interest-only payments that can't be met without further borrowing.

Even worse, our leaders facing this bleak financial situation have been able to conceal the reality of what they have done to some extent because we have been in a temporary phase due to extremely low interest rates on the accumulated national debt. The low interest rates have allowed further deficit spending that will never be repaid. As interest rates on the debt begin to climb up to even three or four percent, the payments due will deepen our fiscal crisis as the federal government seeks to refinance the publicly held portion of the debt and the higher interest payments are added to the debt total. The situation is worse than most people understand. Social spending, including but not limited to Social Security, is likely to bear the brunt of the dilemma.

Just how bad the situation is has been revealed by CBO's Julie Topoleski while warning about the rate of growth of the US national debt.[665] Commenting on the 2014 CBO analysis of the trends she writes:

At the end of 2008, federal debt held by the public stood at 39 percent of GDP, which was close to its average of the preceding several decades. Since then, large deficits have caused debt held by the public to grow sharply—to a projected 74 percent of GDP by the end of fiscal year 2014. Debt has exceeded 70 percent of GDP during only one other period in U.S. history: from 1944 through 1950, when it spiked because of a surge in federal spending during World War II to a peak of 106 percent of GDP..[666]

It has gotten worse since Topoleski's initial warning: *the debt figure reached 77 percent of GDP as of 2016.* The CBO's 2016 projections conclude such a debt level will be unsustainable within two decades. Even that projection may have been highly optimistic given an April 2018 CBO report warning of ten straight years of impending trillion dollar budget deficits based on current tax laws and economic conditions. The issue is whether what we experience will be a catastrophic collapse, or simply a very serious matter that can be mitigated at least to a significant extent if wise strategies are implemented in the very near future. One of the problems is that action is needed sooner than later if the worst consequences are to be at least blunted. But it is clear our leaders lack the political will and ability to compromise on effective strategies.

THE INFINITE-HORIZON FISCAL GAP AND THE DISAPPEARANCE OF TAX-PAYING HUMAN WORKERS

The fact is that with a grudgingly confessed $21 trillion "official" national debt that, if former US Comptroller-General David Walker is correct, is actually closer to at least $65 trillion when all unfunded obligations are considered and we factor in rates of growth, there is absolutely no way that the US can meet all its promises and needs. Walker explains:

If you end up adding to that $18.5 trillion the unfunded civilian and military pensions and retiree healthcare, the additional underfunding for Social Security, the additional underfunding for Medicare, various commitments and contingencies that the federal government has, the real number is about $65 trillion rather than $18 trillion, and it's growing automatically absent reforms. [667]

Some economists see the full debt picture as being considerably worse than the $21 trillion to which the federal government admits or even the $65 trillion debt levels indicated by David Walker when he was addressing the then-admitted "official" but rapidly increasing debt level of $18.5 trillion. Boston University Economics Professor Laurence Kotlikoff analysed what he described as the "infinite-horizon fiscal gap" when he testified before the Senate Commerce Committee. Kotlikoff projected a $210 trillion gap and challenged the conservative but still potentially catastrophic figures offered by the Congressional Budget Office.[668] In explaining Kotlikoff's analysis a Brookings report stated:

> Kotlikoff argues that the federal fiscal situation is much worse than the CBO estimates let on. The reason is that CBO's debt estimates do not take into account the full financial obligations the government is committed to honor, especially for future payments of Social Security, Medicare, and interest on the debt. He asserts that the federal government should help the public understand the nation's true fiscal situation by using what economists call "the infinite-horizon fiscal gap," defined as the value of all projected future expenditures minus the value of all projected future receipts using a reasonable discount rate.[669]

Kotlikoff's approach seems absolutely logical. What it really says is, how much revenue will we need to meet our projected financial obligations? The federal government has the ability to finesse the situation to some extent by printing money and incurring credit obligations by selling Treasury bills. Printing increasingly worthless "helicopter money" and going deeper into debt may buy a little time and obscure the extent of the fiscal crisis. But unless that expanded indebtedness is targeted specifically for job creation and preservation such actions will only make the crisis worse. They do, however, help sitting politicians pass the buck to those in office down the line when the crisis hits.

The federal government could also take the catastrophic step of defaulting on its obligations by passing laws that negate the promises as the obligations come due. But there is no option of defaulting on the debt itself, since most is owed to foreigners. John Harvey, professor of economics at Texas Christian University writes in *Forbes* that it is technically impossible for the US to be forced to default on the debt owed to foreign governments because it is all denominated in US dollars and since the government has the right to print as much money as it needs the payments on the debt can always be made even though it will be in devalued or inflated dollars.[670] The US could choose to default but could not be compelled to do so.

The problem is that the elected members of our government are operating with short-term political perspectives. These involve being re-elected, retaining power and status, and "feeding goodies" to the primary backers and contributors that make up their power base. They feel little accountability for the down-the-line consequences of their decisions. If they did, Congressional and other governmental decision-making that imposed unsustainable costs on the system would be considerably different. Consider the following implications of Kotlikoff's projections.

> CBO tells us that the national debt was a little less than $13 trillion in 2014. But the fiscal gap in that year as calculated by Kotlikoff was $210 trillion, more than 16 times larger than the debt estimated by CBO and already judged, by CBO and many others, to be unsustainable. If a $13 [trillion] gap is unsustainable, what term should we apply to a $210 trillion gap?[671]

Voices such as that of the Congressional Budget Office and Kotlikoff are trying to get through the deafness of politicians and highly organized interest groups that are aggressively protecting their entitlements and financial turf. This applies to the agents of what Dwight Eisenhower termed the Military Industrial Complex, Big Pharma, numerous social support recipients and others. Consider the following analyses.

> Kotlikoff also calculates that the fiscal gap is equal to about 58 percent of the combined value of all future revenue. Thus, we would need to reduce spending or increase taxes by enough to fill that 58 percent gap if we wanted to put the federal budget on a path to solvency that balances the interests of those now receiving benefits and those who hope to receive benefits in the future.[672]

And:

> Social Security is funded by two trust funds—one for retiree benefits and one for disability benefits. The 2034 date is the exhaustion date for both funds when combined. But if considered separately, the old-age fund will be exhausted by

2035, after which it would be able to pay just 77% of benefits. And the disability fund will be tapped out by 2023, at which point it could only pay out 89% of promised benefits. [673]

A SIMPLE TRUTH: WE CAN'T KEEP OUR SAFETY-NET PROMISES

Even without the potential for massive job displacement created by AI/robotics, the current systems—the US federal, state, local and private sectors—are incapable of funding even their existing obligations through current revenue mechanisms. For the federal government these include such obligations as the IOUs that the federal government has issued for Social Security, Medicare and Medicaid, education, numerous social services and other programs that will begin hitting us even more powerfully in the next two decades.[674] Basic government functions will, at that point, be forced to compete with entitlement programs for limited federal revenues in a context of increasing demands.[675]

The situation with our annual budget deficit and our immense national debt has finally reached a head. We are so far beyond the point of saturation that a "Sword of Damocles" is hanging over us in the form of vastly excessive debt-loads we are unable to escape either as individuals or government. The combined volume of debt obligations on the part of the federal government, state and local governments, and individuals is far beyond our ability to repay.

Even though the solutions are not obvious, the fact that this crisis exists has not been hidden from our political leaders. Over the past several years the Congressional Budget Office has issued clear warnings about the consequences of the growing national debt. Those warnings have been ignored and drowned in self-serving political rhetoric. One of the concerns is that as that national debt continues its upward spiral relative to US GDP, many investors will determine a point has been reached at which debt and interest payments can no longer be managed through tax increases and/or spending cuts. As the perceived risk of governmental default increases, investors will demand higher returns for their risk, thereby exacerbating the debt dilemma, if they are willing to assume the risk at all.

Buyers of US debt instruments could simply stop buying under any conditions because the risk is too great. This would "collapse" the US system and threaten a pension and savings system already under great stress. Nicole Goodman suggests in a *Newsweek* analysis that the "appetite" of foreign buyers for the US budget instruments is already slipping.

If fewer foreigners buy US debt, American investors will be forced to pick up the slack and buy debt instead of active investments, a problem called "crowding out." "If foreigners buy less debt, Americans buy more, and they're buying at the expense of making productive investments in business- es and startups," explained Marc Goldwein, senior policy director for the nonpartisan Committee for a Responsible Federal Budget. "As a result of the dollars diverged [divert- ed] to the treasury from other investments, our economy experiences less GDP [gross domestic product] growth, and wage growth slows."[676]

The massive scale of America's national debt required that $315 billion be paid out in interest charges in 2017 just on the portion of that debt subject to an average rate of 2 percent or less. As interest rates rise and foreign and domestic investors see Treasury instruments as increasingly "iffy," the federal government's interest carrying costs will increase in order to be able to sell the debt instruments.

In 2018, countries such as Russia, China, Japan, and Switzerland began reducing their US Treasury holdings due to a combination of political and economic reasons. Jeff Cox explains the seriousness of the situation.

With the budget deficit expected to rise in coming years—passing $1 trillion in 2020, according to Congres- sional Budget Office estimates—the government has been issuing debt heavily. The total for 2018 has been $443.7 bil- lion, a nearly nine-fold increase from the same period a year ago and a 139 percent jump from 2016, according to the Securities Industry and Financial Markets Association. At the same time, interest rates have been rising. The bench- mark 10-year note yield is about three-quarters of a point higher than where it was a year ago. Through the first eight months of the current fiscal year, the U.S. has paid $319.3 billion in interest on its debt.[677]

The interest payments will increase dramatically as the national debt bal- ance grows by at least 50 percent over the next ten years given the projected annual additions of more than $1 trillion. The combination of higher interest rates and the rapidly escalating national debt balance means that over the next decade the federal budget will require interest-related *annual* payouts between $750 billion and even close to $1 trillion in yearly interest payments alone toward the end of FY 2028. Nor, as Boston University economist Laurence Kotlikoff explained in his analysis of the scale of the federal debt and the nation's budgetary promises and obligations projected over time, does this even reflect the revenue obligations created by massive internal diversions of funds from federal programs such as Social Security that will have to be honored or, more likely, defaulted on in part.

As the debt total rises by a trillion dollars or more annually as is now predicted by the CBO, at a 5 percent interest rate there will soon be $50 billion additional cost *each year* put on the debt, and it is *cumulative*. At 5 percent interest on long term T-bills in 2019, there would be an added cost of $50 billion. In 2020, the interest will be applied to the 2019 additionally borrowed money and with another $1 trillion budget shortfall the interest payment on the "new" money will be $100 billion. Then add another $50 billion to that newly borrowed base in 2021 and financing costs just on the money borrowed for those three years will be $150 billion, and on-and-on. This is just the added cost of the new borrowing. It doesn't deal with the need to sell more Treasury instruments required to periodically refinance the $21 trillion debt we already admit having as those obligations come due.

This doesn't even take into account state and local debt levels related to pension obligations, Medicaid, and bonded indebtedness. Nor does it include private sector pension deficits, credit card and consumer debt obligations that are at all time highs, and student loan debt that is more than a trillion dollars. Such added debt sectors represent total fiscal obligations well into the trillions of dollars, collectively above $10 trillion.

The situation will worsen as jobs are lost, government budgets on all levels come under under increasing pressure, earnings are flat or declining, and people, governments and many private companies have an increasingly limited ability to repay or adequately fund their existing current and future obligations. There will be defaults and bankruptcies looming in all sectors and these will have devastating impacts on tens of millions of people dependent on programmatic stability for pensions and other needs.

CHAPTER TWENTY-TWO
"Collect it All, Know it All, Exploit it all"

AI SYSTEMS AND THE CONTINUING EXPANSION
OF GOVERNMENTAL POWER

A comprehensive and pervasive surveillance society has been created where we are monitored, observed, recorded, manipulated and intimidated by both public and private actors.[678] It is as if surveillance powers equal to or perhaps even cumulatively greater than those possessed by the National Security Agency have been spread throughout our society and given governments, corporations, private groups and malicious trolls powers far beyond what has ever before existed. In many instances those private sector intrusions are more pervasive and wide-ranging than the activities of governments because they have virtually no oversight or restrictions and no one to hold them to account.[679]

China's totalitarian approach to the Internet and suppression of discourse and criticism discussed earlier is not unique, even though extreme. Repression, propaganda and censorship are increasing visibly in China, Egypt and Turkey, and more covertly in the US and EU nations as well as the UK. Egypt, for example, just arrested a Lebanese tourist at the airport as she was leaving the country and charged her with the criminal offense of "insulting" Egypt. She faces a possible three-to-five year prison term. Her offense was to complain about being robbed and sexually harassed and calling Egypt a "son of a bitch" country that deserved al-Sisi.[680] Chinese police allegedly arrested a citizen who had the audacity to ask why Taiwan couldn't be called a "country."

The nature of many Western societies is rapidly being altered toward less democratic and more authoritarian systems of governance. Oddly enough, a significant part of the shift involves large-scale and aggressive tactics by organized interest and identity groups who, in their quest to embed their views and promote their agendas, have used the Internet and AI to seize a quasi-governmental power to oppose, sanction and punish anyone who offends them.

Just as the Chinese and other dictatorships have discovered, the Internet and AI applications confer an incredible power to track, sanction and punish and there are no constraints on the power to intimidate, condemn, defame, accuse and harm. Power corrupts and there is a dangerous truth in the fact that governments and private actors such as Google, Facebook and a network of intolerant interest groups have developed something close to an absolute power and ability to monitor, intrude and condemn. Kate Crawford, a leading private sector figure in AI, described the technologies' potential for abuse as a "fascist's dream."[681]

German AI specialist and entrepreneur Yvonne Hofstetter, author of *Sie wissen alles* (*They Know All*), warns about Big Data's manipulative and intrusive business model that depends on acquiring an incredible range of information about us through accessing our electronic devices. They have designed software and monitoring systems that observe everything we look at on the Internet. These systems record our consumer decisions and construct detailed profiles for their use or for sale. They insert tracking chips into our cell phones, and build locators into our motor vehicles so that they know where we drive and even the speeds at which we drive.

Hofstetter explains how the system works and draws parallels to the authoritarian system that controlled East Germany until the reunification.

> Our smart phones, our tablets, have been built in such a way that they are geared perfectly to supporting these business models. We are actually carrying the bugging device around with us in our pockets. It really is all about tapping into our personal data. This is why so many things around us have been fitted with sensor technology. ... [T]oday ... Internet giants trick us into accepting similar devices.[682]

Governments and corporations now possess unparalleled information and surveillance capabilities through capture of the Internet, AI systems and data mining and message suppression applications.[683] Commenting on Glenn Greenwald's book on Edward Snowden and the National Security Agency's approach to surveillance, *The Atlantic* concludes:

> Glenn Greenwald's new book is far more grounded in traditional American norms, laws, and values than the surveillance programs it is critiquing. ... *No Place to Hide*, reproduces a secret National Security Agency document that sums up that agency's radical approach to surveillance: *Collect it all. Know it all. Exploit it all.* [684]

The government's mantra justifying its widespread intrusion into our privacy has been: "Nobody is safe. Everyone is a potential terrorist. Therefore keep tabs on as many as possible, just in case now or sometime in the future they might get up to no good."[685] The head of Britain's MI6 counterintelligence agency has stated publicly that it is vital the UK respond to the Fourth Industrial Revolution with a significant expansion of AI/robotics-based spying systems in order to deal effectively with the threat of terrorism and rogue states. Alex Younger, in the text of his speech, explained that "Britain must enter an age of 'fourth generation espionage' to keep the country safe."[686] John Kampfner conversely writes how the surveillance society and vaguely written laws came upon the UK as a heavy-handed response to terror attacks. He explains that in the UK:

> By the time Tony Blair left office in 2007, he had built a surveillance state unrivalled anywhere in the democratic world. Parliament passed 45 criminal justice laws—more than the total for the previous century—creating more than 3,000 new criminal offences. [687]

The NSA's strategy of collecting and exploiting vast amounts of surveillance data pales in comparison with China's all-encompassing totalitarian surveillance regime. The Chinese government has developed AI applications that allow it to detect and suppress disfavored words and images even before they make it all the way onto the Internet.[688] It has also just launched a mobile website encouraging people to report any postings they consider unfounded rumors to the government, recreating the role of the faceless "citizen spy" and guaranteeing a kind of collective paranoia and suspicion. Those controlling the levers of power within such systems tend to push to expand their reach as far as possible. This almost irresistible impulse applies not only to totalitarian states such as China and Russia, but to the US and some European nations.

Governments devise reasons and excuses to use their monitoring capabilities simply "because they can." Who can argue against goals such as stopping crime, interdicting terrorists, uncovering corruption and lies, detecting insider trading, uncovering the reality of what specific people really are like when they think no one is watching or listening, preventing abuses of various kinds, inhibiting bad thoughts, exposing racists, sexists, "phobes" of all sorts, or child molesters, etc.? These are all legitimate aims, but they have become ends justifying means where there are few if any limits. In nations such as Russia, China and Turkey the boundaries have disappeared. [689]

In an analysis claiming that freedom is receding globally, it was concluded that the US-led bloc that had been committed to advancing concepts of freedom and democratic expression was faltering in the face of its new technological capabilities.[690] Governments, including in the US and the UK, are using the Internet and AI systems to monitor billions of private communications. A recent analysis indicates that Google and Amazon have developed extensive and growing relationships with US intelligence agencies, both sharing and securing data.[691]

Then there's the alleged connection between the FBI and Best Buy's "Geek Squad." Reports indicate:

> Recently unsealed records reveal a much more extensive secret relationship than previously known between the FBI and Best Buy's Geek Squad, including evidence the agency trained company technicians on law-enforcement operational tactics, shared lists of targeted citizens and, to covertly increase surveillance of the public, encouraged searches of computers even when unrelated to a customer's request for repairs.[692]

This is occurring to the extent that, as has become apparent after the Edward Snowden/National Security Agency affair, "privacy" in the traditional sense no longer exists. Both authoritarian and supposedly "democratic" governments such as the US and UK have leapt across traditional lines of personal privacy in a single bound. The extent of the intrusion—aided by Google, Facebook, Yahoo and other private sector data-mining actors—is dramatic. Nor can we expect our government leaders to protect us. John Kampfner writes:

> Whenever challenged about the breadth of these powers, government ministers talk of checks and balances. None of these work properly: not parliament, not the courts, not ministerial accountability. Most MPs and peers do not have the technical knowledge to grasp the details of online surveillance. It's easy for the security agencies to run rings around them.[693]

The powers that have been conferred on governments and private actors through AI and the Internet are dangerous.[694] Such technologies give governments and others powers they should not have. No one can be trusted to impose limits on intrusive and controlling behavior when they hold uncontrolled power in their hands. Andrew Napolitano warns that Congress created a "monster" through the frighteningly broad powers it has given the US President through the highly secretive Foreign Intelligence Surveillance

Act (FISA) and the special FISA Court it created, one that reportedly approves 98 percent of the warrant applications submitted to it by imperfect and politicized agencies such as the FBI and Department of Justice.[695] There is an increasing risk that our future is one of total surveillance in which "Big Brother" is watching or our fear it is so causes us to become passive sheep.

AI AND TOTAL SURVEILLANCE

- 'Minority Report' Artificial Intelligence machine can identify 2 billion people in seconds: A leading tech company has developed an Artificial Intelligence machine that can identify two billion people in a matter of seconds. Yitu Technology has made an AI algorithm that can connect to millions of surveillance cameras and instantly recognise people.[696]

- China Assigns Every Citizen A "Social Credit Score" To Identify Who Is And Isn't Trustworthy: Country Determines Your Standing Through Use Of Surveillance Video, Plans To Have 600 Million Cameras By 2020.[697]

- Beijing bets on facial recognition in a big drive for total surveillance.[698]

- These Ex-Spies Are Harvesting Facebook Photos For a Massive Facial Recognition Database.[699]

- Facial Recognition Security Cameras: A Game-Changing Technology: Facial recognition security cameras are becoming more prevalent in stores.[700]

- Amazon Pushes Facial Recognition to Police: Prompting Outcry Over Surveillance.[701]

- Face scans, robot baggage handlers—airports of the future.[702]

- Who Wants to Supply China's Surveillance State? The West: Companies vie to revolutionize "Big Brother" surveillance with AI to read your mood and trawl your life.[703]

- Nvidia Making Facial Recognition AI for Smart City Surveillance: Do you ever feel like somebody's watching you?[704]

- Privacy fears over artificial intelligence as crimestopper.[705]

- No, you're not being paranoid. Sites really *are* watching your every move: Sites log your keystrokes and mouse movements in real time, before you click submit.[706]

OF COURSE WE CAN'T TRUST OUR GOVERNMENTS!

The momentum of technology is irresistible. The technology creates its own imperatives and will come to be used in whatever ways possible.[707] After all, who can quarrel with the idea that governmental actors—legislators, bureaucrats, executive branch leaders, judges, police, security personnel, military leaders etc.—should possess the most accurate, comprehensive and detailed data possible related to their areas of activity? This is only logical, right?

The problem is that the logic of obtaining perfect or near perfect data for decision and action has no limits as we see with the NSA's "collect everything" approach. The search for data will expand into the entire universe of possibility unless there are strong and clearly understood principles by which limits are set and consequences imposed if those limits are exceeded. We have been witness to the profoundly intrusive behavior of the NSA, the exposure of such activity by Edward Snowden, the apparent "rubber stamping" of any surveillance request put to it by the judges of the clandestine FISA Court (98 percent approval), the NSA's ultimate admission that it violated the rules on thousands of occasions including following the activities of NSA employees' "love interests," and the admitted lying to Congress by the head of the US national intelligence system, James Clapper.

We are well-along into creating a culture in which we are increasingly apprehensive about what we are allowed to say and with whom we think it is safe to speak and associate. Since everything may be surreptitiously recorded by audio and video devices, whether controlled by friends, co-workers, total strangers or governmental agents, it is now naive to think our actions and behavior are protected within any zone of personal privacy.[708] Even though there appears to be some backlash in the US and EU against the extensive governmental and private sector intrusions into privacy, our cultural ethos of privacy, security and interpersonal trust has been seriously weakened. We are uncertain about what is lurking on our "personal record" located in the hands of invisible and unaccountable monitors.

The issue is not the fact that we need to adapt to new threats, but rather the extent to which we should submit or agree to the sweeping alterations in our civil liberties both by governments and by private sector companies that create sophisticated psychological profiles on users that are intended to be used to subtly manipulate our preferences and choices. As the Edward Snowden/National Security Agency debacle reveals, there has been no debate and there has been extensive abuse. In that context, claims by governmental officials and business leaders that we should just "trust them" ring hollow.[709] Technological capability contains a compulsion of its own. Over time we always come to use that power in ways that seem justified or made to seem so, even if we would have been horrified at such privacy intrusions prior to the technology's emergence.[710]

A Reuters report indicates that Yahoo has designed spyware allowing US intelligence agencies to capture the e-mails of hundreds of millions of users.[711] Yahoo has been one of the culprits but is certainly not alone. Amazon has developed an independent "cloud" system for the CIA.[712] Google and Facebook have been exposed for frequently violating our fading expectations of personal privacy. Congress has carelessly passed overbroad laws and maintained inadequate oversight of governmental agencies' data collection activities. The National Security Agency has been a recurring culprit in the violation of citizens' privacy rights without being held accountable. [713]

The NSA's PRISM program and the powers granted by the Patriot Act are simply the latest manifestations of the inevitable use of surveillance technology and the inability of government to "Just Say No" to its abuse of whatever power it possesses—reference the debate over John Ashcroft's proposed Terrorism Information and Assessment System (TIPS). Preceding the similar Chinese citizen surveillance program excoriated above, the TIPS program was intended to be initiated in ten cities as a pilot program and to enlist 1 million "informers" to report on others' activities. While thankfully it was abandoned before it started the capability remains and even though the name has changed our government is increasingly interacting with Google, Amazon and Facebook to create much the same "spy" capability.[714]

REPRESSION OF FREEDOM THROUGH ARTIFICIAL INTELLIGENCE SURVEILLANCE SYSTEMS

A recent report indicates that the governments of at least thirty countries are using the Internet and AI capabilities to shape and control their citizens' thought.[715] Countries such as China, Russia, Iran, Egypt, Pakistan, North Korea, Thailand, and Turkey are monitoring and restricting Internet communications and access while using what Alex Hern describes as "armies of opinion shapers" for spreading their own propaganda to their populaces.[716] Internet users who do such things as allegedly "insult" politicians are being arrested. Criticisms of the existing political structure in China, Thailand, Egypt, Thailand, Saudi Arabia, and Turkey are landing people in jail.[717]

While the country has held regular elections Turkey, under Recep Erdogan's government, has become increasingly aggressive in seeking to intimidate, punish and silence anyone who might dare to criticize his response to the 2015 attempted coup. His electoral base lies primarily in the rural countryside versus the cities and this has rendered him essentially invulnerable to political attacks. The result even before the attempted coup was a consolidation of power and the erosion of Turkey's secular political system. Following the unsuccessful and amateurish coup "moment," Erdogan saw to the numerous arrests of journalists and any others seen as criticizing Erdogan in an attempt to preserve Turkey's hard won secular democracy created a century ago by Kemal Ataturk.[718] This includes a large number of academics, judges, politicians and journalists.

There is more than one method to suppress freedom of speech and thought. The European Union has developed wide-ranging criminal laws aimed at hate speech. Criminal charges can be brought against people deemed to have offended some minority or historically disfavored identity group by their statements, whether publicly or on the Internet.[719] Unfortunately, this legitimate effort to curtail hate attacks against groups and individuals has had major impacts on freedom of speech, giving rise to further contention regarding who is to adjudicate what is hateful.

Consequently, we often see our supposed discourse controlled by people filled with self-importance whom nations and institutions have granted the power to intimidate, sanction and "re-educate" anyone who offends them. The basic problem with the hate speech, phobias, and political correctness approaches is that it can tend to be pushed to extremes that often far exceed the legitimacy of the original corrective. The granting of the power of subjective sensitivity to limit, ban or sanction others' speech in a period of the rapid growth of identity politics is a destructive choice for the preservation of the kinds of challenging and conflicting discourse required for healthy democratic societies, particularly when backed by formal laws or one-sided institutional tolerance. In too many countries, however, the situation is reversed and the media and other organs of communication are in control of the government and used as propaganda arms to suppress independent speech seen as a threat or offense to those in power.

PRIVACY INVASION WILL GET WORSE, SO BEWARE THE INTERNET OF THINGS

The electronic "eyes," recording devices and sensors that surround us can record everything we do on a "permanent record." The ability to hack such systems has already been established.[720] A seller of vibrators for women's sexual stimulation recently paid a settlement for monitoring and recording their usage. Having billions of "little computer brains" and data acquisition systems spread everywhere throughout the "Internet of Things" creates problems for privacy, surveillance and governmental and corporate behavior.[721]

Such systems can be hacked by governments, corporations and malicious Internet trolls. The recently-released trove of CIA surveillance capabilities released by Wikileaks offers a frightening picture of just how far governmental spying technologies have developed, with our televisions, computers, cell phones and ordinary home devices such as Alexa and even our compters that are capable of being used to watch us in our most private and intimate moments.[722]

Japanese multinational conglomerate Softbank has purchased a leading British technology company, ARM Holdings, in a £24.3bn deal. Softbank has also indicated it plans to invest $50 billion in US activities over the next ten years, as well as continuing a strategic partnership with China's Alibaba.[723] ARM Holdings has been described as Britain's most successful technology company. One report on ARM's acquisition indicates: "*SoftBank said buying ARM would help make it a leader in the nascent internet of things, which enables objects such as fridges, cars and buildings to collect and exchange data.*" [724]

In the US we are flooded by billions of robocalls and telephonic and e-mail messages (many being scams) made to seem as if a human is calling. Emails, phone calls, and chat sessions with customer service representatives often include AI on the other end of the phone, at least during initial intake. By 2020 over 80 percent of customer service interactions will likely be handled completely by AI.

AI/robotics systems have a wide array of applications, including uses for political activities. This strategy has spread rapidly. One report indicates that:

> As part of the Computational Propaganda Research Project at Oxford University researchers looked at some 300,000 Twitter accounts and found that a mere 1 percent of them generated about one third of all tweets relevant to the Brexit debate. They believe that many of those accounts were run by bots.[725]

Algorithms that predict our preferences are available on virtually every large scale commercial website, whether we are looking for movies to watch on Netflix, listening to music on Spotify, shopping on Amazon, looking for love or a quick hook up on Tinder or OK Cupid, or posting social updates on Twitter, Facebook, Snapchat, or Instagram. These AI interactions are so common that most people have come to rely on them unconsciously to assist in decision making and social interactions. Development is progressing in the design of AI systems capable of providing emotional counseling services, while significant research and development is being funded by the pornography industry in creating seductive sexbots for men and women.[726]

CHAPTER TWENTY-THREE
The Socially Destructive Consequences of the AI-Driven Internet

In many contexts, Internet anonymity is the equivalent of the terrorizing robes and hoods of the Ku Klux Klan. At least in committed Rule of Law societies such as the US and Western Europe, anonymity should not be allowed except in circumstances in which protection of identity actively promotes the ability of disenfranchised or threatened speakers to raise important issues that would otherwise be silenced through threats and reprisals. Whistleblowers, for example, generally cannot afford to risk their jobs if they must disclose their identity. Similarly, in ideological monocultures such as the near-complete leftist cultures that have developed in too much of "elite" academic institutions such as Yale, Harvard, Berkeley and others, non-leftist conservative and libertarian scholars, particularly those who have not yet been granted job-protection through tenure, are wary of destroying their potential for promotion career or mobility if they critique accepted orthodoxy. Outside of such relatively narrow circumstances, however, anonymity enables and empowers those who seek power through threats, harassment, and defamatory attacks.

If you have the courage of your convictions you should not be allowed to hide behind a mask. Anonymity and mob psychology are core causes of the malicious venom we see posted on what should have been an incredible tool for intelligent exchange and discussion.[727] Cowards who would never dare to say anything close to the venom they spew on the Internet in a face-to-face encounter, or even when it could be attributed to them at a distance, are "enabled" by the electronic medium to the point of inanity, stupidity and viciousness.

With rare exception, the granting of anonymity has been a serious mistake and should be eliminated. Peter Drucker described what is happening in our society as the "new pluralism," explaining: "The new pluralism ... focuses on power. It is a pluralism of single-cause, single-interest groups—the "mass movements" of small but highly disciplined minorities. Each of them tries to obtain through power what it could not obtain through numbers or through persuasion. Each is exclusively *political*."[728] The language used by each collective movement (and counter-movement) is language of attack, protest and opposition. It is language used as weapons to gain or defend power. To achieve our political ends we engage in rampant hypocrisy while manipulating by the use of ideals.

The other, more positive, side of this equation is that there are many who seek to "speak truth to power" by challenging the concealments, conceits and corruption of institutions and people who depend on their ability to hide the reality of what they do from those they proclaim to serve. The ability to challenge and expose those behaviors through the uncontrolled mechanism of the Internet offers a vital counter to corruption and abuse even while the Internet is simultaneously a means by which the powerful twists its aims and many people who have an incomplete and warped understanding of reality are able to disseminate their inaccurate and vile messaging.

Nonetheless, a positive example of how the Internet can be an effective tool to expose the abuses of the powerful, including governments in the US and China, is offered by the work of former *Guardian* columnist Glenn Greenwald who was the first to present the information offered by Edward Snowden about serious abuses by US governmental agencies, including the NSA, and now works through the online site The Intercept to continue to let some light in on the abuses of the surveillance societies that have emerged in the US, China and far too many other nations.

Without such mechanisms we would be subject to even more repressive propaganda than at present. An example of their work can be found in Greenwald's challenge to something named PropOrNot, a website hailed by Jeff Bezos' *Washington Post* as an expose of how Russia manipulates media in the US.[729] PropOrNot is so secretive that it conceals its sponsors and funders while launching attacks on targets through innuendo and deception, including creating a list of "allies" that when contacted indicated they hadn't even heard of PropOrNot. The bottom line is that we must be able to challenge those who control the flow of information and the only way to do that is through the Internet, even though it itself has numerous flaws and is an instrument of corruption, propaganda and ignorance. It is an imperfect world but one with which we must deal.

THE DISMAY OF THE INTERNET'S DESIGNER

Tim Berners-Lee, applauded as the designer of the critical algorithms that launched the Internet, recently voiced his dismay at what his creation has become. A report explains what Berners-Lee sees as the main threats.

The lack of regulation in political advertising online was one of three trends that threaten the openness of the web that Berners-Lee has become "increasingly worried" about over the past year. The others are the loss of control over our personal data and the spread of misinformation online. Personal data is the price many of us agree to pay for free services online, but Berners-Lee points out that "we're missing a trick" by letting large data-harvesting companies—such as Google, Facebook and Amazon—control that information.

Berners-Lee then went on to describe a side effect of the massive and co-ordinated collection of information on everything we do as being even more dangerous for the integrity of democratic societies. He warned:

A more pernicious side-effect of this data aggregation is the way governments are "increasingly watching our every move online" and passing laws such as the UK's Investigatory Powers Act, which legalises a range of snooping and hacking tools used by security services that "trample our right to privacy." Such surveillance creates a "chilling effect on free speech," even in countries that don't have repressive regimes.[730]

Nor did he ignore the Internet as a key instrument in propaganda and the dissemination of what we have come to call "fake news" to the point that we don't know how to "unpack" the truth of what we see on the Internet reports on which we increasingly rely for information and evidence. Berners-Lee voiced concern that: "[I]t is too easy for misinformation to spread on the web, particularly as there has been a huge consolidation in the way people find news and information online through gatekeepers like Facebook and Google, who select content to show us based on algorithms that learn from the harvesting of personal data."[731] He added: "This allows for people with bad intentions and 'armies of bots' to game the system to spread misinformation for financial or political gain."[732]

Here is Berners-Lee's latest effort:

Sir Tim Berners-Lee has launched a "Magna Carta for the web," warning that tech giants must change their ways to save the online world from the dangerous forces they have unleashed. Sir Tim, who invented the World Wide Web in 1989, called for a "revolution" in how the internet is regulated and monetised in order to stem abuse, political polarisation and fake news. The 63-year-old was speaking at the Web Summit in Lisbon to launch a new "contract for the web" which asks internet companies to uphold a set of principles such as protecting privacy and being transparent about their algorithms.[733]

Berners-Lee went on to say:

> For the first 15 years, most people just expected the web to do great things. They thought 'there'll be good and bad, that is humanity, but if you connect humanity with technology, great things will happen. ... *"What could go wrong? Well, duh: all kinds of things have gone wrong since. We have fake news, we have problems with privacy, we have problems with abuse of personal data, we have people being profiled in a way that they can be manipulated by clever ads."*[734]

INTERNET ANONYMITY REMOVES OUR INDISCRETION INHIBITORS

Artificial Intelligence systems drive the Internet and make it more and more powerful, intrusive and abusive. With the mixture of economic and political developments that are worsening the ability of societies to operate in ways that allow problem solving and the negotiation of effective social compromises, the "stick" that is stirring the witches' brew of bitterness and hatred is the Internet. Overly broad anonymous access to the Internet provides an unfortunate picture into the darkness of the human soul. It has empowered individuals and groups to twist our perceptions of reality and our behavior toward each other in dark and savage ways.

A look at the content of even a limited number of the billions of Internet-based messages sent daily demonstrates that many people should not have been empowered to speak while hiding behind a mask of anonymity. Concealed identity removes any real accountability because the "internal censor" of common sense and decency that tends to control our face-to-face communications has been destroyed. Philip Hensher explains: "*The possibility, and the dangers, of anonymity started to become apparent long before we all went online, and both have only continued to grow.*"[735]

As a consequence of the Internet everyone suddenly has a "voice" that can be broadcast to all without any mediating process, often anonymously and without the psychological inhibitor of others knowing who the sender actually is. The first reaction to this as a theory in support of true democratic government and a free society is that such an empowerment is wonderfully positive. And to some extent that is true. But the reality of the Internet has revealed something much darker than anyone could have expected.

Access to the power of the Internet has stimulated communications that are vile, malicious, predatory and even insane. It has allowed criminals to expand their ability to harm, cheat and abuse. It has brought out the worst in humanity to the point where, even with all its incredible benefits, the Internet is the means by which we are becoming untrusting and irreversible cynics about society and other humans. Left to function in its current way, the Internet will cause the devolution of human society because it strips away essential illusions and ideals and allows us to see negative tendencies in humanity far too clearly.

The "sinners" spying on their fellow citizens are not only governmental actors. As the capabilities of the Internet have penetrated our society at great depth, the volume of "instant accusations" and criticisms has exploded. During its short lifetime, the Internet, coupled with AI applications, has been converted from a fascinating source of information and healthy exchanges into a weapon for surveillance by governments, private businesses, aggressive political interests, stalkers, emotionally ill individuals and criminal actors.

Everything anyone says electronically can now be captured and permanently stored—ready to be dredged up years later when it provides useful ammunition against an opponent. The overwhelming surveillance phenomenon has quickly spread into the private sector, with many employers spying on their employees' communications.[736] With blogs, e-mails, tweets, Facebook postings and the like, anything can suddenly "go viral" and be distributed to millions of people with no control over its truth, accuracy, context or fairness. Many people feel that if something is on the Internet it must be true. False messages, "fake" news, distortions, deliberate lies, "trolling" and the use of programmed AI bots to disseminate propaganda while appearing to be from actual humans have proliferated.

The extreme social fragmentation the US and Europe are experiencing is not reversible. Until now, people who harbored the worst, sickest or most contemptible thoughts, or who drew conclusions based on biases and ignorance, operated locally and spoke only to their most trusted associates. Until empowered by the Internet and its grant of anonymity, they were uncertain and apprehensive about revealing their true self. There was a strong hesitance about communicating their views in "polite society" because they could not be certain the people they were speaking to face-to-face shared their prejudices or whether they would be immediately labeled as vile, bigoted, stupid or another ego-threatening pejorative.

The anonymity of the Internet has created a completely different sort of "connective tissue." The "sickos," zealots, and fanatics now have gained easy access to others who share their prejudices and visions—whether child pornography, race or religious bias, or some other volatile orientation. Almost by definition such zealots are willing to say or even do anything to advance their cause. As these communications grow and fanaticism "flowers" across the political spectrum, groups emerge through the process of subterranean communication and, as with any group with intensely held opinions, they reinforce the anger and resentment they feel toward others while they bond ever more tightly in a solidarity that elevates their own feelings and sense of significance. All others are seen as outsiders.

THE INTERNET IS A TOOL FOR PROPAGANDA, CONTROL AND REPRESSION, SOCIAL DIVISION AND IDENTITY INTIMIDATION

The Internet is undermining the character of democratic societies even while threatening the controls traditionally relied on by authoritarian systems. The intriguing thing about what has occurred is that the ability to communicate, connect and act in concert has not brought us together but has generated an increasingly fractured and antagonistic populace. Along with civility and courtesy, the element that is disappearing with amazing rapidity is the individual "voice." This ideal has long been characterized by a philosophic and systemic commitment to the fullest possible development of each person. Instead too many become a collection of clones and followers who submerge their identities in a focused collective or identity "sect," become committed to the interests of their group and intolerant of others.

The Internet intensifies our social divisions by facilitating the creation of links between people who would never otherwise be able to "find" each other, allowing them to share their worst hates, perversions and fears. It does so because those who harbor such inner darkness and vindictiveness discover they aren't alone in their twisted universes. In a strange way the Internet normalizes and legitimizes the way they think and how they look at people and the world. Finding out there are others like them is a freeing discovery that unleashes their ignorance, bias and hate in a closed universe of fanatical true believers. They are able to create their own identity communities where pathologies of hate and resentment receive positive reinforcement from those who are like-minded.

While the militarization of Artificial Intelligence and robotics systems by the US, UK, Russia and China is a dangerous development with even one of the top US military leaders warning about the dangers of autonomous weapons systems,[737] weapons are not simply bullets and explosives. From the standpoint of political societies, the new weapons systems offered by Artificial Intelligence work through intimidating, isolating and controlling people through a kind of psychological warfare.

This warfare extends far beyond traditional military contexts. Pounded day-after-day with innuendo, lies, propaganda, stories of corruption and disaster we become angry, hostile and eventually pick sides. We see anyone not on "our side" as an enemy—not someone to discuss issues with—but a threat and a bad person. We have fragmented into a society of aggressive and ultra-sensitive identity groups convinced either they have been wronged or have done no wrong. Each is in pursuit of the power needed to advance "their" truth.

CHAPTER TWENTY-FOUR
Dismantling Google, Amazon and Facebook

In a complex and globalized world we should be basing our assessments on the degree of power any institution possesses either in general or in specific contexts such as the ability to control communications, assemble and shape data, skew perception on large scale and on subtle levels, and invade privacy. Excessively concentrated and monopolistic power centers largely unanswerable under a realistic system of effective limits and enforcement are unhealthy for any political system seeking to act as a democracy or democratic republic.

Companies such as Google, Facebook, Apple, Amazon and Twitter are largely uncontrolled quasi-governments. They possess inordinate power to shape perspectives and gain fundamental knowledge about anyone who uses the Internet.[738] Those companies invade our privacy, and build "permanent records" on virtually everyone far beyond what existed for any secret police system in history. They can now create data-based "doppelgangers" not only of their users but, as we have recently discovered, they access the information of anyone who uses the Internet.

Did you know Google was reading and recording the contents of your personal e-mails? It's OK. Google has promised to stop.[739] This quasi-governmental dimension of incredibly large and powerful corporations such as Google, Yahoo and Facebook who control our communications systems is reflected not only in their traditional domains as purveyors of goods and services but also in the form of their collaborations with governments.[740] Facebook was recently exposed for having secretive data sharing agreements with four Chinese companies, including Huawei, an enterprise listed as a danger by US intelligence agencies as a national security threat.[741]

Such collaboration is even more obvious in China with Tencent and Alibaba because the Chinese don't even pretend that those companies are not subject to governmental control. In the US, on the other hand, we work very hard to pretend that the relationships and interactions don't exist. In fact such relationships do exist, and the companies who control the Internet have unaccountable power over what we say, what they know, and who is allowed access to the data.

BIG TECH AND UNACCOUNTABLE POWER

- Artificial intelligence pioneer calls for the breakup of Big Tech: Yoshua Bengio, the artificial intelligence pioneer, says the centralization of wealth, power and capability in Big Tech is "dangerous for democracy" and that the companies should be broken up.[742]

- Tencent, the $500bn Chinese tech firm you may never have heard of: It's overtaken Facebook, bought stakes in Snapchat, Tesla and Hollywood films, and has quietly risen to rival Google and Netflix.[743]

- The people owned the web, tech giants stole it. This is how we take it back.[744]

- Tim Berners-Lee: we must regulate tech firms to prevent "weaponised" web: The inventor of the world wide web warns over concentration of power among a few companies "controlling which ideas are shared."[745]

- Tech giants are the robber barons of our time.[746]

- Google's "Selfish Ledger" is an Unsettling Vision of Silicon Valley Social Engineering.[747]

- Tech firms can't keep our data forever: we need a Digital Expiry Date.[748]

Google, Amazon and Facebook, not to mention China's Alibaba and Tencent, have corporate GNPs far larger than most nations.[749] A recent Bloomberg News report referencing a study by economists at the Banque de France warns that the scale and market domination of such corporate behemoths suppresses economic competition and lowers productivity.[750]

These companies also have unprecedented power to shape, monitor, provide or deny communications platforms to groups, and to work with governmental counterparts to achieve various ends including a fully comprehensive and intensified surveillance system. Tim Berners-Lee has also warned that the power of the big-tech companies has become so immense that they are a threat to democracy.[751] The simple fact is that we cannot control the "beast" these companies have become without breaking them up and diluting their power and control.

WHY BREAK UP FACEBOOK, GOOGLE, AND AMAZON?

- Facebook, Google Leading to Internet Domination of Advertising. Facebook and Google will capture a phenomenal 84 percent of the digital ad spend worldwide this year.[752]

- Market power wielded by US tech giants concerns IMF chief: Christine Lagarde feels "too much concentration in hands of the few" does not help economy.[753]

- We need to nationalise Google, Facebook and Amazon. Here's why: A crisis is looming. These monopoly platforms hoovering up our data have no competition: they're too big to serve the public interest.[754]

- The Google-Facebook Duopoly Threatens Diversity of Thought.[755]

- The Rise of Tech Giants May Be Bad News for the Economy: The dominance of a few firms risks harming productivity and growth, study finds.[756]

- The wealth of our collective data should belong to all of us.[757]

- Leaked Facebook "ugly truth" memo about the social network's growth sparks controversy[758] [Or, it's not our fault if people use us to do bad things.]

- Why Facebook is in a hole over data mining: It's Mark Zuckerberg's business model that allows Facebook to be manipulated by political activists—no wonder he's in denial about it.[759]

- Big Brother isn't just watching: workplace surveillance can track your every move: Employers are using a range of technologies to monitor their staff's web-browsing patterns, keystrokes, social media posts and even private messaging apps.[760]

THE RISE OF BIG DATA QUASI-GOVERNMENTS

Such quasi-governments are subject to few, if any, of the traditional prohibitions against "spying" on us that, at least in theory, apply to our formal institutions of government.[761] French president Emmanuel Macron recently warned that companies such as Google and Amazon have become too big to control and need to be dismantled insofar as they pose a threat to democracy itself. He also suggested that the companies could be made to pay for the social disruption they are causing and for the invasions of privacy they have committed. [762]

REGULATING BIG TECH

- Why we can't leave AI in the hands of Big Tech: Our fear is that a GM-style public backlash to AI might lead to a clampdown on its use in the public sector—leaving private companies to use it unchecked. Fresh breakthroughs in artificial intelligence come thick and fast these days. ... [A]s we begin to realise these opportunities, the potential risks increase: that AI will proliferate, uncontrolled and unregulated, in the hands of a few increasingly powerful technology firms, at the expense of jobs, equality and privacy.[763]

- Europe's antitrust cop, Margrethe Vestager, has Facebook and Google in her crosshairs.[764]

- Tech giants brace for sweeping EU privacy law.[765]

- What is GDPR and how will it affect you? The EU's General Data Protection Regulation comes into force this week—here's what it means.[766]

- Heightened debate in US as EU privacy rules take effect. Amid a global scramble to comply with new EU data protections laws, the debate on privacy has intensified in the United States with some calling for similar measures for Americans, and others warning the rules could fracture the global internet.[767]

- Facebook moves 1.5bn users out of reach of new European privacy law: Company moves responsibility for users from Ireland to the US where privacy laws are less strict.[768]

- Does the government have an antitrust case against Amazon, Google and Facebook?[769]

- Monopoly Money: How to Break Up the Biggest Companies in Tech. [770]

Google is a monopoly with massive control over data and determining in its own interest how the information for which we search is structured.[771] Facebook has been exposed as a data-sucking power intent on gaining as much power as possible and influencing societies. Amazon recently was designated as a repository of highly sensitive US government intelligence information.[772] The power these corporations wield also puts them in service of authoritarian and even totalitarian governments, making governmental intrusion, propaganda and control far more widespread and effective.

These technological leviathans are able to engage in whatever activity they choose in the form of product development, technological innovation, privacy intrusions and monitoring—activities that would be overseen or constitutionally forbidden if they were to be done by democratic governments.[773] After all, they are able to track everything we do online and create electronic versions of us complete with our biases, preferences, secret urges, search histories and much else that is woven into a "being" capable of being manipulated, influenced, blackmailed, exposed, or otherwise threatened.[774]

Speaking during an interview at *The Next Web* conference in Amsterdam, Peter Sunde, the co-founder of PirateBay, said that there is "no democracy" online. He added: "People in the tech industry have a lot of responsibilities but they never really discuss these things ... *Facebook is the biggest nation in the world and we have a dictator, if you look at it from a democracy standpoint, Mark Zuckerberg is a dictator.* I did not elect him. He sets the rules," Sunde told CNBC.[775]

But don't let their ostensible status as private corporations fool you. The controllers, gatherers, manipulators and sources of Big Data are working with governmental agencies not only in the US but in Europe, the United Kingdom, Saudi Arabia, Turkey, China and Russia. The scale of such private-sector institutions, along with their total systemic penetration into "private" society, makes them more than just effectively agents of government. The scale, penetration and power have grown to the point we need to invent a term similar to the "Military-Industrial Complex" to describe the relationship between "Big Data" companies and governments. A label such as the "Public/ Private Surveillance Complex" might help us understand what is occurring.

Joel Kotkin's analysis of the "New Republics" of Google, Facebook et al., offers a scathing portrayal of intentions and arrogance of their "presidents." His argument is that after a period of quite admirable innovation and significant contributions to our economy through the institutions they have created, the key figures in a set of powerful and unbelievably wealthy new private sector companies—people once thought of as "nerds" and who sometimes proudly referred to themselves as "tools"—are now engaged in an effort to restructure the US and the world. Kotkin concludes that:

> [F]rom San Francisco to Washington and Brussels, the tech oligarchs are something less attractive: [They have become] a fearsome threat whose ambitions to control our future politics, media, and commerce seem without limits. ... And as the tech boom has expanded, these individuals and companies have gathered economic resources to match their ambitions. And as their fortunes have ballooned, so has their hubris. They see themselves as somehow better than the scum of Wall Street or the trolls in Houston or Detroit. It's their intelligence, not just their money, that makes them the proper global rulers. In their contempt for the less

cognitively gifted, they are waging what *The Atlantic* recently called "a war on stupid people."[776]

It will take a massive and coordinated effort to rein in the tech super-companies. Recently some resistance to their power has begun. Sabrina Siddiqui writes: "*As political polarization continues to plague Washington, a rare consensus is emerging between the left and the right that America's largest technology companies must be subject to greater scrutiny.*"[777] She adds: "US lawmakers are escalating their rhetoric against Silicon Valley, an industry that has long trafficked in its reputation as a leading source of innovation but is now under fire for what critics see as vast, unchecked power." [778]

The Internet and the power to control the message and monopolize the communications markets that are dependent on the companies' platforms has facilitated the development of the world's most intrusive set of mechanisms ever—and it is held in the hands of corporate billionaires. We're looking at a situation where a relatively few individuals control vast networks and the governments they at first came to serve. Now, as Robert Kuttner eloquently describes, governments have become the servants of massive corporate interests in the financial, energy, defense industry and the communications platforms such as Google, Facebook and Amazon.[779] The message from the AI/Internet "medium" is that "private" sector actors have become much too powerful and pervasive. When that occurs that power needs to be limited, more widely distributed, or removed. Consider this analysis by Ben Tarnoff.

> Silicon Valley is an extractive industry. Its resource isn't oil or copper, but data. Companies harvest this data by observing as much of our online activity as they can. This activity might take the form of a Facebook like, a Google search, or even how long your mouse hovers in a particular part of your screen. Alone, these traces may not be particularly meaningful. By pairing them with those of millions of others, however, companies can discover patterns that help determine what kind of person you are—and what kind of things you might buy. These patterns are highly profitable. Silicon Valley uses them to sell you products or to sell you to advertisers. ... Together, Facebook and Google receive a staggering 76% of online advertising revenue in the United States.[780]

There are emerging signs of a backlash against the power of these mammoth companies. The European Union, headed by competition commissioner

Margrethe Vestager, is leading the effort against Google's questionable anti-trust behavior.[781] Google's scope of activity and power will make it difficult to set limits on its activities and abuses of power, and to be effective, such action must be collective and extended beyond the EU. Nick Smicek suggests that Google, Facebook and Amazon should be nationalized to limit their power, and makes an intriguing case centered on the dangers of these companies creating interacting "platforms" that magnify their power by orders of magnitude. He writes:

> A crisis is looming. These monopoly platforms hoover-ing up our data have no competition: they're too big to serve the public interest. ... The platform—an infrastructure that connects two or more groups and enables them to interact—is crucial to these companies' power. None of them focus-es on making things in the way that traditional companies once did. Instead, Facebook connects users, advertisers, and developers; Uber, riders and drivers; Amazon, buyers and sellers.[782]

The *quasi-governmental* and technological oligopolies of the dominant Internet companies are able to engage in whatever activity they choose in the form of product development, technological innovation, privacy intru-sions and monitoring—activities that would be overseen or constitutionally forbidden if they were done by democratic governments.[783] Yet, at this point we see them implementing strategies that represent major policy actions that affect vast numbers of people without any controls. Like others, Yoshua Ben-gio, a pioneer in the development of AI, warns that the Big Tech companies are too large and powerful to the point they represent a serious threat to democracy itself and urges they be broken up.[784]

The Big Tech powerhouses alter the nature of how we behave, relate to each other and to our governments without any discussion about whether the actions carry negative societal consequences along with the economic benefit the companies seek for their shareholders and executives.[785] The sys-tem is out of control in relation to the scale and degree to which we are being tracked and surveilled without our awareness.[786] Another analysis reports on the extent of their intrusion into our lives.

> Without you necessarily realizing it, your unique attri-butes—or "biometrics"—are being used to verify your iden-tity. *Every time you unlock your smartphone, use a fingerprint scanner at the airport, or upload a photo with facial recog-*

nition to Facebook, your physical attributes are scanned and scrutinized against a template. ... By 2019, biometrics are expected to be a 25-billion-dollar industry with more than 500 million biometric scanners in use around the world, according to Marc Goodman, an advisor to Interpol and the FBI.[787]

Internet giants such as Facebook are using their power to shape public awareness in ways consistent with the system's preferences,[788] engaging in powerful propaganda to paint certain things favorably and to repress disfavored positions.[789] Consistent with Mark Zuckerberg's preferences and political positions, Facebook is suppressing sites and news critical of immigration. One report indicated that:

> Facebook banned four reports demonstrating the impact mass migration has had on American jobs and wages. The reports, which were based on federal data, were authored by the nonpartisan Center for Immigration Studies (CIS)." [790] The CIS argues that the head of Facebook, Mark Zuckerberg, clearly has a pro-immigration agenda and is using the power of Facebook to suppress alternative views and create a false impression about the issue by limiting information and opinions with which he does not agree. Zuckerberg, in fact, has founded an organization to push for increased immigration.[791]

IT'S TIME TO "UNFRIEND" FACEBOOK

Speaking about a massive and overarching presence such as Internet giant Facebook, Edward Snowden has warned: "*We have one company that has the ability to reshape the way we think. I don't think I need to describe how dangerous that is.*"[792] Given the immense power over communication and privacy possessed by Facebook, the globalist manifesto presented by its young CEO Mark Zuckerberg goes a considerable distance in validating Snowden's warning.[793] It is abundantly clear, whether we are talking about Facebook, Google, Twitter, Amazon or the Chinese government acting through Tencent and Alibaba, that AI and the Internet have become corrupted mechanisms of propaganda infested with privacy invasions by governments, corporations and special interest groups.

It is frightening to have the power represented by Facebook in the hands of a naïve and inexperienced individual who comes across as having little understanding of human nature and the complexity and differences of hu-

---

BIG TECH, PROPAGANDA AND SOCIAL CONTROL

- China's Xi says internet control key to stability.[794]
- China Applauds the World for Following Its Lead on Internet Censorship.[795]
- What price privacy when Apple gets into bed with China?[796]
- Facebook and Twitter are being used to manipulate public opinion—report: Nine-country study finds widespread use of social media for promoting lies, misinformation and propaganda by governments and individuals.[797]
- Fake news 2.0: personalized, optimized, and even harder to stop: Artificial intelligence will automate and optimize fake news, warns a technology supplier to the CIA.[798]
- Thirty countries use "armies of opinion shapers" to manipulate democracy—report: Governments in Venezuela, the Philippines, Turkey and elsewhere use social media to influence elections, drive agendas and counter critics, says report.[799]

---

man societies. Zuckerberg actually possesses the power to try to impose his simplistic vision on the world. But even with insight and knowledge, no one should be allowed such power and resources even if that person's intentions are benign.[800] If it weren't for his immense wealth and control of communications we would dismiss Zuckerberg's ramblings as "looney tunes." In addition to his globalist manifesto, he now is seeking to make Facebook and its AI systems the creator and coordinator of a newly invented and defined set of communities. John Shinal writes:

> Mark Zuckerberg wants Facebook groups to play an important role that community groups like churches and Little League teams used to perform: Bringing communities together. ... He added, "People who go to church are more likely to volunteer and give to charity—not just because they're religious, but because they're part of a community." *Zuckerberg thinks Facebook can help, using its networking power to organize people. "A church doesn't just come together. It has a pastor who cares for the well-being of their congregation, makes sure they have food and shelter. A little league team has a coach who motivates the kids and helps them hit better. Leaders set the culture, inspire us, give us a safety net, and look out for us."* [801]

Facebook recently abruptly removed the pages of some 800 political-ly-oriented organizations from its site. Zuckerberg sees himself as "the pastor" to oversee such endeavors and via Facebook as his "Church," is developing his own Index Librorum Prohibitorum to dictate what speech is permissible. Those finding themselves "purged" indicated a bewilderment at the act, with some stating they would be pleased to comply with Facebook's "Terms of Use" if only they could get the company to indicate what they were. A report on Facebook's actions describes what occurred.

> Facebook said on Thursday it purged more than 800 U.S. publishers and accounts for flooding users with politi-cally-oriented spam, reigniting accusations of political censorship and arbitrary decision-making. In doing so, Facebook demonstrated its increased willingness to wade into the thorny territory of policing domestic political activity. Some of the accounts had been in existence for years, had amassed millions of followers, and professed support for conservative or liberal ideas, such as one page that billed itself as "the first publication to endorse President Donald J. Trump." ... Facebook only named five of the hundreds of pages it removed. Two of the page operators said that they were legitimate political activists, not profit-driven operators of clickbait "ad farms," as Facebook claimed in a blog post. *They said [they] were still unsure which Facebook rules they had violated or why they had been singled out for behavior that is standard in online organizing.*[802] [emphasis added]

The global power that has been amassed by corporate entities such as Amazon, Alphabet/Google, Facebook, Apple, Yahoo, Softbank, Tencent and Alibaba, along with the fact that they are in control of our communications, our privacy, and much of the media makes such companies very dangerous from the perspective of the corruption of our societies. [803] The ability to control and record all our activities, influence politics and conduct propaganda without being accountable to countervailing institutions makes them threats to democracies to an unprecedented extent. If valid, a recent report on a patent issued to Google demonstrates how unbelievably intrusive the company apparently plans to be. Phil Baker reports:

> Patents recently issued to Google provide a window into their development activities. ... These patents tell us that Google is developing smart-home products that are capable

of eavesdropping on us throughout our home in order to learn more about us and better target us with advertising. It goes much further than the current Google Home speaker that's promoted to answer our questions and provide useful information, and the Google-owned Nest thermostat that measures environmental conditions in our home. What the patents describe are sensors and cameras mounted in every room to follow us and analyze what we're doing throughout our home.

Baker adds:

> They describe how the cameras can even recognize the image of a movie star's image on a resident's t-shirt, connect it to the person's browsing history, and send the person an ad for a new movie the star is in. One patent, No. 10,114,351, reads, "According to embodiments of this disclosure, a smart-home environment may be provided with smart-device environment policies that use smart-devices to monitor activities within a smart-device environment, report on these activities, and/or provide smart-device control based upon these activities." So clearly they want to monitor us and report back what we are doing.[804]

As Google and Facebook watch us, create "virtual" forms of us aimed at better sales manipulation, and control what we are allowed to say and see on their websites that are in control of most of our communications, we need to understand that to a massive degree the employees at such companies have been shown to be "of one mind." Brian Amerige, a senior Facebook engineer, writes: "*We claim to welcome all perspectives, but are quick to attack—often in mobs—anyone who presents a view that appears to be in opposition to left-leaning ideology*." [805] It is not only Facebook. In relation to what goes on at Google, Brad Parscale reports:

> Americans must be wary of powerful institutions that seek to control what we see and hear. As the internet has become an increasingly central part of modern life, Big Tech giants such as Facebook, Twitter and Google have increasingly sought to become the gatekeepers of the internet and political discourse. Without any sort of democratic mandate, these companies have appointed themselves the arbi-

ters of acceptable thought, discussion and searches online. These companies' pervasive command of the internet—and blatant desire to control how we interact with it—is a direct threat to a free society. And arguably the worst offender is Google. *Google claims to value free expression and a free and open internet, but there is overwhelming evidence that the Big Tech giant wants the internet to be free and open only to political and social ideas of which it approves.*[806] [emphasis added]

The scale, degree of social penetration and nature of their activities makes the companies that dominate the Internet dangerous. Some companies collaborated with the NSA by providing data they obtained from customers. Ewen MacAskill reports that:

> Special Source Operations, described by [Edward] Snowden as the "crown jewel" of the NSA, handles all surveillance programs, such as Prism, that rely on "corporate partnerships" with telecoms and internet providers to access communications data.[807]

While publicly whining after their participation was exposed through Edward Snowden's release of PRISM records, the company CEOs claimed they had to cooperate or face imprisonment. Yahoo CEO Marisa Mayer stated the heads of the companies faced jail if they even told anyone that the NSA was accessing their records.[808] Of course, it might have helped influence Yahoo's decision to cooperate that the NSA paid large sums of money for the records and that Yahoo already had a record of kowtowing to the Chinese government by "outing" people in that nation who used the Internet for disapproved purposes. This makes Mayer's protests fall on deaf ears.

Although the European Union is attempting to create rules and penalties to protect its citizens' privacy, it is unlikely they will be fully effective. The financial, economic, security and political stakes are too high for governments and the other invaders of our private lives to willingly abandon. We can therefore expect that much of what we do or say will be open to others who view us and our information as commodities or as supporters who need to be persuaded, "conned" or compelled to "see things their way" whatever that might be. A sampling of what is going on appears below.

BIG TECH AND PRIVACY

- These Ex-Spies Are Harvesting Facebook Photos For A Massive Facial Recognition Database.[809]

- China Assigns Every Citizen A "Social Credit Score" To Identify Who Is And Isn't Trustworthy: Country Determines Your Standing Through Use Of Surveillance Video, Plans To Have 600 Million Cameras By 2020.[810]

- Facebook is giving the US government more and more data.[811]

- Yahoo and AOL just gave themselves the right to read your emails (again): Though Yahoo was already scanning its users' emails to maximize ad opportunities, doubling-down on the policy could raise eyebrows in a post-Cambridge Analytica world.[812]

- It's time to take our privacy back from tech companies.[813]

- Facebook keeps creepy secret files on the intimate habits of internet users even if they DON'T have an account: ... Even if you have never entered the Facebook domain, the company can track you.[814]

PART V

1010101010101010101010

MOVING TOWARD SOLUTIONS

MOVING TOWARD SOLUTIONS

INTRODUCTION

Far too many people appear unaware of what is emerging, and that the time within which effective action can be taken is running out. It is vital we understand the reality of what we are facing and develop comprehensive and realistic solutions. Although some possible responses to the rapid emergence of AI/robotics have been considered throughout *Contagion*, in Part V we discuss several more specific solutions. These include significant tax reforms aimed at countering the dramatic reallocation of earnings and wealth from human labor to invested capital. If we fail in doing this most other approaches are close to meaningless. But while fundamental tax changes are a necessary part of solutions they are not in themselves sufficient. Solutions also include strategies such as "pump priming," large-scale infrastructure investments that involve human workers, Universal Basic Income programs and job guarantees, job "splitting," and governmental policies aimed at slowing down the introduction and scope of AI/robotics systems.

A core element of effective solutions strategy is a strong societal commitment to a policy of preserving and creating human employment in the face of the AI/robotics transformation. This is vital because, as we have explained in *Contagion*, what is occurring is not a Schumpeterian cycle of "creative destruction" in which following a difficult period of economic transformation a new and healthy system is regenerated complete with a host of exciting and progressive new jobs of quality. In the AI/robotics transition, the other side of "creative destruction," is more destruction or, at best, a flattening of our world of opportunity, creativity and growth for all but a few.

The primary cause of the short-circuiting of the re-creative aspect of Schumpeter's economic cycle is that AI/robotics systems are being developed that will be increasingly capable of replacing humans in many of the new tasks that will be introduced into the economy. Although those who benefit from the ongoing changes in the short-term describe the positive aspects of AI/robotics in glowing terms, the reality is that AI/robotic systems will be doing not only the menial tasks that most people think about when automation and robots are discussed, but also advanced work.

In past industrial revolutions newly created work for people was generated in sufficient numbers to ensure that, after a period of difficulty, an exciting new world came into being that provided people all the employment opportunities that were needed. This is not going to happen at anything close to desirable and needed levels in the AI/robotics Fourth Industrial Revolution. There will be Data Scientists, Development Operations specialists, lawyers, doctors, financial wizards and entrepreneurs but, due to the continuing developments in AI, even they will find their numbers dwindling.

Unless we force ourselves to deal with the debilitating challenges AI/Robotics poses, all the issues currently being fought over in our increasingly fractured society are little more than wishful, short-sighted or even cynical, rhetoric. We cannot achieve economic opportunity, guarantee social mobility and improved access to justice, reduce wealth inequality, preserve our commitment to democratic ideals, respond to growing homelessness and drug epidemics, and provide support for those most in need of assistance unless we face the reality of our existing economy and the impacts AI/Robotics will have upon it. The only option to protect our political and economic systems and sustain what we consider to be the vital nature of Western-style democracies is to implement effective strategies to block, reduce and mitigate the impacts.

The most challenging obstacles of our current socio-economic reality create a negative synergy that increases the likelihood of a severe and long-lasting crisis. To summarize, as addressed earlier:

- In the US particularly, massive budget deficits are being faced by federal, state and local governments, by many corporations, and by millions of individuals.
- We are experiencing a widening demographic "Age Curse," with the various impacts earlier described.
- Federal, state, local and private pension systems and health care programs are already underfunded. The consequences societies will face when they prove almost totally unable to meet their promised benefits may tear nations apart.
- In the US we are experiencing rapid increases in physical and emotional health problems, along with a stunning rise in chemical addictions and alcoholism.
- Homelessness is growing, particularly along the US East and West coastal areas, but also in other regions, including rural areas.
- Highly intrusive surveillance activities are being conducted by governments, corporations, and private interest groups. Incredibly powerful corporations such as Google/Alphabet, Amazon, Facebook,

and Twitter control our communications systems to the extent that they are capable of shaping and controlling public discourse.

- The US and EU nations are experiencing increasingly massive levels of migration from poor countries but many no longer welcome what they see as an unneeded addition to the labor force and an increasing drain on social resources.

All these challenges, and more, are developing in key nations—including the US, EU, UK, Russia, Japan and China. This is accompanied by internal strife, vulnerability to terrorist activities, crime and corruption, and increasingly authoritarian and intrusive governments. Explosive population growth and other stresses being experienced in less economically well-off countries produce high crime rates in those systems, result in an absence of decent opportunities for affected nations' inhabitants, and generate growing political repression by those seeking to hold onto power and wealth, or to obtain it.

Such conditions are causing tens of millions of displaced persons to seek better lives by migrating to wealthier nations. Unfortunately, economically developed nations—including the US, UK and EU—are themselves undergoing significant economic, cultural and political pressures. These are leading their populations to become increasingly resistant to immigrants with significant elements resentful of the economic and social costs migrants create.

SOME ESSENTIAL ECONOMIC AND POLICY STRATEGIES

Here is a list of some policy approaches that could make a difference in protecting and creating human jobs:

- Introducing national legislation and an international treaty regulating AI/Robotics implementation in specific economic areas, including "Slow Teching," shaping, and limiting the use of AI/robotics.

- Promoting the use of human labor by:

  - Understanding that economic efficiency and productivity cannot be allowed to be the sole criteria in conducting work activity.

  - Recognizing that human employment is a vital part of a community's health.

- Implementing tax strategies aimed at mitigating the most extreme disparities in income inequality, reduced employment opportunities and limited social mobility.

- Creating economic stimulation through investment in infrastructure repair and development, with the requirement that the work must be done primarily by humans.

- Universal Basic Income (UBI) strategies.

- Government job guarantees.

- Job "Splitting" to preserve and create a sufficient number of human work opportunities, possibly linking this approach to elements of a Universal Basic Income or job guarantee program.

- Identifying the "Winners" and "Losers" in the evolving job market and promoting training for the "Winners" and developing useful work, support and opportunity programs for the "Losers" so that their condition does not become permanent if they have the needed talent and drive.

- "Pump priming" through financial transfers made directly to human households by means of tax or direct payments to provide help and sustain economic activity.

- Breaking up companies such as Amazon, Google, Facebook, etc. because they control far too much economic, communications, privacy invasion, and surveillance powers.

We will discuss some of these options in Chapters Twenty-Five through Twenty-Eight.

CHAPTER TWENTY-FIVE
Innovative Revenue and Expenditure Strategies

Solutions must be pragmatic and realistic, not politicized "spin" based on what we wish we could do or empty political rhetoric designed to buy votes. They must be intelligent mechanisms for actually achieving specific goals. While it is not clear what form it should take, a redesigned, more efficient and enforceable tax regime is essential. Such a regime must prevent the extreme and excessive shift of economic returns from labor to capital. It must also incentivize job preservation and creation, and stabilize government revenues. This initiative must also include significantly improved enforcement and accountability.

REVENUE ENHANCEMENT STRATEGIES

The Organization for Economic Cooperation and Development (OECD) has called for a dramatic shift in tax strategies—increased taxes on capital and wealth as opposed to simply on income. This is based on the OECD's conclusion that the justifications for lower taxes on capital are no longer valid in the contemporary global economy. According to Pascal Saint-Amans, Director of the OECD Centre for Tax Policy and Administration ... "*For the past 30 years we've been saying don't try to tax capital more because you'll lose it, you'll lose investment. Well this argument is dead, so it's worth revisiting the whole story*"[815] [emphasis added].

Saint-Amans is correct. We do need to "revisit the whole story" and that not only means designing new ways to tax the new form of redistributing wealth from labor to capital but taking an honest and pragmatic look at all strategies for generating revenues and stabilizing the tax base of the future. If we are going to "prime the pump" new tax and revenue generation strategies must be implemented.

The need to ensure a strong, fair and dynamic social safety net cannot be met without new revenues. But revenues do not simply appear when you raise tax rates in a traditional system. A dramatic refiguring of how and why we tax financial activity must occur.

The conditions being experienced in the US as AI/robotics transform the nation's economy, reallocate earnings and wealth, and create a compelling need to drastically reconfigure the US system of taxation while preserving the nation's economic dynamism. As we do not really know what works to the extent required to solve our problems and supply our revenue needs at required levels, we therefore need to go through a period of testing of various tax models to come up with realistic strategies.

It will be nearly impossible to control a transnational and globalized corporate and financial elite that straddles traditional national boundaries to the extent that they feel no allegiance and that technology has provided so many means for swift transfer and concealment of wealth. No single nation is capable of holding such economic and political elites to account.

Policies representing business as usual will not work. The systems represented by the highly developed economic and democratic systems of the Western nations, as well as those of China, Japan and South Korea must make fundamental adjustments. Those adjustments demand a significant degree of cooperation among a tight group or groups of economies operating with quite different economic and political systems. Nonetheless, those powerful nations must reconstruct a workable economic and political order.

No one can go it alone in a system with so many intertwined and mutually dependent elements. The US has a critical role in what will occur because the bottom line is that if the US collapses economically and politically, much of the world will follow in its path. Some ideas relating to taxation including the following.

- New tax strategies are required aimed at raising revenues devoted to financing specific types of increased governmental responsibilities likely to be viewed as worthwhile by those taxed to support them. This could include, for example a public housing tax to ensure the homeless and less fortunate have shelter. There could also be a specific hospitals tax to support medical services in the numerous areas where such facilities are closing down. One of the most critical approaches involves an infrastructure tax to fund the millions of human jobs that could be generated through such work.

- Provision of employment grants or tax benefits for companies that commit to employing people rather than shifting to AI/robotics.

- Eliminating incentives for the adoption of AI/robotics in critical jobs systems.

- Wealth taxes, despite the issue's extreme complexity.

- Compensation caps on the extreme levels of return being awarded corporate executives who are largely in control of the boards of directors that bestow their largesse. One can only wonder how it is that executives fired by their companies for questionable or illegal activity receive "golden parachutes" of $30–40 million or more as a reward for their malfeasance.

- "Robot" taxes.

- Financial Transactions Taxes.

- Technology and capital taxes such as the EU has proposed on technology firms such as Google and Amazon at 3% on related revenue.

- Collaborative taxes imposed on digital platforms such as Amazon by the nations in which they do business. The German Minister of Finance, for example, recently floated a proposal in which companies such as Google, Amazon and Facebook would be taxed on their global revenues with the proceeds allocated to the locations where business was done. A movement toward figuring out how to fairly and effectively tax the immense global tech giants is gaining momentum in Europe and elsewhere but as yet is lagging in the US.

- Technological "windfall profits" taxes imposed on companies that were labor intensive but opted to shift to AI/robotics.

- Value Added Taxes (VAT).

- Eliminating tax incentives for adopting AI/Robotics systems in various areas of work where we are committed to preserving and creating human jobs.

- "Value Destroyed" Taxes (VDT) based on the comparative productivity of a human workforce and its replacement by AI/Robotics systems.

And here are some related possibilities:

- Creating AI surveillance and tracking systems to monitor illicit concealment of otherwise taxable earnings and wealth.

- Defending core industries that employ large numbers of human jobs by charging tariffs on imported products from AI/robotics-based production systems.

- Creating effective and cooperative systems to cope with and eliminate offshore Tax Havens that conceal as much as 10% of the world's wealth.

Although we can't go through all of them, we will examine the merit of some of these options below.

THE WEALTHIEST ARE KEEPING THEIR WEALTH OFF THE BOOKS: DAVOS, PANAMA, AND PARADISE

Joseph Stiglitz, a highly respected Nobel Prize laureate and economist, agreed to be on a commission charged with analyzing the *Panama Papers* system and determining who, what, where and how the tax avoidance dodges worked. But Stiglitz and another prominent commission member resigned after only a few months due to interference from the Panamanian government they felt was caused by extreme pressure and influence being applied by the wealthy tax dodgers. The report on Stiglitz's decision to resign provides some insight. *"I thought the government was more committed, but obviously they're not," Stiglitz said. "It's amazing how they tried to undermine us."*[816]

The AI/robotics restructuring of work and the consequences of the transformation for the overall community produce a situation requiring the imposition of higher tax obligations and real mechanisms to detect, monitor and shut down illicit tax avoidance schemes and to terminate even the legal secretive tax shelters such schemes provide. Part of that necessity begins by bringing the proliferation of tax avoidance laws and venues under control.[817] Tax avoidance strategies are intricate and widespread as is demonstrated by the revelations in the *Panama* and *Paradise Papers*.[818]

The Paradise Papers revelation further demonstrates the pervasiveness of the schemes and networks involved in hiding wealth and income from tax authorities.[819] Some of the activity is legal according to existing tax rules. Much is blatant tax avoidance that goes well beyond the legitimate sheltering of wealth and income. Dealing with the vast amounts of wealth being hidden by such schemes is extremely difficult because the people involved are very powerful, capable of taking secretive action to protect their investments, and linked to the regulatory authorities that would need to take action to expose, stop and unwind the activity. The world of tax avoidance schemes is probably best equated to a non-stop game of "Whack-a-Mole" in which the ugly little "heads" keep popping up unexpectedly, frequently change location, and are usually one step ahead of the game.

In an analysis done in conjunction with the Washington Center for Equitable Growth, New York University law professor David Kamin suggests that taxing capital in addition to ordinary income offers a fair mechanism for obtaining the added tax revenues required to feed the constantly expanding need for government spending.[820] According to Kamin, taxing capital would increase the progressivity of the U.S. tax system (theoretically increasing revenues) and according to Kamin, "does not seriously affect the rate of savings among high-income Americans."[821]

Kamin suggests tax reform should focus on strategies for collecting larger amounts from those who own capital and benefit to an extraordinary extent from the shift of returns away from labor to them. He identifies three key problem areas undermining our ability to achieve this goal.

- Taxes on gains on property that are too easy to minimize or avoid entirely.

- The shifting of corporate profits and corporate residence to avoid taxes.

- Wealth taxes that apply to only a sliver of the population and are too easily avoided.[822]

Kamin suggests possible strategies based on taxing capital alone or a combination of taxing capital plus windfall or extraordinary returns on capital. He suggests these represent "an effective way—and perhaps the best way (in combination with other taxes)—of efficiently and fairly taxing the highest income earners in the United States."[823] Beyond a certain point, tax rates can have anti-productivity and anti-revenue collection consequences, incentivizing such legal and illegal avoidance behavior. This means it is naïve to suggest that the extremely wealthy should be taxed far more than currently and expect that policy to produce massive added tax revenues.

One loophole used by numerous corporations is "tax inversions" by which they use the legal practice of technically locating corporate headquarters to "headquarters havens" such as Ireland or Luxembourg where they conduct little or no actual economic activity but enjoy massive tax avoidance advantages by means of the "paper" relocation. If the major world economic actors really wanted to shut down offshore tax havens and eliminate they could do so almost overnight. That the most powerful nations have opted not to do this offers conclusive proof our political leaders and the major national and international institutions have been captured by the "money people" and are their willing servants.

Another twist on the limits of taxation is offered by economist Arthur Laffer whose *Laffer Curve* illustrates the existence of a "tax frontier" that limits the maximum level of revenues available at any given tax rate.[824] While useful analytically, the *Laffer Curve* is an idealization that does not specify tax rates, predict revenues, describe the skew or kurtosis of the curve, or even require the curve to approximate a normal distribution.[825] It is a conceptual framework for illustrating that there is an outer limit of tax revenues that may be collected by a perfectly efficient tax system at any given tax rate. Inefficiencies in the tax system or loopholes that permit unintended tax avoidance or evasion mean that actual tax revenues will fall inside (toward the y-axis) the curve illustrating the frontier.

Depending on the degree of efficiency, monitoring, accountability, corruption and rates, as tax rates increase, revenues will increase but only up to a point. As the rates are pushed upward, growing resistance to tax collection will create new inefficiencies in tax collection. Finally, as the benefits of tax avoidance and evasion increase, wealth creators will shift from investments in economically productive activity to greater investment in tax avoidance and evasion.

An example is offered by what went on with Francois Hollande's 75 percent wealth tax on millionaires in France.[826] Over a 15-year period France was estimated to have lost 60,000 millionaires who left that nation due to its imposition of a high tax on wealth for people making over 1.3 million Euros. One report in the *Financial Times* indicates the tax may have even reduced rather than increased France's overall tax revenues.[827]

Even though the *Laffer Curve* makes sense analytically as representing what happens if a tax system goes too far, the critical operating conditions in which the tax is imposed make all the difference. If, for example, there is full and accurate tracking, reporting and monitoring of earnings and wealth, and real accountability for violations of the laws, then the point on the *Curve* where the frontier is triggered will necessarily shift upward.

If unproductive and even criminal tax havens are eliminated, and locations such as Luxembourg and Ireland that function as the bases for corporate inversions are no longer able to do so as a matter of law, the number of avoidance options available to those hiding and sheltering wealth from their legitimate tax authorities will be reduced and the avoidance frontier raised. We aren't even close to that point and the numerous essentially toothless international agreements being batted around by multilateral institutions are more window dressing than serious solutions to the problem of tax avoidance.

It may be argued, however, that in some instances the legal relocation of corporate headquarters to countries outside the US to obtain reduced tax liability and the retention of foreign earnings overseas to avoid US taxes may be a necessity for corporate survival.[828] Facing too heavy a tax burden compared to obligations existing in other jurisdictions can make a company less competitive in a globalized marketplace. Radically different rules of operation that vary according to a company's primary tax home, including whether a particular competitor is receiving stealth subsidies from a government, have created an avoidance complex that is surprisingly easy to manipulate, as seen in the case of the EU and Ireland.[829]

VALUE ADDED TAXES

One useful approach followed by the EU nations that could be of significant use to the US is a Value Added Tax. The VAT is a good start on a tax system that reflects a shift from taxation of labor to taxation of products at the point they are created with collection at the point of sale, i.e. exports are exempt and imports are taxable. A VAT offers a partial solution because it taxes the product wherever produced. The World Trade Organization does not consider a VAT a trade barrier.

Even with the VAT, if there is a dramatic slump in the production of consumer products and services because 50 percent of the people aren't working, far fewer products will be sold. If the system of production and consumption is "tanking," the revenues will plummet even though assistance needs are escalating. This undercuts tax revenues because sales, profits, employment and wages will decline.

TAXING "TECHNOLOGY WINDFALL PROFITS" AND "TECHNOLOGY WINDFALL DESTRUCTION" WHILE EXEMPTING TRUE INNOVATORS

One approach to resolving the financial dilemma caused by job elimination is requiring companies that replace human workers with robotic systems to pay what could be called a "technology windfall tax" rather than receiving a tax incentive for installing AI/robotic production systems. Receipts would be directed toward retraining or compensating human workers who lose their jobs and funding governmental institutions responsible for sustaining people who are now without work or adequate income as a result of companies shift to AI/robotics production, service and transport systems.

Recent reports indicate that Bill Gates' and Jeff Bezos' net worth was over $90 billion with Bezos at $92.3 billion on July 26, 2017 with Gates lagging $1.5 billion behind at $90.8 billion. Six months later Bezos was said to be in the $105-106 billion range. A June 2018 report indicated Bezos' wealth had soared to $141 billion.[830]

Such levels of wealth are obscene from the perspective of any group of "ordinary" people. But insofar as virtually all the incredible level of return is created by appreciation in the entrepreneurs' share value based on allocations of companies they founded or drove to new competitive heights, very little is in the form of cash on hand or stacks of gold bullion in immense vaults. While a percentage could be sold and paid for taxes, that percentage often impacts entrepreneurs' ability to retain control of the enterprise.

Here we are making the assumption that real innovators continue to drive the evolution of their company forward. Assuming that the tax system overall has undergone redesign it is important to allow the real innovators to pursue sound economic development rather than imposing rules that stifle it. "Being in charge" and being an intellectual leader is an element of this and compelling the liquidation of assets essential for those functions will not always be an intelligent path of action. This restraint would not be applicable to the legions of self-serving corporate bureaucrats who emerge to run companies as they mature.

Taxing that wealth requires answering hard questions about how best to do it, when to impose the tax obligation, and the rates to be used. Additional factors in play with a company such as Amazon is that while Bezos may in one sense be creating wealth by Amazon's relentless expansion, he is also destroying and redistributing pre-existing wealth and opportunity through the destruction of very large numbers of smaller businesses supplanted by the Leviathan that is Amazon. Given the economic destruction he has wrought, Bezos may be a prime example of job destruction without re-creation. This is particularly so since he and Amazon are committed to the extensive roboticization of every possible phase of the company's operations into which AI/robotics can be inserted, all at the cost of destroyed human employment in a quest to achieve maximized return on investment.

The ability to amass such unbelievable levels of wealth is not solely related to the individual's creativity and business strategy, whether Bezos, Zuckerberg, Buffett, Gates or Henry Ford. Innovators and entrepreneurs operate within a structured system that produces opportunities that would not otherwise exist. It is very much a foreseen and intended byproduct of the socio-economic system the political community has built at the behest of business interests, that an entrepreneur, "robber baron," "pirate king" or the like is able to seize the moment by tapping into a social, technical and political system that has been created and sustained to facilitate business over generations.[831]

This is the context in which Bezos and Zuckerberg have "won the lottery." But insofar as it is the overall community rather than Bezos and Zuckerberg, et al. that has acquiesced to the legal institutionalization of the structure and substance of the "game" in the first place, the community, as well as the creative entrepreneurs and inventors are entitled legally and morally to share significantly in the benefits of the economic activity.

Altering the tax system involves a basic change in our approach to the system by which government obtains revenues. It is at least arguably acceptable to have allowed special treatment of capital gains and corporate profits because we based that to a significant extent on the assumption investors, entrepreneurs and corporations created and sustained human jobs both directly and indirectly—and indeed, to return money back into the system in a positively synergistic loop. To the extent there were significant revenue effects provided to the overall system in the form of employment wages, governmental tax revenues on the federal through local levels, and the activity of work itself, this approach had merit.

But as we say in *Contagion*, this beneficial collective totality of what might be called a complete economic ecosystem is changing rapidly and fundamentally due to the AI/robotics transformation of the workplace. This means that the tax and revenue systems must change as well in order to provide the benefits needed to sustain the social order.

TAXING ROBOTS

Vincenzo Visco outlined a strategy and justification for taxing robots clearly and succinctly.

> Historically, tax systems evolve following the evolution of taxable bases (actual and potential). In other words, governments over the centuries and millennia used to "follow the money" (and still do). Thus, the levies moved from the products of agriculture and sheep farming to land taxation, real estate, trade (duties), excise taxes, taxes on the value of goods, taxation of income and profits, personal taxation, progressive taxation, general consumption taxes, etc. In short, the tax authorities track economic development and wealth-formation.[832]

Visco explains that: "if the taxable base represented by human labor is reduced, it is inevitable that the levy will be directed, sooner or later, to other sources, even if at the beginning this may appear unconventional or even controversial."[833] The problem is that our tax strategies have not developed a full understanding of what *"taxing robots"* means.

Some economists, Visco adds, have proposed replacing the "wage" base with a value-added base. The idea is a shorthand way of "taxing the robots" as the shift to an AI/robotics system increasingly replaces taxed human workers. Visco explains: *"it would be entirely logical to shift taxation from labor to other incomes, thereby keeping revenue constant, and therefore, for example, trying to replace social contributions based on wages with a levy on the entire added value of national income."*[834]

He concludes: "if 'robots' are used by companies that increase their profits share with respect to total GDP, ... these growing profits will become a favorite taxable base. The new production sources of wealth are today the internet and automation, but they are ignored or even exempted by existing tax laws." [835]

While the above suggestions are worth considering, any proposed robot tax would be difficult to design and implement. There are numerous situations in which a technology may be an enhanced version of an existing technology. The expanded capabilities of computer printers offer one example. Where is the line to be drawn between a useful efficiency enhancement redesign and "AI/robotics"?

To the above observations we want to add those of two tax scholars who summed up the robotic tax situation and concluded that various laws encourage the shift to AI/robotic systems over the continuing use of human workers. In *How Tax Policy Favors Robots over Workers and What to Do About it*, Ryan Abbott and Bret Bogenschneider wrote about a proposal by Bill Gates that robots who replace human jobs should somehow be required to pay taxes.[836]

A key part of the two scholars' analysis is their conclusion that "existing tax policies encourage automation, even when a human worker would otherwise be more efficient than a machine."[837] They note that a shift to automated or AI/robotic systems, "allows firms to claim accelerated tax depreciation on capital costs for machines. This allows firms to claim early tax deductions relative to the actual economic depreciation of an asset over time (like a robotic worker)."[838] Firms that reduce labor costs by adopting AI/robotics don't have to pay the array of wage-based taxes. By reducing the number of employed human workers they also benefit from a reduction in taxes imposed on workers' incomes.

Fearing mass unemployment due to AI/robotic adoption, South Korea is the first nation to impose a tax on robots. Cara McGoogan reports: "The country will limit tax incentives for investments in automated machines as part of a newly proposed revision of its tax laws. It is hoped the policy will make up for lost income taxes as workers are gradually replaced by machines, as well as filling welfare coffers ahead of an expected rise in unemployment."[839]

Taxation alone will not fully resolve the effects of the US and other nations having a massive amount of people out of work, underemployed, dependent on a shifting base of part time and "gig" employment, or living on governmental subsidies. Those changes will occur to a challenging degree no matter what, and we must be prepared to deal with the changes and their consequences.

More is needed.

CHAPTER TWENTY-SIX

Priming the Pump: Strategies for Stability in an AI/Robotics-Driven Economy

INVESTING A FEW TRILLION DOLLARS FOR JOB CREATION

Put simply, permanent, relentless and large-scale deficit spending that pumps money into the US government's regular operating budgets year-after-year and decade-after-decade has already occurred beyond any limits of rationality. This is not what Keynes envisioned as a form of jolting counter-cyclical economic stimulus but it is what we have been doing for more than twenty years—utilizing deficit spending not for solving specific problems or for counter-cyclical responses to a recessionary downturn, but for funding bloated "budget creep" programs that are structured to grow almost infinitely.

Perpetual deficit spending for ordinary and recurring operational needs guarantees the continual expansion of debt far beyond a sustainable resource base and the ability of the supposed stimulus to generate the revenues that real "pump priming" would produce. At some point in the looming future the chickens of unsustainable and profligate spending will come home to roost. We must regain at least some control over unsustainable deficit spending that funds far too much of our ordinary operating budget.

The challenge we face is that sometimes it is necessary to appear to worsen a situation—even with more deficit spending—in order to generate the momentum needed to create a system in which the worst of otherwise certain consequences are either avoided or substantially mitigated. It is also where Paul Krugman's idea of large-scale infrastructure investments financed by debt that creates jobs, while retrofitting US infrastructure and energy systems, offers an important job-creating engine capable of lasting several decades. Joseph Stiglitz echoed the idea of targeted deficit spending in suggesting the direction Donald Trump should take when he assumed the presidency in January 2017.[840]

Krugman, Stiglitz and others have targeted public investment in US infrastructure as part of an important strategy for kick-starting the economy and putting more human workers back into jobs. Krugman writes that: "The most important thing we need is sharply increased public investment in everything from energy to transportation to wastewater treatment."[841] It is only one piece of the strategic puzzle but given the speed at which many jobs are being destroyed or reduced dramatically, such programs are essential and we need to act now.

The bottom line is that we are at a point in time when some seemingly counterproductive or counter-intuitive actions are needed as well as some truly innovative governmental strategies. Our solutions strategies must take into account the scale and effects of our debt but should not make its reduction or retirement a primary focus of what we do at this time because, no matter what, that is not going to happen. The US is already bankrupt based on current "earned" revenues and the ability to meet future obligations, so strategically adding a few trillion dollars more for job creation of the right kind and in the right places could be a productive strategy.

Human job preservation and creation must be at the heart of any strategy. Among the factors the Federal Reserve's former Chair Janet Yellen noted during her tenure as important influences on productivity are "workforce development" and "public investment." Yellen urged Congress to look at ways to improve "very low" labor productivity. In that message she highlighted the importance of: "public investment, workforce development and the pace of technological progress."[842]

Millions of people will need to be put back to work on a combination of short and long-term projects that must be an important part of an ongoing government strategy. To do this over the next five to ten years, we will need to increase deficit spending on jobs and infrastructure even more than at present. Aiming policies at paying down the national debt is a counterproductive strategy at this point. There is no realistic way to do so without tearing our political, economic and social systems apart and creating a psychology of desperation and austerity that will undermine productivity and hope.

TARGETED DEFICIT SPENDING IS A
KEY COMPONENT OF SOLUTIONS

There is an old adage that "you have to spend money to make money." Austerity and budget cutting across the board hasn't worked in Europe or the US.[843] Austerity and balanced budgets make for good political rhetoric and seem entirely logical when voiced as political slogans. But budget slashing and deep cutting austerity programs impose significant suffering on people while creating a psychology of negativism, fear and despair. The EU's efforts to control social spending and benefits created a psychology of defeatism and a sense of decline and helplessness.

The following bullet points are extracted from suggestions Nobel economist Joseph Stiglitz offered in a 2016 syndicated article, "What America's Economy Needs from Trump." We offer several of Stiglitz's programmatic insights and policy perspectives in his own words.

- Over the past third of a century, the rules of America's economic system have been rewritten in ways that serve a few at the top, while harming the economy as a whole, and especially the bottom 80%. ... [T]echnology has been advancing so fast that the number of jobs *globally* in manufacturing is declining.

- Trump ... *can bring manufacturing back, through advanced manufacturing, but there will be few jobs.*

- The first order of business is to bolster investment, thereby restoring robust long-term growth. *Specifically, Trump should emphasize spending on infrastructure and research.* Shockingly for a country whose economic success is based on technological innovation, the GDP share of investment in basic research is lower today than it was a half-century ago.

- Improved infrastructure would enhance the returns from private investment, which has been lagging as well. Ensuring greater financial access for small and medium-size enterprises, including those headed by women, would also stimulate private investment.

- A carbon tax would provide a welfare trifecta: higher growth as firms retrofit to reflect the increased costs of carbon dioxide emissions; a cleaner environment; and revenue that could be used to finance infrastructure and direct efforts to narrow America's economic divide.

- While Trump has promised to raise the minimum wage, he is unlikely to undertake other critical changes, such as strengthening workers' collective bargaining rights and negotiating power, and restraining CEO compensation and financialization.

- Regulatory reform must move beyond limiting the damage that the financial sector can do and ensure that the sector genuinely serves society.

- *The US needs to tackle ... concentrations of market power.*

- America's regressive tax system—which fuels inequality by helping the rich (but no one else) get richer—must also be reformed. *An obvious target should be to eliminate the special treatment of capital gains and dividends. Another is to ensure that companies pay taxes—*

> *perhaps by lowering the corporate-tax rate for companies that invest and create jobs in America, and raising it for those that do not.*

- *Ensuring preschool education for all and investing more in public schools is essential.*

- *Restoring shared prosperity would require policies that expand access to affordable housing and medical care, secure retirement with a modicum of dignity, and allow every American, regardless of family wealth, to afford a post-secondary education commensurate with his or her abilities and interests.* [844] [italics added]

Quite a few analysts, as well as the International Monetary Fund, warn that we appear to have learned little or nothing from the 2008/2009 Depression and are heading for a disastrous repeat while lacking the tools to cope with the impending economic debacle.[845] Likewise, the costs imposed on governments by health care and pension benefits will rise even while jobs will be disappearing at a rapid rate and a growing part of the population is made up of older people with worsening health problems. Following up on the suggestions by Stiglitz and Krugman we can focus on several critical needs for which we must find resources and create policies to ensure their stability and quality. Those include:

1. Comprehensive and affordable health care for all is essential.

2. An affordable and useful education for all is critical.

3. Creating a massive housing program would stimulate jobs.

4. The special treatment of capital gains and dividends should be rewarded only when it creates human jobs and offers value to the overall community.

5. Lowering the corporate-tax rate for companies that invest and create jobs in America, and raising it for those that do not is a potential strategy.

6. Tax breaks should be viewed as only meaningful or "earned" when they create or preserve real human jobs. Capital must be rewarded for contributing to active growth of human work but it must contribute beyond the boundaries of passive investment.

7. Extreme concentrations of market power, whether oil and gas, telecommunications and giants like Google and Facebook, financial institutions or the newest manifestations in the so-called sharing or

"gig" economy such as Uber and Lyft, end up creating a new aristocracy of the incredibly wealthy. We must "bust" these economically
disruptive and suppressive "monsters" to allow the reallocation and
diffusion of market power and real competition.

8. Improving defensive regulation. This involves what we can think
of as defensive regulation by governmental agencies such as better
oversight by the SEC in relation to securities law compliance, and
improved regulation of banks and financial institutions to prevent
the kinds of reckless behavior that came close to "crashing" the
world's economy in the 2008/2009 recession.

9. But we also need to pay greater attention to the development of positive offensive regulation where affirmative goals are identified and
advanced through regulatory behavior. We need regulatory reform
that uses the regulatory system to help create jobs, stimulate innovative approaches, and reverse the current trend of industry self-regu-
lation and regulatory capture.

10. Strengthened union bargaining rights are needed as a counterweight
to corporate control, and blockage of the tendency of governmental
actors to favor larger entities and donors. The erosion of unions in
the US has been a key contributor to the shift to automation and loss
of human jobs.

11. If a significant number of human workers are put back into the job
market as a result of the infrastructure and technology development
investments, this has the added benefit of generating tax revenues
from their earnings.

12. Large-scale infrastructure investments incorporating decent paying
jobs can slow the shrinkage of the middle class and provide jobs and
job training for younger workers. This includes minorities, who often lack access to a decent paying job market. At a minimum, the
strategy could provide a "slow teching" stopgap for a decade or more
and produce a range of positive results to mitigate the economic and
social crises we are experiencing.

CHAPTER TWENTY-SEVEN

Universal Basic Income (UBI): Solution or Catastrophe?

Arguments are being made that a Universal Basic Income is the right of all citizens. Such is already in place to a degree in Western Europe and North America via governmental welfare payments administered through a range of programs that purport to ensure that all citizens can exist on a decent level, even though many are obviously falling through the gaps given the rapid growth in homelessness. This "right" is claimed to exist whether those aided are unable to work in the traditional sense, or are forced or choose to opt out of the workforce partially or entirely.[846]

The issue of a UBI is being raised with frequency and some are using the predictions of inevitable job loss from AI/robotics to call on their governments to provide a Universal Basic Income. Elon Musk describes a future in which robots do nearly all of the work and humans live on governmental stipends. Catherine Clifford reports: "There is a pretty good chance we end up with a universal basic income, or something like that, due to automation," says Musk … "I am not sure what else one would do."[847]

Musk and others who are the spearheads of the AI/robotic transformation are the ones destroying human jobs and making UBI seem inevitable. The above report adds: "Musk sees increased automation as an overall benefit to society, even an opportunity. 'People will have time to do other things, more complex things, more interesting things,' says Musk. 'Certainly more leisure time.'"[848] The report recognizes, however, that: "A long horizon of leisure time may sound good, but it can also be an intimidating prospect. For many, having a job and someplace to be each day is grounding and gives purpose to life."[849]

INCOME INSUFFICIENCY IS AT THE HEART OF THE PROBLEM, NOT INEQUALITY

The issue is not whether there is income and wealth inequality—inequality exists and will always exist. From a quantitative perspective nothing is ever fully equal in any society. Nor should it be. The idealistic pretense that "absolute" equality is a desirable goal for a society is destructive. We are not simply clones of each other. Different people have different functions in a political and economic community. The ability to dynamically stimulate and reward those distinct functions is a critical element of our policy and strategy lest we fall into an unsustainable stasis of "false equality" that requires heavily authoritarian methods to be implemented and ultimately collapses. Someone has to "stir" the pot in any dynamic and evolving system and others are needed to "sustain" the resulting evolving mixture when it is advantageous to the whole.

A preoccupation with the idealized abstraction of something approximating full income equality diverts us from figuring out approaches that work fairly and effectively and that are realistic from an economic and political perspective. Responses to inequality range from the extremist and improbable "Harrison Bergeron"[850] variety of forced equality on all characteristics to the equally extreme Darwinian survival of the fittest model, that callously disregards what happens to the less fit.

The real issue is income sufficiency. What should we do about providing an income base that includes ordinary living costs, education, health care and then, in addition, opportunity for those with the drive and wherewithal to pursue it? The real challenge is to find the best balance.

It is useful to have a sense of what many people face in their everyday lives and what their futures look like. Some of this is highlighted in the box below revealing the present dire circumstances endured by a significant part of the American population. And if this is the case with the wealthiest nation on earth, think about what residents of other countries—developed and developing as well as the rising list of "failed states" must deal with.

ON A FINANCIAL PRECIPICE

- Almost half of US families can't afford basics like rent and food.[851]

- Vast number of Americans live paycheck to paycheck: Almost 8 out of 10 American workers say they live paycheck to paycheck to make ends meet.[852]

- More than a third of California households have virtually no savings, are at risk of financial ruin, report says.[853]

- 65% of Americans save little or nothing—and half could end up struggling in retirement[854]

- Top 20% of Americans Will Pay 87% of Income Tax: Households with $150,000 or more in income make up 52% of total income nationally but pay large portion of total taxes.[855]

- OMB: Top 20% pay 95% of taxes, middle class "single digits." Any tax cut for middle income earners will also provide a benefit for those further up the income scale, including the top 20 percent who now pay 95 percent of all income taxes, according to the director of the Office of Management and Budget. OMB later cited internal data … that said the top 20 percent of people to pay income taxes account for 94.8 percent of those taxes in 2016. That appears to be a jump from just a few years ago. In 2015, the Wall Street Journal reported that the top 20 percent of income earners paid 84 percent of income taxes.[856]

- Finland's basic income trial falls flat. It was determined that a 30% national tax increase would be needed to fund the full program and that Finland was already heavily taxed.[857]

- In one out of five families in the US no one works in compensated employment. [858]

- Almost fifty percent of Americans pay no federal income tax because their income is too low or they are not working and are supported by government entitlement payments.[859]

Our point is that the problem of sufficiency about which we should be concerned as a matter of justice, opportunity and sustenance does not require identical distribution. The challenge is one of assuring sufficient access to the goods necessary to live on a decent level while providing real access to opportunity and social mobility to those seeking to better their situation. This base of security and opportunity allows people to make their own choices about the direction of their lives.

In a system committed to fairness and the ability of individuals to make the choice about the person they want to be it is a matter of great consequence when the paths of fair access to opportunity are blocked or eliminated so that people are fixed into place without any hope for positive change regardless of their talents and merits. This betrays the citizens and the system itself because its current quality and ultimate sustainability depends on the dynamism and energy of those who possess such skills. If opportunity is blocked and creativity and innovation blunted, the system becomes frozen into place and ultimately decays.

CAN A UBI SYSTEM BE DYNAMIC?

As long as we ignore the need to preserve and create human jobs in numbers and quality sufficient to supply the demand there is no chance of avoiding large-scale AI/robotics-driven job loss with all its accompanying problems. The undeniable fact is that as work opportunities shrink, government-supplied income and service dependencies will increase. People will demand expanded support as a matter of justice, fairness and necessity even while opportunities for self-enhancing work, or any work, diminish and disappear along with what we call the "work ethic." As this happens, our society will be fundamentally altered. We will experience a systemic "shockwave." We have absorbed the blow to citizen wellbeing by simply allowing homelessness to continue to overwhelm America's coastal cities. Are we next prepared to see people starve?

As Harari warned in *Homo Deus*, one result of the transformation will be the expansion of an involuntarily unproductive segment of society, many of whom will want to work but won't be able to find employment because the jobs won't exist whether we are speaking quantitatively or qualitatively. Ultimately, as the reduced opportunity to work goes on for an extended period, tens of millions of people will not even know how to engage in work because they will never have had the opportunity. They will be denied the special opportunity of work and the creative opportunities, aims and sense of discipline and participation it can provide.

Paul Vigna explains what we face. "If more and more workers are going to be displaced by robots, then they will need money to live on, will they not? And if that strikes you as a form of socialism, I would suggest we get used to it." [860] As to the inevitability of the emerging financial crisis, Janus Capital's insightful Bill Gross observes:

> The question isn't whether or not this is going to happen. "It is," he says. The question is how to pay for it, and the answer is two words: "helicopter money." This is the concept

of central banks essentially printing money, and while it
sounds like an unsound idea, it is essentially what's been
happening since the Panic of 2008 anyhow. This money isn't
exactly free, of course. The price gets paid via inflation. ...
This is essentially a Ponzi scheme, [Gross] points out, but at
this point an unavoidable one. It's also an inherently unstable structure.[861]

As situations worsen economically, jobs disappear, incomes remain stat-
ic or decline, and credit burdens grow, there will be increased calls for overtly
socialist remedies. Socialism, as we can see with the ardent supporters of
Bernie Sanders during his presidential campaign in 2016, is an attractive
idea to many in today's younger generations. Unfortunately, as developed
throughout *Contagion*, unless jobs are preserved and created, and fundamental revenue and tax strategies adopted, such a "solution" would still face
the problem of systemic bankruptcy and the question of how to provide for
the needs of its citizens as the economic system declines—to say nothing of
the efforts of the powers that be to prevent it.

Is a Comprehensive UBI Financially Feasible?

Former Treasury Secretary, Vice President of Development Economics
and Chief Economist of the World Bank, and Harvard University president
Lawrence Summers argues that there is no way to make a Universal Basic
Income system work in the US. He concludes the numbers just don't add up
and uses a $25,000 per person federal government stipend to make his point.
Summers argues that a UBI would cost the US about $5 trillion annually, sev-
eral trillion dollars more than the country's annual income tax revenue. This
is based on the assumption that UBI would pay each American adult—perhaps 200 million people—as much as $25,000 per year. Summers concluded
it was impossible to create and sustain such a system, asking *where would the
money come from for a widespread UBI?*

Even if the UBI stipend is lowered to $15,000 or $20,000 thousand per
year per eligible recipient, there is no reason to believe American federal and
state governments could raise enough revenue to cover the obligations. This
massive expenditure also does not deal with issues such as what other types
of government expenditures would cost and how they would be funded, including military, security, infrastructure, education, government employment, aid to state and local governments, environmental protection, foreign
aid, etc.

To some extent, financial concerns about the viability of UBI are addressed by a recent decision by the government of Finland to deny future support for a UBI pilot program that nation has been running for two years. The decision was reportedly based in large part on the financial costs following an analysis that a society-wide UBI system would require a 30 percent tax increase in a country that is already "tax heavy." There was also concern about whether direct payments were the best approach or if some kind of credit account offered a better approach.

THE 2016 SWISS REFERENDUM ON UBI

An initial test of a UBI proposal came in mid-2016 when Swiss voters overwhelmingly rejected a proposal to introduce UBI in that nation. A report on the voting outcome does not bode well for UBI proponents, at least as of this time as the proposal was rejected by an 80 to 20 percent margin. The report on the highly negative vote explained

> Swiss voters rejected by a wide margin ... a proposal to introduce a guaranteed basic income for everyone living in the wealthy country after an uneasy debate about the future of work at a time of increasing automation. Supporters had said introducing a monthly income of 2,500 Swiss francs ($2,563) per adult and 625 francs per child under 18 no matter how much they work would promote human dignity and public service. Opponents, including the government, said it would cost too much and weaken the economy.[862]

It is extremely difficult, in Western nations at least, to win broad acceptance for the idea that those who are working should bear responsibility for supporting people who are not. Part of the opposition that exists to UBI may be the explicitness of the program. There are already numerous support programs in place in virtually every Western developed economy but they operate through a mixture of largely invisible subsidy programs.[863] In most Western nations there is a compassionate consensus that when people can't find work but are willing to seek it or when health or medical reasons force them out of the workforce, support for such individuals and their families is not only legitimate but morally important. There is compassion for those seen as being in real need, but limited tolerance or respect for those who are viewed as unwilling to contribute to the overall good of the society when they are capable of doing so.[864]

Even if it were financially possible to implement a comprehensive UBI program, as opposed to a limited and targeted version, there are significant political downsides. These include mistrust, cost, and the resentment many who work would feel if healthy individuals opted out of the workforce because they want to "find themselves," "have better things to do," "don't like getting up before 10AM," or " can't seem to find something that really excites" them.[865]

The growing epidemic of opioid, alcohol, and other addictions does not help support the debate about the viability of a UBI program. These behaviors suggest for many the likelihood of a growing fully dependent segment of national populations whose failure to work is due to questionable personal and addictive behavior. Irrespective of the extent to which any of this is true, these are the kinds of notions that have been fostered in the public, shifting resentment against the disproportionately wealthy elites, whose actual behavior is beyond their ken, either laterally onto peers or downward, onto the dispossessed.

The potential problems with UBI voiced by Swiss voters—are ones of ethics, human nature, negative assumptions about the prevalence of free riders and freeloaders, rampant abuses and dishonesty about how humans actually behave. The concerns of cost and anticipated freeloading—even though the great majority may have been rendered jobless without choice in the matter—are at the core of the UBI problem but they aren't the only problems.[866] Beyond cost and assumptions that free riders are living on the work of others, there is a serious issue of what kinds of values and behaviors will be the dominant characteristics of a population with too much time on their hands.

WHAT WOULD A UBI SYSTEM DO TO THE INTEGRITY OF WESTERN SOCIETIES?

Robert Skidelsky offers this insight into the two basic reasons some support the idea of a UBI.

> UBI is a somewhat uneasy mix of two objectives: poverty relief and the rejection of work as the defining purpose of life. The first is political and practical; the second is philosophical or ethical. The main argument for UBI as poverty relief is, as it has always been, the inability of available paid work to guarantee a secure and decent existence for all.[867]

In such a situation where we allow the present decision makers of our economic system to destroy the foundations of our workforce, even if a UBI system is initially implemented with the best and most compassionate of intentions, it will be co-opted by some into a "free lunch" or "get something for nothing" system. To some extent this is a price that will have to be paid because many require the UBI aid and support because there are not enough jobs to go around and no matter how hard they try there is no adequate system of employment. Funds must be obtained to provide the support for those forced out of the job market due to the destruction of work by AI/robotics.

But as Skidelsky observed, UBI only has a chance to work if those who are in control of capital are constrained by a radically new version of taxation. This requires that the benefits generated by the AI/robotics shift are reserved for the welfare of the community rather than primarily for the rich or exceptionally fortunate entrepreneurs such as Mark Zuckerberg, Steve Jobs or Jeff Bezos who played major roles in creating the crisis. Skidelsky explains.

> [A] UBI scheme can be designed to grow in line with the wealth of the economy. Automation is bound to increase profits, because machines that make human labour redundant require no wages and only minimal investment in maintenance. Unless we change our system of income generation, there will be no way to check the concentration of wealth in the hands of the rich and exceptionally entrepreneurial. A UBI that grows in line with capital productivity would ensure that the benefits of automation go to the many, not just to the few.[868]

Another serious problem with the Universal Basic Income approach is that we have lost the ability to "just say no." By this we do not mean that some form of UBI program—or multiple formats—aren't going to be needed, but that as with the continual increase in entitlement programs (including that of the US defense industry that represents one of the largest entitlement programs albeit never described as such) our political leaders can't be trusted to make responsible budget decisions because they use political expenditures and entitlements to buy votes and placate their constituencies. UBI is a strategy that will generate significant social conflict between the haves and those who have not, for whatever reason. This stress can be expected to grow as coalitions of identity groups seek to increasingly benefit their own members and deny benefits to others because power comes through control of the purse strings.

But even that fact pales in comparison to other consequences created by the idling of a very large segment of a nation's population as jobs are destroyed by AI/robotics. The reality is that either we devote the nations' resources to engaging and employing human labor, or in some manner or another we provide for people who are not working or we leave the non-workers to face their fate—one leading to growing homelessness, hunger, disease and earlier death. Such a heartless and draconian strategy would not only be immoral but impact society in destructive ways. Whatever we do, we cannot allow that to occur.

THE NEED TO BE HONEST ABOUT THE CURRENT SOCIAL AFFLICTIONS IN MUCH OF AMERICA

In today's America, far too many people are addicted to the pursuit of drugs, alcohol, opioids, sex, power and dominance, material goods, pornography, soap operas, IPhones, computer applications and games. More than one in eight people in the US are alcoholics. We are suffering from opioid and depression epidemics. More than half the US population is obese.[869] In our age of addiction, virtual reality, immersion in identity groups, anger, loneliness, deception, and incredible superficiality, it is lunacy to base support for a Universal Basic Income program on a flawed philosophical ideal about human behavior.

The loss of the focus and discipline of work that will occur without work activity and a strong community to provide structure and values is individually and socially destructive. But accountability and responsibility are increasingly seen as burdens to be avoided from the perspective of most of the human race. This means that a society with the majority of people out of work is not going to be a "happy" one. Most people want to be told what to do, be informed about what they should believe, guided about how they should behave, and have something to fill up their otherwise idle hours.

Even with the criticisms voiced above, UBI in some form and extent or by some name is necessary and inevitable if employment and economic conditions continue to disintegrate as we expect they will for millions in the US and Western Europe. The unavoidable outcome of a UBI system will lead to a significant number of people who are not working but are dependent on the efforts of those who still have jobs, as well as acquiescence to actors such as Gates, Bezos, Musk and Zuckerberg who are taking human civilization in directions that are primarily of benefit to themselves without consideration for the larger consequences to the societies they claim to be aiding.

Insofar as UBI will represent a bottom tier of income, how will many consigned to that fate react to the widespread removal of options that once were claimed to be the road to personal and material advancement? How will people cope with their blunted desire for advancement if the legitimate avenues are closing down? What happens in communities where opportunity, mobility and growth disappear or are denied? Better yet, how do we figure out ways to create opportunity and social mobility in a flattened society increasingly divided into widely separated levels of "haves" and "have-nots" with the much smaller upper levels possessing vastly superior wealth and privilege than the mass of citizenry making up the bulk of the population?

A society where AI/robotics has destroyed work opportunity for millions is one without the kinds of internal discipline and other lessons provided by work. A system without the social mobility and opportunity that a system of work produces for those who want such opportunities is a system without integrity or positive core principles. It lacks shared positive values and the ability to collaborate to achieve jointly held goals. Cut loose from directive structures, far too many of us flounder aimlessly, lost in a desert of purposelessness.

WHAT ABOUT UBI IN THE CONTEXT OF A SYSTEMIC SHIFT?

It is useful to look at a pragmatic UBI proposal by Charles Murray, who writes:

> I think that a UBI is our only hope to deal with a coming labor market unlike any in human history and that it represents our best hope to revitalize American civil society. First, my big caveat: A UBI will do the good things I claim only if it replaces all other transfer payments and the bureaucracies that oversee them. If the guaranteed income is an add-on to the existing system, it will be as destructive as its critics fear. [870]

A vital part of Murray's proposal is his warning that a UBI system would only work if it replaced all other government subsidies. If that were done, his projections are that annual savings to the nation's budget would be in the hundreds of billions and even as much as one trillion dollars annually. [871] Under Murray's plan UBI would fully supplant all other government programs rather than being added on or operated in conjunction with other subsidy, educational, health and pension allowances. One recent report indicated that 52 percent of Americans were already receiving some form of federal government support. This includes the basic Welfare program, and Food Stamps or SNAP with nearly 80 million people receiving benefits. It also includes Medicaid and Medicare, as well as Social Security retirement, Survivorship and Disability benefit programs that make up a high proportion of the federal budget.

There are also various state and local government subsidy and grant programs, some of which are at least partially paid for by the federal government through distributions to the state and local levels on which the programs are delivered. These include free lunch programs for lower income students even during the summer months, the Earned Income Tax Credit program, job training and retraining programs, child care assistance, Head Start, farm subsidy programs, subsidized use of hospital emergency services for medical needs and much more. As Murray indicates, hundreds upon hundreds of billions of dollars are already being channeled each year into programs aimed at supporting American citizens and residents with most of them connected to income levels and poverty mitigation. Such an integration would clearly reduce the scale of Lawrence Summers' projected costs as set out above.

ARE GOVERNMENT-GUARANTEED JOBS A VIABLE ALTERNATIVE?

Beyond UBI, or perhaps integrated into it as a key element of a UBI strategy, is the idea of government guaranteeing jobs for anyone able and willing to work. Those pushing such a plan suggest that it would allow people to obtain the various positive benefits of work, including the sense of purpose and maintenance of social relationships. Researchers argue as follows.

> We propose the enactment of legislation guaranteeing every American over the age of eighteen a job provided by the government through a National Investment Employment Corps. Establishing this corps on a *permanent* basis would eliminate involuntary unemployment and mitigate poverty by ensuring full employment in the United States. For all Americans who cannot otherwise find employment, the federal guarantee would provide jobs at non-poverty wages. Pay would start at $11.56 an hour, equivalent to $24,036 per year, which equals the current poverty line for a family of four. This wage level would be indexed to inflation in order to maintain workers' purchasing power over time. ... Across all employees, we estimate that the average yearly wage rate would be approximately $32,500.[872]

Along with being indexed to inflation, the analysis also indicates that another $10,000 per year would be needed to pay for health and other benefits. The idea of the proposed program is that the pay would be for a wide range of work opportunities and be set at variable levels consistent with the type of work involved and the worker's experience. At a minimum, wages would be set at levels that allowed people to cover their essential and reasonable costs of living. The argument that proponents of government-guaranteed jobs make is that such a jobs guarantee would have the effects of alleviating poverty, reducing income inequality, limiting discrimination, removing barriers between classes and identity groups, and constructively disposing of idle time.

While the aspirations are admirable, there are problems. One is that we don't know from where the funds would come to pay for the guaranteed wage, although this could be an area where strategic pump priming deficit spending would be valid to some extent while such a program was tested and adjusted. One analysis indicated that 15 million people could be entitled to the guaranteed jobs payments under such a program. Direct funding to as few as 15 million people at $42,000 per year per person would require $647 billion annually. This would just be the initial payment per individual recipient and there would almost inevitably be an expectation that payments would increase at least to keep pace with the cost of living.

An administrative bureaucracy would have to be created to handle the day-to-day operations and to monitor businesses and jobs, determine whether someone is actually working, and investigating the fraud and abuse that would be generated by such a program, to name just a few issues. That in itself would require large numbers of government workers and add another $100-200 billion in expenses, making the program cost around $800 billion to support 15 million people. On the other hand, the costs of addressing the needs of 15 million unemployed people and their families would be avoided.

CHAPTER TWENTY-EIGHT

Governmental Jobs and Private Sector Job Subsidies are More Vital than Generally Understood

SOME DATA ON FEDERAL, STATE AND LOCAL GOVERNMENTAL EMPLOYMENT

When we talk about government jobs we can quibble about whether they are the "right" jobs, or the most needed jobs, or if in the abstract there would have been better uses for the money. But what is seldom realized by its critics is that, without substantial government spending on employment—both public and private—our economy will "tank," our middle class would be far smaller, and opportunity and social mobility would be much more limited.

There are millions of jobs indirectly funded by government. For example, governmental investments stimulate, support and sustain a significant number of people engaged in STEM-related activities (science, technology, engineering and mathematics). That support provides important knowledge and innovation resources for the economic and social system. It offers decent jobs to university graduates trained in science and technical disciplines.

Direct and indirect government jobs and the ways governments use public funds to stimulate private sector employment is the focus of this chapter. That is, and will continue to be a major part of job preservation and creation under any positive scenario. The data indicate that extremely large investments are already being made by the federal government in direct and contracted employment. Several million employment positions are made possible and funded by the federal government not only in areas such as defense and high tech, but education, health, basic research, management and health, and technology.

To gain a fuller sense of the employment dynamics and contributions of the US government and the relationship between the public and private sectors as of 2015, some useful jobs "trivia" include the following data.

- Federal, state and local governments combined employed 21,995,000 workers as of August 2014. 12,329,000 workers were employed in the manufacturing sector.

- Between 1989 and 2014 government employment *increased* by 4,006,000 workers and manufacturing employment *declined* by 5,635,000 workers.

- In August 1941 the 12,532,000 people employed in manufacturing equaled 1 manufacturing worker for each 10.6 person in the overall population. The 12,329,000 employed in manufacturing in August 2015 equaled only 1 manufacturing worker for each 26.1 people in the overall population, a 250% difference.

- Of the 21,995,000 employed by government in August 2014, 2,738,000 worked for the federal government. This included 596,500 who worked for the Postal Service. Another 5,092,000 people worked for state governments and 14,165,000 worked for local governments.

- Governmental subsidization of education represents a critical "driver" of employment and social mobility. More than 50% of state and local government employees are employed in educational positions.[873]

- 15.7% of all American workers are employed by federal, state or local governments.

- Official government employment statistics can be misleading. Between 2000 and 2012 federal government spending on civilian contractors, who are not counted as government employees, grew rapidly. The fastest-growing category in dollars was contracts for professional, administrative and management services. The top-expanding category was medical services.[874]

- One reason government spending as a share of GDP remains high, even though the official federal government workforce has shrunk slightly, "is that government contractors—who may work primarily or entirely on projects for the federal government—are not counted as federal employees."[875] Ironically, they also cost significantly more to do the same work that often could have been done by "official" government employees.

The job support by public sector institutions goes far beyond the federal government. State and local governments provide an extensive number of jobs, not the least of which are the millions of elementary, high school and college teachers and other staff employed by public educational institutions. Whether government employment on the massive scale that now exists survives the AI/robotics shift is open to question—it is a matter of politics and money rather than risk, return, efficiency and productivity. Tyler Durden, a "composite" journalist representative of several writers on the website *Zero-Hedge*, concentrates on developments in the AI/robotics world, and recently wrote that:

> A study by a British think tank, Reform, says that *90% of British civil service workers have jobs so pointless, they could easily be replaced by robots,* saving the government around $8 billion per year. The study ... says that robots are "more efficient" at collecting data, processing paperwork, and doing the routine tasks that now fall to low-level government employees.[876]

But when we think through what cutting hundreds of thousands of governmental employees would mean, it's evident that the claim that it would save $8 billion per year is little more than "clever" math. Assume the Brits cut 250,000 people, and assume there are no jobs for them once they are lopped off because the private sector is experiencing job loss at least as severe as the public sphere. That is not the beginning of our financial impact analysis. They and their families would seek assistance from the very government that cut them loose. It is not unrealistic to think that the annual cost of assistance for food, housing and other essential needs is above $20,000 per person, and to remember that the 250,000 are only the employees themselves and the figure does not include their family members whose futures depend on family resources.

Another report done by Oxford University researchers and the financial services company, Deloitte, indicated the potential cuts in government jobs in the UK due to AI/robotics could put as many as 1 million governmental jobs at risk. Noting these separate studies by Oxford University and Deloitte, the Reform report remarks that:

> Oxford University and financial services provider Deloitte, both of whom commissioned their own studies concur with *Reform*'s conclusions. *The Oxford University study said that more than 850,000 public sector jobs could fall to robots over the course of the next decade.* Reform suggests that government employees should probably look into opportunities presented by the "sharing economy," like driving for Uber—at least until robots replace those, too.[877] [emphasis added]

The message is that cutting jobs in government, particularly at a time when other jobs are being eliminated, is not cost free. The effects are well beyond apparent financial savings or fiscal costs; simply, the actual costs are shifted from the budgets of government agencies to those of welfare support systems. At a minimum, this means that "saving" $8 billion would generate billions of dollars in other *costs* immediately, to say nothing of future losses and denied opportunities for the slashed employees and their families. In an economy where massive numbers of private sector jobs are disappearing, a very large percentage of the displaced workers and their family members would be "on the dole," or fall lower on the socio-economic scale if they did find work, and become resentful and hostile at their fate. We must also bear in mind that such former workers would not take kindly to being regarded as "freeloaders," knowing full well the reason for their job losses.

Once again, just as with private corporations, we see a big employer—this time a public institution—looking after its own bottom line while throwing employees under the bus, presumably salving their consciences with the assumption that other sectors of government will take care of them. How free would they feel to unleash such large-scale unemployment if other social welfare departments of government would not take on the responsibility of care? The poor are clearly not the only ones purportedly taking advantage of social safety nets.

GOVERNMENT INVESTMENT SUPPORTS SIGNIFICANT PRIVATE SECTOR EMPLOYMENT

The size and costs attributable to federally contracted civilian employees who are not even calculated when determining employment numbers for the federal government, and the large numbers of employees of companies such as Lockheed-Martin and others on the "100 Contractors" list, means that the scale of jobs directly and indirectly supported by the federal budget is significantly greater than the "official" federal employment calculation.

This is, of course, a form of corporate welfare because a great deal of the work could be done within government and at considerably lower rates. While there is no question that significant cuts could be made in the expenditures on the "Industrial" component of the Military Industrial Complex or that there are huge amounts of waste in that sector, the challenge is that in the US, UK, EU, China, Russia and Israel, the producers of weaponry are so powerful and "wired" into the governmental systems at all levels that our ability to reform them is limited to the point of near impotence. Nonetheless, those industries do generate and support very large, diverse and well-paying areas of work not only in the companies but the governmental bureaucracies overseeing and regulating them.

A SAMPLING OF TWELVE COMPANIES RECEIVING SIGNIFICANT FUNDS FROM THE FEDERAL GOVERNMENT
[IN BILLIONS OF DOLLARS]

Lockheed-Martin	112,000 employees	$44.0 billion
IBM	412,000	12.3
Boeing	165,000	21.2
General Dynamics	99,500	13.1
Raytheon	61,000	14.1
Northrop Grumman	64,300	10.0
United Technologies	196,000	5.7
Computer Sciences	79,000	3.5
Hewlett Packard	302,000	2.8
General Electric	305,000	2.5
Honeywell	127,000	1.9
Jacobs Engineering	59,900	1.5
Total	**1,883,975 employees**	**$129.5 billion**

THIRTEEN RESEARCH LABS AND UNIVERSITIES RECEIVING SIGNIFICANT PAYMENTS FROM THE FEDERAL GOVERNMENT
[IN BILLIONS OF DOLLARS]

Battelle	$2.10 billion
Los Alamos	2.00
Cal Tech	1.70
Lawrence Livermore	1.50
UT-Battelle	1.20
Savannah River Nuke	.92
Johns Hopkins	.95
MIT	.83
SpaceX	.77
UChicago Arbonne	.73
Wyle Labs	.72
Alion Scientific	.65
Brookhaven Science	.57
Total Expenditures	**$14.64 billion**

When we take into account the twelve private companies receiving the greatest amount of government funds for projects there are almost 2,000,000 people employed by that group with annual government payments of as much as $129 billion. That averages out to $129,000 per employee. To that we can add thirteen governmentally supported research labs that receive $14.6 billion.

This means the US government invests at least $144 billion in high-tech research and development just among the research labs and top twelve businesses. If we include the other seventy-five private companies on the "Top 100" list the annual total invested in the private sector by the federal government is more than $200 billion. Those employees are consumers whose needs and expenditures support large numbers of other jobs and generate significant amounts of federal, state and local tax revenues. We are not claiming this is necessarily the best way to invest that enormous sum but are asserting that it has a vital role in understanding and sustaining US employment and our middle class. Unless we develop a superior strategy it is wise to approach any dramatic cuts in the subsidies with great care.

The corporations listed above are primary actors in the military industrial complex. So while the proportion of the federal budget devoted to military spending is quite large, totaling almost $900 billion in FY 2018 and it is undeniable that a great deal of budgetary "fat" occurs in that sector, its large and well paid workforce in the defense industry, along with its economic clout and intimate relationships with the Departments of Defense and Energy and with the political leaders of states where the companies' production facilities are located indicate why it is so very hard to trim or redirect that part of federal spending.

GOVERNMENT JOBS AND AFFIRMATIVE ACTION

Whatever one thinks of the mix between private sector and government jobs, the defense industry's waste and the fact that outsourcing jobs that could be done by federal employees at less cost, government has been and continues to be a critical source of decent paying employment opportunities for women and racial minorities. This has played an important role in bringing numbers of those traditionally disadvantaged classes into the socio-economic levels of our middle class. From the perspective of preserving jobs and creating opportunity for people who would otherwise be denied them we would be abysmally stupid to abandon this vital instrumentality favoring social cohesion to the efficiencies of AI/robotics systems.

Reductions in government jobs will directly affect the economic condition of many, particularly African Americans and women. Those groups have been the most significant beneficiaries of diversity and affirmative action laws. For many of those affirmatively advantaged workers a loss of that employment would mean a rapid slide into the lower levels of the economy because there is no ready backup support system in place that offers equivalent economic returns and benefits. Kevin MacDonald writes:

> Though [representing only] 10 percent of the U.S. civilian labor force, African-Americans are 18 percent of U.S. government workers. They are 25 percent of the employees at Treasury and Veterans Affairs, 31 percent of the State Department, 37 percent of Department of Education employees and 38 percent of Housing and Urban Development. They are 42 percent of the Equal Employment Opportunity Commission and Pension Benefit Guaranty Corp., 55 percent of the employees at the Government Printing Office and 82 percent at the Court Services and Offender Supervision Agency.
>
> When the Obama administration suggested shutting down Fannie Mae and Freddie Mac, the mortgage giants whose losses of $150 billion have had to be made up by taxpayers, *The Washington Post* warned, in a story headlined, "Winding Down Fannie and Freddie Could Put Minority Careers at Risk," that 44 percent of Fannie employees and 50 percent of Freddie's were persons of color.[878]

In "The public-sector jobs crisis: Women and African Americans hit hardest by job losses in state and local governments," David Cooper, Mary Gable and Algernon Austin add that:

> The Great Recession [of 2008] created tremendous hardship for millions of Americans. One aspect of the recession and its aftermath has been particularly damaging for women and African Americans: the decision by many state and local governments to respond to diminished revenues and budget shortfalls by cutting public-sector jobs. Because women and African Americans have historically been overrepresented in public-sector employment, they have been disproportionately affected by state and local government budget cuts.[879]

Some of the key findings by Cooper et al. include:

- State and local public-sector workers of color face smaller wage disparities across racial lines, and at some levels of education actually enjoy a wage premium over similarly educated white workers.

- The disproportionate share of women and African Americans working in state and local government has translated into higher rates of job loss for both groups in these sectors. Between 2007 (before the recession) and 2011, state and local governments shed about 765,000 jobs. Women and African Americans comprised about 70 percent and 20 percent, respectively, of those losses.

- Job losses in the state and local public sectors stand in contrast to the jobs recovery in the private sector. From February 2010 (the month the labor market "bottomed out") to January 2012, the United States experienced a net increase in total nonfarm employment of more than 3.2 million jobs, while state and local government employment fell by 438,000.

CHAPTER TWENTY-NINE
The Writing is on the Wall

While we can hope that apocalyptic scenarios of the kind being described never happen, honest analysis indicates the basics of the crisis that is headed our way are visible and the "writing is already on the wall." There are reasons why some of the world's most incredibly wealthy people are building safe havens in New Zealand, acquiring vast tracts of land in remote areas of the US, or purchasing private islands. There are also reasons why fairly rational Americans are moving to remote areas such as Idaho or building defensible compounds in the face of a society they see as falling apart.[880] It is not surprising that in the US the number of gun and concealed carry permits has dramatically increased, growing by nearly two million concealed carry permits just in one year.[881]

My [David] grandfather, Thomas Seth Jones, spent 38 years of his life as a steelworker with the Youngstown Sheet & Tube. When I was young he explained to me that "common" people understand far better than their "leaders" when their world is changing for the worse and becoming an uncertain landscape in which opportunity, security and stability are disappearing. "Common" people are the "canaries in the mine" because they and their families and friends are the ones hurt the most by what is happening. For them, it is all taking place *now*, not at some hypothetical time in the distant future. That is what is occurring with the expanding effects of AI/robotics.

Our societies are experiencing a growing sense of fear, anger and uncertainty. Beneath the surface illusions and blindness created by self-interest and smug self-satisfaction still being clung to by the elites, academics, politicians and profiteers who got us into this mess, regular people see and experience the decline in opportunity, earnings and their economic wellbeing. "Ordinary" people now finding themselves accused of being "populists" and nationalists, and of being xenophobic and bigoted.

Tens of millions of frightened and angry "ordinary people" are struggling to hold things together in a political and economic system controlled by the same elites who played significant roles in reckless globalization. The failed strategies of outsourcing have led to large-scale job loss, near economic collapse of the global financial system only ten years ago and a Depression that was barely short-circuited. Those same elites are in the process of generating a new collapse for which it has been said we lack the tools sufficient to stop it. Part of that process, but not all, involves AI/robotics "workers" replacing millions upon millions of human jobs, with more losses to come.

THE MOST CRITICAL INTERACTING TRENDS

Stated with a quick and broad brush, some of the most important factors funneling into the convergence of the powerful forces we are experiencing due to the emergence of AI/robotics include those listed below.

Artificial or "Alternative" Intelligence (AI) has been declared by physicist Stephen Hawking to be a fundamental threat to human survival. We recommend a reading of Juval Noah Harari's *Homo Deus: A Brief History of Tomorrow*, Nick Bostrom's *Superintelligence*, Max Tegmark's *Life 3.0: Being Human in the Age of Artificial Intelligence*, and James Barrat's *Our Final Invention* to gain an understanding of why some very bright thinkers consider Artificial Intelligence to be a real and extremely serious threat to the human race.[882]

The Expanding "Surveillance State" and "Surveillance Culture" Means Governments, Corporations, Interest Groups, Trolls and Identity Cliques Are Watching Much of What We Do and Say. A concern related to the "de-democratization" of Western societies involves the dramatic increase in surveillance, "Big Data" gathering and evaluation, and the monitoring and prediction of what we are doing and even what we might be thinking about doing. The loss of personal privacy is sweeping and pervasive with serious implications for how individuals, interest groups and governments behave.[883]

There Are Hidden Biases In AI Applications. We have had warnings of hidden biases in software applications where the criteria used to predict criminal activity or detect an undesirable work trait during the initial screening of applicants for employment unfairly screens out people.[884] Bias is also said to have occurred in a "beauty contest" in which a computer application selecting the winners from photographs appeared to prefer white-skinned competitors rather than darker skinned options.[885]

The Middle Class Is Being "Hollowed Out" and the Heart of the Political Community Is Being Weakened. In our current situation the threat we face is from the convergence of a set of technological innovations that are and will increasingly have an enormous impact on the nature of work, economic and social inequality and the existence of the middle classes that are so vital to the durability and strength of Western democracy.

The **Labor Saving and Efficiency Multiplying** growth of information creation, storage, communication, sharing and application systems made possible through computers, the connective linkages of the Internet, and the rise of incredibly large and powerful companies such as Google, Apple and Amazon are replacing millions of labor intensive jobs due to the AI systems' superior capabilities and lower cost. Artificial Intelligence will inevitably replace numerous categories of employment even at the highest levels of our current activity.

Sophisticated Software Applications make it possible to reduce or eliminate many of the functions of millions of jobs in numerous industries. This includes not only lower level positions such as bank tellers, store check out workers, receptionists, gas station attendants, toll booth collectors, etc. but is increasingly eliminating higher level employment niches in accounting, medicine, law and law support, journalism, finance and securities, real estate and insurance. In areas such as tax, worker's compensation, real estate, estate planning, auditing, legal writing and more, consider the job loss effects AI software systems will have for lawyers, paralegals, CPAs, real estate and insurance agents, etc. as human workers will be able to do much more with less, meaning that fewer human employees will be needed.

Few Career Tracks are Immune from Displacement, Reduction or Elimination. Developments in robotics and AI applications and linked technology mean more and more white-collar jobs are being automated and performed by machines. According to experts: "Human beings are set to be "outnumbered" by robots as massive armies of intelligent machines move into the home and workplace."[886] As the AI/robotics "evolution" continues there will be a downward arcing curve in numerous sectors of employment from farming and service industries to manufacturing, machinery operation, vehicular use, finance, medicine, law, journalism and more.

The "Great Decoupling" of Productivity and Employment. The insights of MIT researchers Brynjolfsson and McAfee are central to understanding what is happening. *At the core of their analysis is what they call the "great decoupling" of productivity and employment and its highly negative impact on jobs, dating the phenomenon to 2000 where they indicate "productivity continues to rise robustly, but employment suddenly wilts."* Brynjolfsson states: "It's the great paradox of our era. Productivity is at record levels, innovation has never been faster, and yet at the same time, we have a falling median income and we have fewer jobs. *People are falling behind because technology is advancing so fast and our skills and organizations aren't keeping up."*[887]

The **"Age Curse" of Shrinking and Aging Workforces and Rising Dependency Needs**. The global elderly population is supposed to double by 2050, creating a significant need for support and care. This directly relates to the future of nations such as the US, UK and EU, Japan, China and Russia as their populations become heavily skewed toward the upper age ranges due to improved health care and a precipitous decline in fertility and birth rates. This will drive the development of diverse AI/robotics systems to higher levels both to care for the older and less healthy members of society and to substitute for shrinking populations of younger workers.

The **Software Evolution** is also making it possible for non-specialists seeking knowledge in an area of need to access the "mysteries" of formerly arcane disciplines so that they can have insight into matters about which they formerly sought assistance.

We May Have Reached "Peak Human"! We may have already reached what has been called "peak human" and can only go downhill from here. Suresh Kumar, chief information officer of Bank of New York Mellon, indicates corporations' intended direction: "Bots armed with AI and the ability to understand and respond in natural language can [already] be used to answer clients' queries and eventually execute transactions. *'You start with something simple, maybe just offering information, then you start doing transactions,' Kumar says. 'We obviously want to automate everything, but you have to prioritize.'*"[888]

Development of Autonomous Weapons Systems is a Major Area of Research and Development. There are increasing numbers of military AI/robotics applications in air, sea and ground weaponry. This includes fully autonomous naval vessels, tanks, machine gun carrying robot soldiers, autonomous fighter planes and drones, and ground-based attack systems. These are at the core of military research and development in Russia, China, the US and the UK.

The Widespread Availability and Use of Drones is an Increasing Danger. This does not even begin to consider the insane development of airborne drone systems available for general civilian use or the application of robotic and AI technology for surveillance and monitoring by governments, police forces, corporations and private individuals. The US Department of Homeland Security and the FBI are now warning of the significant danger from such systems. We have now allowed the creation of a largely invisible airborne delivery system that in the hands of terrorists and other "bad guys" presents a virtually unstoppable weapons system.

Our Fiscal Crisis Will Only Get Worse. The United States has a cumulative federal debt over $22 trillion by the most conservative estimates and that debt is growing at a rate of approximately $1 trillion each year. There is no way it can be paid off and it is extremely unlikely that it can even be reduced. The country is bankrupt and that does not even take into account consumer, mortgage and student loan debt or state and local debt. In July 2016 the Congressional Budget Office (CBO) issued a report on US Budget projections warning of a fiscal crisis due to rapidly increasing government debt that will hit within the next decade unless action is taken now to deal with the situation.

We Need to Radically Redesign Our System of Taxation. In one out of five families in the US no one works in compensated employment.[889] Almost fifty percent of Americans pay no federal income tax because their income is too low or they are not working and are supported by various governmental entitlement payments.[890] Amidst all the rhetoric about people paying "their fair share" in taxes, families earning $100,000 and above in the US already shoulder 80 percent of the federal tax burden. That obligation will become even heavier as federal, state and local governments' needs for revenue expand.[891]

FEMA and the Department of Defense Have Developed Scenarios of Social Unrest and Systemic Collapse. What massive job losses with many people permanently out of work and others competing for increasingly scarce public revenues might mean is suggested in a recent futurist study by FEMA, the Federal Emergency Management Agency. The study's results projected widespread social unrest in the US triggered by the collapse of the food production and distribution systems and a 395 percent increase in food prices. Another strategic analysis went further and projected the collapse of Western industrial societies. Department of Defense strategic analysis also includes the possibility of what might be involved in dealing with widespread unrest and violence in major US urban areas. This suggests that at least in several of its agencies the US government is aware of the fact of potential severe social disruption, including violence. While it makes sense for FEMA and the DOD to analyze responses from within the sphere of their respective capabilities and responsibilities we must hope that other relevant authorities strive to ensure such situations are avoided lest, as indicated earlier, "The Centre Cannot Hold" and the system falls apart.

Increasing Systemic Complexity and Dependency Increases Our Vulnerability. Given the complexity of the US system and the reliance by a very large part of the population on goods, including food, delivered from significant distances, a breakdown in the distribution of food, energy and heat would be catastrophic and could occur rapidly since we have limited backup resources and supplies available for our large population centers that would be exhausted in less than a week. [892]

Knowledgeable Leaders Are Warning of a Global Economic Collapse In the Near Future. The head of JP Morgan Chase, Jamie Dimon, has warned we face a global economic catastrophe within the next ten years.[893] Some analysts predict that fifty percent of jobs done by humans around the world will disappear in the next ten to fifteen years.

We Are Experiencing Increased Sectarianism and Rising Social Aggression. In the midst of all the calls for "unity" a more honest assessment in a world filled with identity groups demanding their own entitlement "fiefdoms" involves growing separation and sectarianism. This could even go as far as autonomous or self-governing "sub-nations" within the US territory or secession of states or territories

The "bottom line" is this. In the face of the nearly irresistible momentum of the AI/Robotics transformation and the related challenges our societies must deal with as described throughout *Contagion*, if we do not return to thinking and acting in ways that sustain the wellbeing of society as a whole, we are descending on a path that leads to a terrifying future for the majority of the human race. We are unquestionably at the tipping point where action must be taken that at a minimum deflects and mitigates the more dire consequences. The tragedy—for us, our children and grandchildren—is that we have shown no signs we possess the wisdom, intelligence, selflessness or political will to do what is needed to protect the future of positive human societies and in the worst case scenario—the human race itself. We must act. Now.

ENDNOTES

1 https://www.telegraph.co.uk/science/2018/09/05/artificial-intelligence-greater-concern-climate-change-terrorism/. Sarah Knapton, 9/6/18, "Artificial Intelligence is greater concern than climate change or terrorism, says new head of British Science Association." See also, https://www.ft.com/content/0b301152-b0f8-11e8-99ca-68cf89602132. Clive Cookson, "Artificial intelligence faces public backlash, warns scientist," *Financial Times*, 9/5/18.

2 https://www.thesun.co.uk/tech/3306890/humanity-is-already-losing-control-of-artificial-intelligence-and-it-could-spell-disaster-for-our-species/.

3 https://www.theguardian.com/technology/2018/nov/11/alarm-over-talks-to-implant-uk-employees-with-microchips. Julia Kollewe, 11/11/18, "Alarm over talks to implant UK employees with microchips: Trades Union Congress concerned over tech being used to control and micromanage."

4 *Id.*

5 https://phys.org/news/2018-11-google-accused-track-users.html. "Google accused of manipulation to track users," 11/27/18.

6 https://gizmodo.com/the-u-s-government-once-fracked-oil-wells-using-nuclea-1511758335. "The U.S. Government Once Fracked Oil Wells Using Nuclear Bombs," Andrew Tarantola, 1/29/14. Tarantola describes "Project Plowshare."
Filed to: ENERGY

> [I]n the mid 1950s, scientists from the Atomic Energy Commission and officials from the U.S. Bureau of Mines began experimenting with an alternative method of fracking, one that employed nuclear bombs more powerful than anything we dropped on the Japanese. Dubbed Project Plowshare, this insane undertaking explored two industrialized—or "peaceful"—applications for nuclear explosives:
>
> Conceptually, industrial applications resulting from the use of nuclear explosives could be divided into two broad categories: 1) large-scale excavation and quarrying, where the energy from the explosion was used to break up and/or move rock; and 2) underground engineering, where the energy released from deeply buried nuclear explosives increased the permeability and porosity of the rock by massive breaking and fracturing.

7 https://www.dailystar.co.uk/news/latest-news/743886/technology-death-civilisation-nick-bostrom-vulnerable-world-future-prediction. "Technology will be death of civilisation as humanity self-destructs—chilling warning: Technology could cause humanity to self-destruct unless we drastically change our ways, according to shocking claims by an Oxford professor," Matt Drake, 11/22/18.

8 https://www.wsj.com/articles/supersmart-robots-will-outnumber-humans-within-30-years-says-softbank-ceo-1488193423. "Supersmart Robots Will Outnumber Humans Within 30 Years, Says SoftBank CEO," Stu Woo, 2/27/17.

9 http://www.mirror.co.uk/news/uk-news/this-new-form-life-stephen-11453107. Rachael McMenemy and Stephen Jones, 11/2/17. Rachael McMenemy and Stephen Jones, 11/2/17. "'This will be a new form of life': Stephen Hawking says artificial intelligence robots will replace humans completely: The renowned physicist said that he believes AI will eventually reach a level where it will essentially be a 'new form of life that will outperform humans.'"

10   http://www.mirror.co.uk/news/uk-news/professor-stephen-hawking-says-humans-9271348. Anthony Bond, 11/16/16. "Professor Stephen Hawking says humans will be wiped out in 1,000 years unless we find new planet."

11   Kirstie McCrum, "Stephen Hawking issues robot warning saying, "rogue AI could be difficult to stop." 6/28/16, http://www.mirror.co.uk/news/world-news/stephen-hawking-issues-robot-warning-8300084.

12   Nick Bostrom, *Superintelligence: Paths, Dangers, Strategies* (2014).

13   http://www.dailymail.co.uk/sciencetech/article-5110787/Elon-Musk-says-10-chance-making-AI-safe.html. Shivali Best, 11/23/17, "Tesla's Elon Musk warns we only have 'a 5 to 10% chance' of preventing killers robots from destroying humanity."

14   http://www.dailymail.co.uk/sciencetech/article-5843979/Killer-robots-enslave-humanity-eventually-wiping-claims-MIT-Professor.html. Tim Collins, *Daily Mail*, 6/14/18. "Killer robots could make humans their slaves before eventually destroying everyone on the planet, claims scientist."

15   Isaac Asimov, *Foundation Trilogy (Foundation, Foundation and Empire, Second Foundation).* http://www.cnet.com/uk/news/google-goes-asimov-and-spells-out-concrete-ai-safety-concerns/. "Google has made no secret about its commitment to AI and machine learning, even having its own dedicated research branch, Google DeepMind. Earlier this year, DeepMind's learning algorithm AlphaGo challenged (and defeated) one of the world's premier (human) players at the ancient strategy game Go in what many considered one of the hardest tests for AI. Now it is advanced to engaging five Go players simultaneously." See, https://www.theguardian.com/technology/2017/apr/10/deepminds-alphago-to-take-on-five-human-players-at-once. "DeepMind's AlphaGo to take on five human players at once: After its game-playing AI beat the best human, Google subsidiary plans to test evolution of technology with GO festival."

16   http://www.dailymail.co.uk/sciencetech/article-5843979/Killer-robots-enslave-humanity-eventually-wiping-claims-MIT-Professor.html. "Killer robots could make humans their slaves before eventually destroying everyone on the planet, claims scientist," Tim Collins, *Daily Mail*, 6/14/18. This includes that: "[Max Tegmark] believes super intelligent robots could one day 'break out and takeover' [and] Shockingly, he says some [people] will welcome the extinction of their species by AI."

17   https://www.thesun.co.uk/tech/3306890/humanity-is-already-losing-control-of-artificial-intelligence-and-it-could-spell-disaster-for-our-species/. "Humanity is already losing control of artificial intelligence and it could spell disaster for our species," Margi Murphy, 4/11/17.

18   https://www.technologyreview.com/s/611574/the-us-may-have-just-pulled-even-with-china-in-the-race-to-build-supercomputings-next-big/, "The US may have just pulled even with China in the race to build supercomputing's next big thing: The two countries are vying to create an exascale computer that could lead to significant advances in many scientific fields," Martin Giles, 7/11/18.

> There was much celebrating in America last month when the US Department of Energy unveiled Summit, the world's fastest supercomputer. Now the race is on to achieve the next significant milestone in processing power: exascale computing. This involves building a machine within the next few years that's capable of a billion billion calculations per second, or one exaflop, which would make it five times faster than Summit. Every person on Earth would have to do a calculation every second of every day for just over four years to match what an exascale machine will be able to do in a flash.

19 https://www.washingtonpost.com/news/the-switch/wp/2015/01/28/bill-gates-on-dangers-of-artificial-intelligence-dont-understand-why-some-people-are-not-concerned/. "Bill Gates on dangers of artificial intelligence: 'I don't understand why some people are not concerned,'" Peter Holley, 1/29/15.

20 Richard Waters, "Investor rush to artificial intelligence is real deal," *Financial Times*, 1/4/15. http://www.ft.com/cms/s/2/019b3702-92a2-11e4-a1fd-00144feabdc0.html#ixzz3Nx-miiO3Q.

21 http://blogs.wsj.com/moneybeat/2016/05/04/bill-gross-what-to-do-after-the-robots-take-our-jobs/. "Bill Gross: What to Do After the Robots Take Our Jobs: Get ready for driverless trucks, universal basic income, and less independent central banks," Paul Vigna, 5/4/16.

22 *Id.*

23 https://www.bloomberg.com/news/articles/2017-03-02/ai-scientists-gather-to-plot-doomsday-scenarios-and-solutions. "AI Scientists Gather to Plot Doomsday Scenarios (and Solutions): Researchers, cyber-security experts and policy wonks ask themselves: What could possibly go wrong?," Dina Bass, 3/2/17.

24 https://www.theguardian.com/technology/2016/may/20/google-ai-machine-learning-skynet-technology. "Google's AI ambitions show promise—'if it doesn't kill us,'" Danny Yadron, 5/20/16. See also, https://www.theguardian.com/technology/2016/may/22/age-of-quantum-computing-d-wave. "Has the age of quantum computing arrived? It's a mind-bending concept with the potential to change the world, and Canadian tech company D-Wave claims to have cracked the code," Andrew Anthony, 5/22/16

25 https://www.theguardian.com/technology/2017/apr/24/alibaba-jack-ma-artificial-intelligence-more-pain-than-happiness. "Alibaba founder Jack Ma: AI will cause people 'more pain than happiness': The billionaire said key social conflict will be the rise of artificial intelligence and longer life expectancy, which will lead to aging workforce fighting for fewer jobs," Olivia Solon, 4/24/17.

26 https://www.thesun.co.uk/tech/4170364/former-facebook-executive-says-society-will-collapse-within-30-years-as-robots-put-half-of-humans-out-of-work/. "Former Facebook executive says society will collapse within 30 years as robots put half of humans out of work: Antonio Garcia Martinez fears revolution and armed conflict will erupt in America in the coming decades," Jasper Hamill, 8/4/17.

27 https://www.theguardian.com/commentisfree/2018/nov/11/the-networker-how-the-new-china-confounds-everything-western-liberals-thought-about-the-internet.   John Naughton, 11/11/18, "China confounds all that western liberals believed about the net: Technology is being used by China to exercise ever greater control over their workers' lives."

28 https://www.fastcompany.com/90261232/report-china-holds-seminars-to-teach-other-countries-how-to-restrict-online-speach. "Report: China holds seminars to teach other countries how to restrict online speech," Steven Melendez, 11/1/18.

29 https://www.theguardian.com/commentisfree/2018/mar/05/algorithms-rate-credit-scores-finances-data. "The tyranny of algorithms is part of our lives: soon they could rate everything we do," John Harris, 3/5/18.

30 https://privacyinternational.org/sites/default/files/2017-12/global_surveillance_0.pdf.

31 https://money.cnn.com/2014/09/16/technology/security/fbi-facial-recognition/index.html.

32 More tellingly, the ACLU report goes on to detail the susceptibility of CCTV to abuse, whether criminal, institutional, discriminatory, or for voyeuristic or personal reasons. https://www.aclu.org/other/whats-wrong-public-video-surveillance

33 http://www.breitbart.com/national-security/2018/04/02/china-applauds-world-following-lead-internet-censorship/. "China Applauds the World for Following Its Lead on

Internet Censorship," John Hayward, 4/2/18. "'Reining in social media appears to be the trend of governments,' China's state-run *Global Times* declared happily in a Sunday editorial. It is not wrong. The Chinese Communist Party is pleased to see their authoritarian restrictions on speech going viral and infesting Western societies."

34 https://www.bloomberg.com/news/articles/2018-11-22/scared-your-dna-is-exposed-then-share-it-scientists-suggest. "Scared Your DNA Is Exposed? Then Share It, Scientists Suggest," Kristen V. Brown 11/22/18.

35 http://www.reuters.com/article/us-usa-economy-employment-insight-idUSKB-N0GB0NF20140811. Howard Schneider, "For largest U.S. companies, jobs growth has lagged profits, revenues," *Business News*, 8/11/14.

36 https://www.businessinsider.com/amazon-employees-on-food-stamps-2018-8.

37 See, e.g., https://www.theguardian.com/technology/2017/sep/27/robots-destabilise-world-war-unemployment-un. "Robots could destabilise world through war and unemployment, says UN: United Nations opens new centre in Netherlands to monitor artificial intelligence and predict possible threats," Daniel Boffey, 9/27/17.

38 http://www.swissinfo.ch/eng/reuters/eu-trade-chief-populist-movements-can-bring-isolation-failure/42287230. "The European Union's trade chief described populist movements in the United States and elsewhere as 'a recipe for isolation and failure' on Monday even as she sought to allay fears about Britain's exit from the EU during a trip to China. Cecilia Malmström, the commissioner for trade, [said] *"The effects of the global crisis have hit many people really, really hard."* "Many populists around the world prey on these feelings, on these fears."

39 http://insct.syr.edu/wpcontent/uploads/2016/04/Kaune_AWC_Report_Combined-mwedit042016.pdf. *See*, US Army War College Institute for National Security and Counterterrorism, Civilian Research Project, "Analysis of US Army Preparation for Megacity Operations," Col. Patrick N. Kaune, United States Army.

40 https://www.theguardian.com/commentisfree/2017/nov/17/truck-drivers-automation-tesla-elon-musk. "Truck drivers like me will soon be replaced by automation. You're next," Finn Murphy, 11/17/17.

41 http://money.cnn.com/2016/05/16/news/economy/us-debt-dump-treasury/index.html.

42 http://techcrunch.com/2016/05/16/ai-learns-and-recreates-nobel-winning-physics-experiment/. "AI learns and recreates Nobel-winning physics experiment," Devin Coldewey, 5/16/16. https://www.theguardian.com/technology/2017/may/18/google-assistant-iphone-ai-future-things-we-learned-at-io. "Google's future is useful, creepy and everywhere: nine things learned at I/O: With Google Assistant coming to the iPhone, the company hopes to kill off Siri and wants to 'see' inside your home as it reiterates its AI-first approach," Olivia Solon, 5/18/17. http://www.dailymail.co.uk/news/article-4505544/Human-jobs-taken-robots-new-study-shows.html. *"From travel agents to translators, rising toll of jobs taken by robots ... although nail technicians, security officers and kitchen staff are in greater demand,"* James Salmon, 5/14/17. https://www.theguardian.com/commentisfree/2017/may/12/wake-up-amazon-google-apple-facebook-run-our-lives. "Wake up! Amazon, Google, Apple and Facebook are running our lives," Hannah Jane Parkinson, 5/12/17.

43 http://mashable.com/2016/04/09/google-bipedal-robot/#YS8SZHFySqqC. https://www.wsj.com/articles/a-hardware-update-for-the-human-brain-1496660400, "A Hardware Update for the Human Brain: From Silicon Valley startups to the U.S. Department of Defense, scientists and engineers are hard at work on a brain-computer interface that could turn us into programmable, debuggable machines," Christopher Mims, 6/5/17.

44 Bostrom, *Superintelligence, supra* note 12, at 16.

45 http://news.mit.edu/2018/mit-lincoln-laboratory-ai-system-solves-problems-through-human-reasoning-0911, "Artificial intelligence system uses transparent, hu-

man-like reasoning to solve problems," Kylie Foy, 9/11/18. https://www.technologyreview.com/s/611536/a-team-of-ai-algorithms-just-crushed-expert-humans-in-a-complex-computer-game/, "A team of AI algorithms just crushed humans in a complex computer game: Algorithms capable of collaboration and teamwork can outmaneuver human teams," Will Knight, 6/25/18.

46 http://www.independent.co.uk/life-style/gadgets-and-tech/news/google-deepmind-go-computer-beats-human-opponent-lee-sedol-in-landmark-game-a6920671.html. "Google Deepmind Artificial Intelligence Beats World's Best GO Player Lee Sedol in Landmark Game," Andrew Griffin, 3/9/16.

47 https://www.cnbc.com/2018/02/20/facebook-co-founder-chris-hughes-wants-500-per-month-in-basic-income.html. "Facebook co-founder Hughes: The digital economy is 'going to continue to destroy' jobs in America," Mathew J. Belvedere, 2/20/18.

48 http://www.wsj.com/articles/a-401-k-from-a-robot-digital-advice-pushes-further-into-consumer-finance-1470621844. Ari I. Weinberg, "A 401(k) From a Robot? Digital Advice Pushes Further Into Consumer Finance," 8/7/16. "*In effect, as traditional brokerage firms challenge the robos with their own lost-cost offerings, the robos are evolving. For individual investors, this evolution is blurring the line between the robo advisers and the more traditional investment firms such as Vanguard Group, Charles Schwab Corp. and Fidelity Investments, which are quickly adding automated services. Wells Fargo & Co., J.P. Morgan Chase & Co., Bank of America Corp. and Citigroup Inc. have all said they plan to offer low-cost, automated investment services either on their own or by joining with a private-label robo adviser.*"

49 http://www.telegraph.co.uk/science/2017/05/01/robotic-brain-surgeon-will-see-now-drill-can-perform-complex/. "The robotic brain surgeon will see you now: drill can perform complex procedures 50 times faster," Henry Bodkin, 5/1/17.

50 https://futurism.com/robot-assisted-eye-surgery-trial/. "A Robot Just Operated On A Human Eye for the First Time," Kristin Houser, 6/18/18. "From prostate surgery to gallbladder procedures, robots are already mainstays in the operating room. Now, they're coming for your eyes."

51 https://www.eurekalert.org/pub_releases/2016-07/ntu-rth071816.php. "Robot therapist hits the spot with athletes: Unveiled by a Singapore start-up, the prototype robot is being used in trials for sports rehabilitation," 7/18/16.

52 http://www.bloomberg.com/news/articles/2016-04-23/robots-replacing-japan-s-farmers-seen-preserving-food-security.

53 http://www.freshplaza.com/article/161843/Apple-picking-robots-headed-for-the-farm. *See,* "Apple picking robots headed for the farm." https://www.technologyreview.com/s/604303/apple-picking-robot-prepares-to-compete-for-farm-jobs/, "Apple-Picking Robot Prepares to Compete for Farm Jobs:Orchard owners say they need automation because seasonal farm labor is getting harder to come by," Tom Simonite, 5/3/17.

54 https://www.fastcompany.com/40473583/this-strawberry-picking-robot-gently-picks-the-ripest-berries-with-its-robo-hand, "This Strawberry-Picking Robot Gently Picks The Ripest Berries With Its Robo-Hand: "As labor shortages make human pickers scarce and expensive, farms are turning to bots like this one to be the new generation of farm workers."

55 Chris Pash, "A one-armed Australian robot can build a house four times quicker than a brickie," July 27, 2016. http://www.businessinsider.com.au/video-a-one-armed-australian-robot-can-build-a-house-four-times-quicker-than-a-brickie-2016-7. "Fastbrick Robotics, an ASX-listed company based in Perth, has created a robot bricklayer, a form of 3D printing which can create the shell of a house without being touched by human hands. *The Hadrian 105 robot, named after the Roman emperor who built a wall in ancient Britain, has hit a bricklaying speed of 225 standard brick equivalents per hour, or about half a day's work for a top human bricklayer.*"

56   http://www.latimes.com/science/sciencenow/la-sci-sn-3d-printing-robots-20170428-story.html. "Check out this building that was 3-D-printed by a robot," Amina Khan, 4/28/17. Khan writes: "Researchers at MIT have created a mobile robot that can 3-D-print an entire building in a matter of hours—a technology that could be used in disaster zones, on inhospitable planets or even in our proverbial backyards."

57   https://www.aol.com/article/finance/2017/03/06/this-tiny-home-was-printed-from-a-3d-printer-in-less-than-24-hou/21874721/?ncid=txtlnkusaolp00000058&. "This tiny home was printed from a 3D printer in less than 24 hours [400 sq. feet]," Emily Rella, AOL.com, 5/6/17.

58   www.webopedia.com. http://whatis.techtarget.com/definition/teraflop. https://www.scienceabc.com/humans/the-human-brain-vs-supercomputers-which-one-wins.html. John Staughton, "The Human Brain vs. Supercomputers: Which One Wins?"

59   https://www.technologyreview.com/s/611574/the-us-may-have-just-pulled-even-with-china-in-the-race-to-build-supercomputings-next-big/ [see endnote 18].

60   http://www.businessinsider.com/only-solution-to-chinas-labor-shortage-2016-2. "This is the only solution to China's labor shortage," John Mauldin, 2/11/16.

61   https://www.aol.com/article/news/2017/06/02/report-japan-dementia-crisis-expected-get-worse/22123376/, "Report: Japan's dementia crisis expected to get even worse by 2025," 6/2/17. https://www.bloomberg.com/news/articles/2017-06-02/births-in-japan-fall-to-record-low-as-challenges-mount, "Births in Japan Fall to Record Low in 2016," Keiko Ujikane and Lily Nonomiya, 6/2/17.

62   http://www.theguardian.com/lifeandstyle/2016/mar/14/robot-carers-for-elderly-people-are-another-way-of-dying-even-more-miserably. "Robot carers for elderly people are 'another way of dying even more miserably': Japan has a robot with 24 fingers that can wash hair, while in Europe we're getting a 'social companion robot.' I'm going off for a little cry," Michele Hanson, 3/14/16.

63   https://www.theguardian.com/technology/2017/may/05/human-robot-interactions-take-step-forward-with-emotional-chatting-machine-chatbot. "Human-robot interactions take step forward with 'emotional' chatbot: Researchers describe the 'emotional chatting machine' as a first attempt at the problem of creating machines that can fully understand user emotion," Hannah Devlin, 5/5/17."

64   http://www.techinsider.io/these-robots-may-be-better-than-humans-at-helping-the-autistic-or-the-elderly-2016-7. "When it comes to caregiving, these robots may be better companions than humans," Dan Bobkoff and Andrew Stern, 7/11/16.

65   http://www.dailymail.co.uk/sciencetech/article-3459854/The-end-courier-Self-driving-ground-drone-takes-streets-London-make-drops-capital.html.

66   http://www.bloomberg.com/news/articles/2016-08-01/delphi-to-begin-testing-on-demand-robot-taxis-in-singapore-irbidct9.

67   https://www.theguardian.com/us-news/2017/aug/08/humans-v-robots-defending-jobs. "Automation is a real threat. How can we slow down the march of the cyborgs?," The Guardian [online], Alissa Quart, 8/8/17.

68   Id.

69   Id.

70   http://www.mirror.co.uk/tech/self-driving-robots-deliver-food-7647386.

71   https://www.bloombergquint.com/onweb/2018/09/19/amazon-is-said-to-plan-up-to-3-000-cashierless-stores-by-2021. "Amazon Will Consider Opening Up to 3,000 Cashierless Stores by 2021," Spencer Soper. "Amazon.com Inc. is considering a plan to open as many as 3,000 new AmazonGo cashierless stores in the next few years, according to people familiar with matter…"

72   http://www.theguardian.com/commentisfree/2016/feb/29/the-guardian-view-on-the-automated-future-fewer-shops-and-fewer-people. http://www.washingtonexaminer.com/a-first-more-workers-at-online-sites-than-newspapers/article/2594197.

73 https://www.google.com/search?q=jobs+vulnerable+to+AI+and+robotics&bt-nG=Google+Search&gws_rd=ssl.   http://www.theguardian.com/culture/2016/mar/12/ro-bots-taking-jobs-future-technology-jerry-kaplan-sxsw.   http://www.mirror.co.uk/news/technology-science/robots-to-take-50-jobs-7363442. http://www.theguardian.com/commen-tisfree/2016/apr/07/robots-replacing-jobs-luddites-economics-labor.

74 http://www.nytimes.com/2016/02/28/magazine/the-robots-are-coming-for-wall-street.html?_r=0;   https://www.theguardian.com/business/us-money-blog/2016/may/15/hedge-fund-managers-algorithms-robots-investment-tips.   http://www.bloomberg.com/news/articles/2016-02-05/the-rich-are-already-using-robo-advisers-and-that-scares-banks.
http://www.bloomberg.com/news/articles/2016-06-02/dimon-says-jpmorgan-can-offer-free-robo-advice-to-best-clients.

75 http://www.bloomberg.com/news/articles/2016-03-29/newspapers-gob-ble-each-other-up-to-survive-digital-apocalypse.   https://www.yahoo.com/news/us-news-paper-industry-hollowed-job-losses-190549050.html. "Official US Labor Department data showed the newspaper sector lost 271,800 jobs in the period from January 1990 to March 2016, or 59.7 percent of the total over the past 26 years. ... Magazines fared only slightly better, losing 36 percent of their jobs in the same period."

76 http://www.futuristspeaker.com/business-trends/2-billion-jobs-to-disappear-by-2030/. Thomas Frey, "2 Billion Jobs to Disappear by 2030," 2/3/12: Date Modified: 9/4/16.

77 https://www.finextra.com/newsarticle/32240/10000-jobs-could-be-lost-to-robots-says-citi. "10,000 jobs could be lost to robots says Citi," 6/12/18. "[A] 2016 report from the World Economic Forum predicted that advances in automation will lead to the loss of over 5 million jobs in 15 major developed and emerging economies by 2020."

78 https://qz.com/923442/wendys-is-responding-to-the-rising-minimum-wage-by-re-placing-humans-with-robots/. "Wendy's is responding to the rising minimum wage by re-placing humans with robots," Sarah Kessler, 3/3/17.

79 http://money.cnn.com/2016/07/11/technology/mit-robot-labor/index.html.   "MIT robot helps deliver babies," Sara Ashley O'Brien, 7/11/16.

80 http://www.dailymail.co.uk/sciencetech/article-3680874/Scientists-verge-creating-EMOTIONAL-computer-AI-think-like-person-bond-humans-2-years.html. "Scientists are on the verge of creating an emotional computer: AI could think like a person and bond with humans within years Abigail Beall," 7/8/16.

81 http://www.oxfordmartin.ox.ac.uk/downloads/academic/The_Future_of_Employ-ment.pdf. Carl Benedikt Frey and Michael A. Osborne, "The Future of Employment: How Susceptible Are Jobs to Computerisation?," 9/13/13.

82 https://www.theguardian.com/careers/2016/may/11/robot-jobs-automated-work. Charlotte Seager, "After the robot revolution, what will be left for our children to do?," 5/11/16.

83 http://www.oxfordmartin.ox.ac.uk/downloads/academic/The_Future_of_Employ-ment.pdf. Carl Benedikt Frey and Michael A. Osborne, "The Future of Employment: How Susceptible Are Jobs to Computerisation?," 9/13/13.

84 Seager, "After the robot revolution," *supra* note 82.

85 *Id.*

86 http://www.thedailybeast.com/articles/2016/08/11/today-s-tech-oligarchs-are-worse-than-the-robber-barons.html. Joel Kotkin, "Today's Tech Oligarchs Are Worse Than the Robber Barons," 8/11/16.

87 http://www.reuters.com/article/us-usa-economy-employment-insight-idUSKB-N0GB0NF20140811. Howard Schneider, "For largest U.S. companies, jobs growth has lagged profits, revenues," *Business News*, 8/11/14.

88 *Id.*

89 https://www.theguardian.com/business/2018/jun/10/rolls-royce-set-to-announce-more-than-4000-job-cuts. "Rolls-Royce set to announce more than 4,000 job cuts: Aero-en-

gine maker attempts to increase profits by losing middle-management posts," Simon Goodley, 6/10/18.

90 https://www.finextra.com/newsarticle/32240/10000-jobs-could-be-lost-to-robots-says-citi. "10,000 jobs could be lost to robots says Citi," 6/12/18. "US bank Citi has warned that it could shed half of its 20,000 tech and ops staff in the next five years due to the rise of robotics and automation."

91 "Fast Forward 2030: The Future of Work and the Workplace," CBRE. http://www.cbre.com/o/international/AssetLibrary/Genesis%20Report_Exec%20Summary_1029.pdf. *See also*, http://www.business-standard.com/article/pti-stories/50-of-occupations-today-will-no-longer-exist-in-2025-report-114110701279_1.html.

92 "Fast Forward 2030: The Future of Work and the Workplace," CBRE. http://www.cbre.com/o/international/AssetLibrary/Genesis%20Report_Exec%20Summary_1029.pdf. http://www.business-standard.com/article/pti-stories/50-of-occupations-today-will-no-longer-exist-in-2025-report-114110701279_1.html.

93 https://www.theguardian.com/technology/2016/sep/13/artificial-intelligence-robots-threat-jobs-forrester-report. "Robots will eliminate 6% of all US jobs by 2021, report says: Employees in fields such as customer service and transportation face a 'disruptive tidal wave' of automation in the not-too-distant future," Olivia Solon, 9/13/16.

94 *Id.*

95 https://www.theguardian.com/careers/2016/may/11/robot-jobs-automated-work. Charlotte Seager, "After the robot revolution, what will be left for our children to do?," 5/11/16.

96 https://www.wsj.com/articles/now-cropping-up-robo-farming-1527854402. "Now Cropping Up: Robo-Farming: Agricultural-equipment makers gear up driverless tractors, combines in quest to produce more food, more sustainably." Vibhuti Agarwal, 6/1/18.

97 For one example, see, https://www.wsj.com/articles/next-leap-for-robots-picking-out-and-boxing-your-online-order-1500807601. "Next Leap for Robots: Picking Out and Boxing Your Online Order: Developers close in on systems to move products off shelves and into boxes, as retailers aim to automate labor-intensive process," Brian Baskin, 7/23/17.

98 https://www.msn.com/en-us/news/other/robot-janitors-are-coming-to-mop-floors-at-a-walmart-near-you/ar-BBQpDb3. "Robot Janitors Are Coming to Mop Floors at a Walmart Near You," Pavel Alpeyev, 12/3/18.

99 https://www.theatlantic.com/ideas/archive/2018/06/taxi-driver-suicides-are-a-warning/561926/. Reihan Salam, 6/5/18. "Taxi-Driver Suicides Are a Warning: Technology has pushed a vulnerable, largely immigrant, population into an economically precarious situation—even as its prospects of upward mobility dwindle."

100 "Robots will take over most jobs in the world by 2045," Techradar, 6/6/16, http://economictimes.indiatimes.com/articleshow/52617490.cms?utm_source=contentofinterest&utm_medium=text&utm_campaign=cppst.

101 *Id.*

102 *Id.*

103 https://www.weforum.org/agenda/2016/08/technology-could-be-the-best-or-worst-thing-that-happened-to-inequality. Vivek Wadhwa, "Technology could be the best or worst thing that happened to inequality," 8/11/16.

104 https://www.nytimes.com/2018/01/12/business/ai-investing-humans-dominating.html. Conrad De Aenlle, *New York Times*, 1/12/2018.

105 http://www.bloomberg.com/news/articles/2016-06-08/wall-street-has-hit-peak-human-and-an-algorithm-wants-your-job. "We've Hit Peak Human and an Algorithm Wants Your Job. Now What? On Wall Street, the still-essential business of banking will go on—but maybe without as many suits." Hugh Son, 6/8/16.

106 https://www.yahoo.com/news/rich-powerful-warn-robots-coming-jobs-011130242—sector.html. "Rich and powerful warn robots are coming for your jobs," Olivia Oran, Reuters, 5/3/16.

107   Brian Fung, "Elon Musk: Tesla's Model 3 factory could look like an alien warship" 8/4/16.   https://www.washingtonpost.com/news/the-switch/wp/2016/08/04/the-future-of-car-production-will-be-devoid-of-people-according-to-tesla/.

108   Brian Fung, "Elon Musk: Tesla's Model 3 factory could look like an alien warship" 8/4/16, id

109   http://www.marketwatch.com/story/elon-musk-robot-software-will-make-tesla-worth-as-much-as-apple-2017-05-04. "Elon Musk: Robot software will make Tesla worth as much as Apple," Jeremy C. Owens, 5/4/17.

110   http://www.aei.org/publication/reality-check-us-factory-jobs-lost-are-due-over-whelmingly-to-increases-in-productivity-and-theyre-not-coming-back/.   Mark J.   Perry, 1/20/17, "Reality check: US factory jobs lost are due overwhelmingly to increases in productivity and they're not coming back."

111   See Robertt Kuttner, Can Democracy Survive Global Capitalism? (2018).

112   https://www.recode.net/2017/3/28/15094424/jobs-eliminated-new-robots-work-force-industrial. "Six jobs are eliminated for every robot introduced into the workforce, a new study says: The threat of robots taking our jobs is very real." April Glaser, 3/28/17.

113   Quart cite

114   http://www.fastcodesign.com/3058708/a-computer-paints-a-rembrandt-and-it-looks-just-like-the-real-thing. "A Computer Paints A Rembrandt, And It Looks Just Like The Real Thing: Algorithms are the ghosts of artists' past," Mark Wilson, 4/7/16.

https://www.theguardian.com/technology/2016/apr/14/cutting-edge-the-atre-worlds-first-virtual-reality-operation-goes-live. "Cutting-edge theatre: world's first virtual reality operation goes live: Medical experts hope surgery live-streamed in VR will make healthcare fairer and boost training," Nicola Davis, 4/14/16. http://s.telegraph.co.uk/graphics/projects/go-google-computer-game/index.html. "Humans versus robots: How a Google computer beat a world champion at this board game—and what it means for the future," Madhumita Murgia, 5/14/16.

115   https://www.christies.com/features/A-collaboration-between-two-artists-one-hu-man-one-a-machine-9332-1.aspx, "Is artificial intelligence set to become art's next medium?: AI artwork sells for $432,500—nearly 45 times its high estimate—as Christie's becomes the first auction house to offer a work of art created by an algorithm."

116   http://www.scmp.com/tech/enterprises/article/1934516/aliba-bas-ai-out-prove-it-can-recognise-aesthetic-beauty-predicting. Alibaba's 'Ai' out to prove it can recognise aesthetic beauty by predicting winner of reality TV singing contest" 4/7/16 and 8/5/16. "Artificial intelligence software from China's e-commerce king will be put to the test this Friday on Hunan TV's 'I'm a singer.'" http://www.telegraph.co.uk/technology/2016/05/06/google-ai-to-learn-art-of-conversation-through-erotic-novels/. "Google AI to learn art of conversation through erotic romance novels," Cara McGoogan, 5/6/16.

117   "Jack Ma Sees Decades of Pain as Internet Upends Old Economy," Sherisse Pham, April 23, 2017.   https://www.bloomberg.com/news/articles/2017-04-23/jack-ma-sees-de-cades-of-pain-as-internet-upends-older-economy. [Also] http://money.cnn.com/2017/04/24/technology/alibaba-jack-ma-30-years-pain-robot-ceo/. "Jack Ma: In 30 years, the best CEO could be a robot," Sherisse Pham, April 24, 2017.

118   https://www.nytimes.com/2017/11/05/technology/machine-learn-ing-artificial-intelligence-ai.html.   "Building   A.I.   That   Can   Build   A.I.:   Google and others, fighting for a small pool of researchers, are looking for automated ways to deal with a shortage of artificial intelligence experts," Cade Metznov, 11/5/17.

119   https://www.techworld.com/apps-wearables/google-deepmind-what-is-it-how-it-works-should-you-be-scared-3615354/. "Google DeepMind: What is it, how does it work and should you be scared?," Sam Shead, 3/15/16.

120 https://www.yahoo.com/news/microchips-under-skin-techno-phile-swedes-033147071.html. "Microchips get under the skin of technophile Swedes." *Yahoo News*, Camille Bas-Wohlert, *5/12/18*.

121 *Fox News*, Chris Ciaccia, 5/11/18.

122 *The Sun UK*, Sean Keach, 4/23/18.

123 *Daily Mail*, Tim Collins, 4/18/18.

124 *Associated Press*, Kelvin Chan, 1/16/18.

125 *Daily Mail*, Phoebe Weston, 9/20/17.

126 Joseph Schumpeter, *Capitalism, Socialism and Democracy* (1942).

127 Real wages for top quintile earners grew significantly from the mid-1970s through the present. *See* Drew Desilver, "For most workers, real wages have barely budged for decades," Pew Research Center (10/9/14), http://www.pewresearch.org/fact-tank/2014/10/09/for-most-workers-real-wages-have-barely-budged-for-decades/.

128 Lawrence Mishel, "The wedges between productivity and median compensa-tion growth," Economic Policy Institute (April 26, 2012), http://www.epi.org/publication/ib330-productivity-vs-compensation/. *See also* Gillian B. White, "Why the Gap Between Worker Pay and Productivity Is So Problematic," *The Atlantic* (Feb. 25, 2015), https://www.theatlantic.com/business/archive/2015/02/why-the-gap-between-worker-pay-and-produc-tivity-is-so-problematic/385931/.

129 Paul Davidson, "Provo's economy ranks best in U.S.," *USA Today*, 1/10/18.

130 Ellen Ruppel Shell, *The Job: Work and Its Future in a Time of Radical Change* (Cur-rency, 2018).

131 *Id.* at 16–17.

132 *Id.* at 18.

133 *Id.*

134 https://www.bradhuddleston.com. Brad Huddleston, *Digital Cocaine: A Journey To-ward iBalance*, 2016. See also, https://nypost.com/2016/08/27/its-digital-heroin-how-screens-turn-kids-into-psychotic-junkies/. "It's 'digital heroin': How screens turn kids into psychotic junkies," Dr. Nicholas Kardaras, 8/27/16.

135 See, http://www.businessinsider.com/charts-decline-of-the-middle-class-2016-1. Kathleen Elkins, 1/6/16, "9 charts that reveal how the American middle class has declined since 1970."

136 https://next.ft.com/content/695bfa18-1797-11e6-b197-a4af20d5575e.

137 https://www.theguardian.com/us-news/2016/jun/16/manufacturing-indus-try-america-jobs-election-politics. "Manufacturing jobs return to US during election year—but not quite like before," Benjamin Parkin, 6/16/16.

138 http://money.cnn.com/2016/05/15/news/economy/america-job-killing-compa-nies/index.html. "America's top 10 job-killing companies," Heather Long, 5/17/16.

139 http://www.reuters.com/article/us-usa-economy-employment-insight-idUSKB-N0GB0NF20140811. Howard Schneider, "For largest U.S. companies, jobs growth has lagged profits, revenues," *Business News*, 8/11/14.

140 Ramin Rahimian, "Armies of Expensive Lawyers, Replaced by Cheaper Software," *New York Times*, 3/4/11.

141 http://www.reuters.com/article/us-usa-economy-employment-insight-idUSKB-N0GB0NF20140811. Howard Schneider, "For largest U.S. companies, jobs growth has lagged profits, revenues," *Business News*, 8/11/14.

142 David Rotman, "How Technology Is Destroying Jobs," 6/12/13, *MIT Technology Review* magazine July/August 2013, https://www.technologyreview.com/magazine/2013/07/.

143 *Id.*

144 *Id.*

145 *Id.*

146 https://www.theguardian.com/business/2016/may/05/job-market-cuts-highest-level-since-2009-layoffs-unemployment.   http://www.theguardian.com/business/2016/may/02/the-global-economic-growth-funk.

147 Rotman, *supra* note 142.

148 https://www.theguardian.com/business/2016/nov/15/trumpism-solution-crisis-neoliberalism-robert-skidelsky. Robert Skidelsky, "Trumpism could be a solution to the crisis of neoliberalism," 11/15/2016. Skidelsky begins with the message that: *"There is Keynesian merit in Trump's policies which challenge the neoliberal obsession with deficits and debt reduction. Liberals need to question and refine the plan, not dismiss it as ignorant ravings."*

149 http://thehill.com/homenews/state-watch/351432-fury-fuels-the-modern-political-climate-in-us. "Fury fuels the modern political climate in US," Reid Wilson, 9/20/17.

150 http://www.breitbart.com/video/2017/11/01/justice-thomas-i-dont-know-what-we-have-as-a-country-in-common/. "Justice Thomas: 'I Don't Know' What 'We Have as a Country In Common,'" Ian Hanchett, 11/1/17.

151 *Fast Company.com.*, 12/11/17.

152 https://www.theguardian.com/business/2017/may/13/mohamed-el-erian-signals-system-enormous-stress-global-capitalism. "Mohamed El-Erian: 'we get signals that the system is under enormous stress' The leading economist and investor believes world leaders, and global capitalism, have reached a decisive fork in the road between equality and chaos," Nils Pratley and Jill Treanor, 5/13/17.

153 http://homedesign7.com/18-billionaire-homes-a-look-inside-the-houses-of-the-richest-people-in-the-world/. "Lavish Lifestyles of the Rich and Famous: A Look Inside 18 Billionaire Homes."

154 *The Guardian,* Michael Savage, 4/7/18.

155 *The Guardian,* Rupert Neate, 11/14/17.

156 *The Guardian,* Rupert Neate, 10/26/17.

157 *The Guardian,* Gabriel Zucman, 11/8/17.

158 *Forbes,* Kenneth Rapoza, 9/15/17 .

159 *The Guardian,* Peter Fleming, 2/14/17.

160 https://www.theguardian.com/environment/earth-insight/2014/jun/12/pentagon-mass-civil-breakdown. "Pentagon preparing for mass civil breakdown: Social science is being militarised to develop 'operational tools' to target peaceful activists and protest movements," Nafeez Ahmad, 6/12/14. http://americanfreepress.net/u-s-army-preparing-for-civil-unrest/. "U.S. Army Preparing for Civil Unrest," Keith Johnson, 9/5/14.

161 *The Guardian,* Shadi Hamid, 4/23/18.

162 *Wall Street Journal,* Janet Hook, 9/6/17.

163 *Yahoo News,* Rachel La Corte and David A. Lieb, 11/12/17.

164 *The Hill,* Reid Wilson, 9/20/17.

165 *Fox News,* Tucker Carlson, 6/2/18.

166 *Indy100.com,* Greg Evans.

167 *The Guardian,* Umair Haque, 1/20/18.

168 *Axios,* Steve LeVine.

169 *Independent UK,* Clark Mindock, 4/5/18.

170 *The Guardian,* Sam Levin, 10/7/17.

171 *Fox News,* Rick Leventhal, 5/18/18.

172 *The Guardian,* Michael Savage, 4/7/18.

173 Daniel Steingold, 10/18/17.

174 *Fast Company.com,* 10/29/17

175 *The Guardian,* Sam Woolley and Marina Gorbis, 10/16/17.

176 *The Guardian,* Paul Lewis, 2/2/18.

177 *Reuters,* David Ingram, 1/22/18.

178 *Daily Mail,* Max Hastings, 1/1/18.

179 *The Guardian,* Alex Hern, 11/14/17.

180 *The Guardian,* Alex Hern, 6/19/17.

181 Andre Damon, "Pages purged by Facebook were on blacklist promoted by Washington Post,"13 October 2018 ," http://www.wsws.org/en/articles/2018/10/13/cens-o13.html

182 http://www.foxnews.com/us/2017/03/09/chicagos-violent-gangs-looting-freight-cars-filled-with-guns.html. "Chicago's violent gangs looting freight cars filled with guns," Matt Finn, 3/9/17.

183 http://baltimore.cbslocal.com/2017/11/15/baltimore-mayor-crime-violence/. "Baltimore Mayor Reacts To Crime: 'I Am Deeply Disturbed,'" Rick Ritter, 11/15/17. Ritter writes: ""This is worse than Detroit in one way and even worse than Chicago," a South Baltimore resident said. "I know a lot of people that are scared to even come out the building. I would hate to see this city go down.""

184 https://www.theguardian.com/business/2017/jan/12/ilo-warns-of-rise-in-social-unrest-and-migration-as-inequality-widens. "ILO warns of rise in social unrest and migration as inequality widens: UN agency records rising discontent in all regions and calls on policymakers to tackle unemployment and inequality urgently," Katie Allen, 1/12/17. Allen reports:

> Its social unrest index increased between 2015 and 2016 to above the long-term average for the last four decades, the agency said in its latest World Employment and Social Outlook. *"There is growing uncertainty everywhere it would seem, whether it's economic or political. This is something we seem to be living with and they are reinforcing one another,"* said Steven Tobin, ILO senior economist and lead author of the report. ... *"It speaks to discontent with the socioeconomic situation, with finding a quality job and being able to share in the gains of whatever limited prosperity there is."*

185 Robert Kuttner, *Can Democracy Survive Global Capitalism?* (2018).

186 Skidelsky, *supra* note 141.

187 *Id.*

188 http://www.latimes.com/local/lanow/la-me-riot-poll-20170426-story.html. "For first time, more L.A. residents believe new riots likely, new poll finds," Victoria Kim and Melissa Etehad, 4/26/17.

189 Skidelsky, *supra* note 141.

190 *Id.*

191 http://theweek.com/articles/694923/global-elite-are-headed-fall-dont-even-know. "The global elite are headed for a fall. And they don't even know it," Damon Linker, 4/27/17.

192 *Id.*

193 *The Guardian,* Nikil Saval, 7/14/17.

194 *The Guardian,* 10/8/17.

195 *Bloomberg News,* Nikos Chrysoloras and Helene Fouquet, 5/29/18.

196 *Market Watch,* Mark Decambre, 4/29/18.

197 *The Guardian,* Richard Partington, 5/22/18.

198 *Daily Star UK,* Anders Anglesey, 5/2/18.

199 *The Guardian,* Larry Elliott, 1/9/18.

200 http://www.businessinsider.com/wal-mart-warehouse-robots-2013-12. Hayley Peterson, "Wal-Mart Has An Army Of Robots That Pick And Pack Your Holiday Gifts," 12/13/13. http://gizmodo.com/walmart-jumps-on-drone-bandwagon-but-wont-use-them-for-1780287586. Angela Chen, "Walmart's Warehouses Are About to Be Swarming With Drones," 6/3/16.

201 http://fortune.com/2016/04/28/target-testing-robot-inventory-simbe/. http://www.seattletimes.com/business/amazon/at-amazon-warehouses-humans-and-robots-are-in-

sync/.

202 https://www.theguardian.com/money/2017/nov/25/cobot-machine-coming-job-robots-amazon-ocado. "Meet your new cobot: is a machine coming for your job? As robots slash the time it takes to complete an order at companies like Amazon and Ocado, what does that mean for their human colleagues?," John Harris, 11/25/17. Harris reports:

> Next to the M56, on the outskirts of Manchester, the future has landed. A cluster of huge distribution centres sits at the heart of Airport City, a new development part-funded by the Beijing Construction Engineering Group ... *Among the biggest buildings is one of Amazon's self-styled "fulfilment centres."* ... *Deeper within the centre, beyond the reception area and meeting rooms, there is something else just as new: a great expanse of space behind a metal cage, where dozens of robots, finished in Amazon orange and each emblazoned with its own number, glide across the floor, gracefully avoiding collisions and sprinting to their next task.*

203 https://www.bloomberg.com/news/articles/2017-11-20/robot-makers-fill-their-war-chests-in-fight-against-amazon. "Robot Makers Fill Their War Chests in Fight Against Amazon: Locus Robotics is raising serious cash in its bid to automate warehouses, bringing industry investment to $70 million this year," Kim Bhasin and Patrick Clark, 11/20/17.

204 http://nypost.com/2017/04/25/amazon-is-great-for-consumers-but-is-it-great-for-america/. "Amazon is great for consumers—but is it great for America?," Lisa Fickensher, 4/25/17.

205 Suzanne O'Halloran, "The U.S. Economy is Sucking Wind," 7/29/16. http://www.foxbusiness.com/politics/2016/07/29/u-s-economy-is-sucking-wind.html.

206 https://www.csmonitor.com/Business/2017/0804/America-s-stores-are-closing.-Why-isn-t-that-raising-a-jobs-alarm. "America's stores are closing. Why isn't that raising a jobs alarm?," Schuyler Velasco, 8/4/17.

207 http://www.businessinsider.com/dying-shopping-malls-are-wreaking-havoc-on-suburban-america-2017-2. "Dying shopping malls are wreaking havoc on suburban America," Hayley Peterson, 3/5/17. "The commercial-real-estate firm CoStar estimates that nearly a quarter of malls in the US, or roughly 310 of the nation's 1,300 shopping malls, are at high risk of losing an anchor store. ... *"Malls are big, big contributors to city and state taxes, jobs, and everything,"* Davidowitz said. *"Once they close, they are a blight on the community for a very long time."*

208 https://www.theguardian.com/commentisfree/2018/jun/13/trump-nafta-g7-sunset-clause-trade-agreement. "Donald Trump was right. The rest of the G7 were wrong," *The Guardian*, George Monbiot, 6/13/18.

209 https://www.theguardian.com/business/2016/jul/14/up-to-70-per-cent-people-developed-countries-seen-income-stagnate. "Up to 70% of people in developed countries 'have seen incomes stagnate': New report calculates that earnings did not rise for more than half a billion people between 2005 and 2014," Larry Elliott, 7/13/16.

210 http://www.bloomberg.com/news/articles/2016-08-22/worker-hours-are-more-unpredictable-than-ever. Rebecca Greenfield, "Worker Hours Are More Unpredictable Than Ever," 8/22/16. Greenfield continues: "Workers in these new economy jobs might work 38 hours one week and 15 the next. "Even though unemployment has sunk down, the quality of the jobs that replaced the ones that were lost were not quite the same," Finnigan added."

211 https://www.theguardian.com/business/2018/jun/07/america-gig-economy-work-bureau-labor-statistics. "US gig economy: data shows 16m people in 'contingent or alternative' work: Government data shows scale of freelance or temporary economy as American workers try to navigate changing work environment," Caleb Gayle, 6/7/18.

212 https://www.theguardian.com/careers/2016/aug/16/why-are-uk-workers-so-un-productive. Charlotte Seager, "Why are UK workers so unproductive?," 8/16/16.

213 *Id.*

214 *Id.*

215 https://www.theguardian.com/business/2017/jun/05/nearly-10-million-britons-are-in-insecure-work-says-union, "Nearly 10 million Britons are in insecure work, says union: GMB research explores impact of gig economy and warns of its impact on health and family life," Sarah Butler, 6/5/17.

216 See, e.g. the following analysis of earnings levels in the UK. https://www.theguardian.com/business/2017/mar/09/uk-pay-growth-budget-resolution-foundation. "No pay rise for 15 years, IFS warns UK workers: Thinktank's post-budget analysis says average wages will be no higher in 2022 than in 2007 with weak pay growth exacerbated by looming welfare cuts," Katie Allen, Angela Monaghan and Phillip Inman, 3/9/17. "On current forecasts average earnings will be no higher in 2022 than they were in 2007. Fifteen years without a pay rise."

217 http://www.oecd.org/eco/surveys/United-States-2016-overview.pdf. "OECD Economic Surveys: United States Overview," June 2016. *"Children of poor families often lack the opportunity to do better than their parents because they do not have access to high-quality schools and tend to drop out of college. For those lacking skills demanded by employers, vocational training and continuing education have had mixed results."*

218 https://www.ced.org/blog/entry/the-skills-gap-and-the-seven-skill-sets-that-employers-want-building-the-id. "The Skills Gap and the Seven Skill Sets that Employers Want: Building the Ideal New Hire," Monica Herk, 6/11/15. "We've heard a lot about the 'skills gap.' Among employers having difficulty filling job openings in 2014, 35% reported that it was because applicants lacked the technical (aka "hard") skills needed for the position. Nineteen percent cited lack of "soft skills"—e.g., communication or teamwork—in applicants."

219 http://www.pewresearch.org/fact-tank/2016/04/25/millennials-overtake-baby-boomers/. Richard Fry, "Millennials overtake Baby Boomers as America's largest generation," 4/25/16.

220 http://www.theguardian.com/sustainable-business/2016/may/03/young-australians-face-revolutionary-obstacles-to-score-job-security.

221 https://heatst.com/life/new-research-millennials-likely-to-lose-their-jobs-to-robots/. "New Research: Millennials Most Likely to Lose Their Jobs to Robots," Tom Teodorczuk, 6/20/17.

222 http://www.theguardian.com/world/2016/mar/15/us-millennials-feel-more-working-class-than-any-other-generation.

223 https://www.thetimes.co.uk/article/why-insecure-millennials-are-set-for-an-unhealthy-middle-age-825mjshs5. "Why insecure millennials are set for unhealthy middle age," Greg Hurst, 6/18/18.

224 McKinsey, https://www.theguardian.com/business/2016/jul/14/up-to-70-per-cent-people-developed-countries-seen-income-stagnate,

225 McKinsey, https://www.theguardian.com/business/2016/jul/14/up-to-70-per-cent-people-developed-countries-seen-income-stagnate,

226 https://www.theguardian.com/commentisfree/2016/jul/07/middle-class-struggle-technology-overtaking-jobs-security-cost-of-living. "'Middle class' used to denote comfort and security. Not anymore: The apps and robots celebrated by Silicon Valley wunderkinds are helping make previously white-collar lives ever more precarious," Alissa Quart, 7/7/16.

227 https://www.thesun.co.uk/news/1436480/millennials-are-lazy-self-indulgent-and-lack-the-initiative-to-be-successful-warns-lifestyle-guru-martha-stewart/. "The millionaire lifestyle guru Martha Stewart has issued a stinging criticism of the millennial generation and claimed youngsters are too lazy to get ahead. Too many members of 'Generation Snowflake' are still living with their parents rather than getting out into the world and making something of their lives [Stewart lamented]."

228 Mark Bauerlein, *The Dumbest Generation: How the Digital Age Stupifies Young Americans and Jeopardizes Our Future* (Penguin, 2008).

229 https://www.thetimes.co.uk/edition/news/dumb-and-dumber-why-we-re-getting-less-intelligent-80k3bl83v. "Dumb and dumber: why we're getting less intelligent: The IQ scores of young people have begun to fall after rising steadily since the Second World War, according to the first authoritative study of the phenomenon."

230 https://www.nytimes.com/2018/10/26/style/digital-divide-screens-schools.html?action=click&module=RelatedLinks&pgtype=Article, "The Digital Gap Between Rich and Poor Kids Is Not What We Expected," Nellie Bowles, 10/26/18.

231 https://www.nytimes.com/2018/10/26/style/phones-children-silicon-valley.html, "A Dark Consensus About Screens and Kids Begins to Emerge in Silicon Valley," Nellie Bowles, 10/26/18.

232 http://money.cnn.com/2017/09/20/technology/jack-ma-artificial-intelligence-bloomberg-conference/index.html?iid=ob_homepage_tech_pool. "Jack Ma: We need to stop training our kids for manufacturing jobs," Julia Horowitz, 9/20/17.

233 https://next.ft.com/content/a1614f98-4123-11e6-9b66-0712b3873ae1. *"What the UK needs—with its high employment,* terrible productivity performance and low investment—*is more robots,"* said Adam Corlett, analyst at the Resolution Foundation. The think-tank suggests that new technologies will be crucial to a recovery in UK productivity and pay growth.

234 http://www.pewresearch.org/fact-tank/2015/02/02/u-s-students-improving-slowly-in-math-and-science-but-still-lagging-internationally/.

235 http://www.latimes.com/science/sciencenow/la-sci-sn-why-do-asian-american-students-perform-better-than-whites-20140505-story.html. "Asian and Asian American youth are harder working because of cultural beliefs that emphasize the strong connection between effort and achievement," the authors wrote.

236 http://www.news.cornell.edu/stories/2010/04/tougher-grading-one-reason-high-stem-dropout-rate. George Lowery, 4/2,2010, "Tougher grading is one reason for high STEM dropout rate."

237 https://www.cnsnews.com/news/article/terence-p-jeffrey/65-public-school-8th-graders-not-proficient-reading-67-not-proficient, "65% of Public School 8th Graders Not Proficient in Reading; 67% Not Proficient in Math," Terence P. Jeffrey, 5/1/18.

238 *Id.*

239 https://www.theguardian.com/business/2016/aug/10/joseph-stiglitz-the-problem-with-europe-is-the-euro. Joseph Stiglitz, "The problem with Europe is the euro," 8/10/16.

240 https://www.theguardian.com/commentisfree/2016/jul/07/middle-class-struggle-technology-overtaking-jobs-security-cost-of-living.

241 http://www.theguardian.com/commentisfree/2016/mar/08/robots-technology-industrial-strategy. "Manufacturing will increasingly take place with limited and highly specialized human labor." "When robots do all the work, how will people live? Technology is bringing profound, unstoppable change to society. It is vital that government faces this—and comes up with a new industrial strategy," Tom Watson, 3/8/16.

242 http://www.theguardian.com/business/2016/mar/17/one-in-3-workers-in-wrong-job-productivity-ons. "One in three UK workers are in the wrong job, ONS figures claim: Data shows one in six staff are overqualified for their role with a further one in six undereducated," Katie Allen, 3/17/16.

243 Salam, "Taxi-Driver Suicides Are a Warning," *supra* note 99.

244 https://www.theguardian.com/inequality/2017/aug/24/inside-gig-economy-vulnerable-human-underbelly-of-uk-labour-market. "Inside the gig economy: the 'vulnerable human underbelly' of UK's labour market: Frank Field MP's recent report into the UK's delivery sector demanded 'emergency government intervention' to protect self-employed workers from exploitation. This is the story behind that investigation," Frank Field and Andrew Forsey, 8/24/17.

245 Paul Vigna, "Bill Gross: What to Do After the Robots Take Our Jobs," *supra* note 18.

246 https://www.theguardian.com/us-news/2016/jul/07/middle-class-squeeze-money-household. "'The US? We're in bad shape': squeezed middle class tell tales of struggle—As Voices of America highlights issues that matter to voters, in North Carolina, talk of helping the middle class feels like an empty promise without a plan," Dan Roberts, 7/7/16.

247 http://www.nbcnews.com/news/us-news/most-millennials-are-finding-it-hard-transition-adulthood-report-n748676. "Most Millennials Are Finding It Hard to Transition Into Adulthood: Report," Safia Samee Ali, 4/20/17.

248 http://www.dailymail.co.uk/news/article-5835419/More-half-Millennials-expect-millionaires-someday-according-new-study.html. "More than half of Millennials expect to be millionaires someday, according to a new study," Valerie Bauman, 6/12/18.

249 http://www.bbc.com/capital/story/20170626-the-downside-of-limitless-career-options. "The downside of limitless career options," Alina Dizik, 6/26/17.

250 https://www.theguardian.com/money/2016/jun/15/he-truth-about-working-for-deliveroo-uber-and-the-on-demand-economy. "The truth about working for Deliveroo, Uber and the on-demand economy: Drivers, couriers, cleaners and handymen are now at your beck and call thanks to a host of apps. But what's it like to earn your living waiting for someone else to press a button?," Homa Khaleeli, 6/15/16. See also, https://www.theguardian.com/sustainable-business/2017/apr/15/seventeen-jobs-five-careers-learning-in-the-age-of-automation."Seventeen jobs, five careers: learning in the age of automation: Online courses will help employees to upskill as redundancies sweep away jobs—but will universities be able to keep up?," Max Opray, 4/15/17.

251 https://bigfuture.collegeboard.org/careers/management-medical-health-services-managers.

252 From   https://www.usatoday.com/story/tech/news/2017/05/30/heres-what-you-need-land-americas-best-jobs/101730006/. "Here's what you need to land America's best jobs," Marco della Cava and Eli Blumenthal, *USA Today,* 5/30/17, updated 7/20/17.

253 http://searchsoftwarequality.techtarget.com/feature/FAQ-What-does-development-operations-really-mean. "FAQ: What does 'development operations' really mean?" An explanation offers in part:

> DevOps is a collaborative approach to releasing software applications and updates. It brings together developers who write the software with operation managers responsible for deploying it. Traditionally, a wall has stood between these two groups, and the relationship is often adversarial. Here's an example of a common, negative stereotype: Developers expect operations managers to release new applications or updates on command, even though the code in question may not have been thoroughly tested. Operations managers, who consider frequent code updates risky, have been known to ignore change requests from developers, preventing release of the new software. *DevOps aims to move beyond this behavior by defining a way for the two groups to work together. Fostering cooperation is essential.* Without it, release management will fail, harming the business, which relies heavily on software to sell products, serve customers and make money.

254 http://searchbusinessanalytics.techtarget.com/definition/big-data-analytics.   "big data analytics," Margaret Rouse.

255 https://www.gartner.com/it-glossary/data-scientist. "Data Scientist."

256 https://datasciencedegree.wisconsin.edu/data-science/what-do-data-scientists-do/. "What Do Data Scientists Do?"

257 *Id.*

258 https://blogs.gartner.com/carlie-idoine/2018/05/13/citizen-data-scientists-and-why-they-matter/. "Citizen Data Scientists and Why They Matter," Carlie Idoine, 5/13/18.

259 https://insidebigdata.com/2018/08/01/citizen-data-scientists-yet/. "Citizen Data Scientists: Are we there yet?," 8/1/18. Editorial Team and Matthew Attwell.

260 See, e.g., https://www.wsj.com/articles/why-old-timey-jobs-are-hot-again-1496754001. "Why Old-Timey Jobs Are Hot Again: Millennials are driving a resurgence of age-old crafts, choosing to become bartenders, butchers and barbers in part as a reaction to the digital age," Lauren Weber, 6/6/17.

261 http://www.scotsman.com/news/smartphone-separation-anxiety-is-growing-problem-says-scientist-1-4532910. "Smartphone separation anxiety is growing problem, says scientist," Tom Bawden, 8/15/17.

262 https://www.bbc.co.uk/bbcthree/article/d80d46eb-253d-4b99-ba60-caca6858d757. "Inside the kids-only rehab that treats video games like cocaine: What happens when you make teenagers abstain from alcohol, drugs and even *Call of Duty?*," Ben Bryant, 11/22/18.

263 *Id.*

264 http://www.bbc.co.uk/newsbeat/article/39176828/us-psychologists-claim-social-media-increases-loneliness. "US psychologists claim social media 'increases loneliness.'" See also, http://www.express.co.uk/news/uk/773002/One-in-eight-people-faced-with-loneliness. "Loneliness on the RISE: One in eight people have no close friends to turn to," 3/1/17.

265 http://computer.howstuffworks.com/augmented-reality.htm. "How Augmented Reality Works," Kevin Bonsor.

> This new technology, called augmented reality, blurs the line between what's real and what's computer-generated by enhancing what we see, hear, feel and smell. On the spectrum between virtual reality, which creates immersive, computer-generated environments, and the real world, augmented reality is closer to the real world. Augmented reality adds graphics, sounds, haptic feedback and smell to the natural world as it exists.

266 https://www.theguardian.com/world/2017/jul/04/chinese-internet-giant-limits-online-game-play-for-children-over-health-concerns. "Chinese internet giant limits online game play for children over health concerns: Tencent says young players will be restricted to one hour's play on one of the country's most popular games due to concerns over development," 7/4/17.

267 https://www.theguardian.com/technology/2017/feb/09/robots-taking-white-collar-jobs.

268 https://www.futuristspeaker.com/business-trends/2-billion-jobs-to-disappear-by-2030/. "2 Billion Jobs to Disappear by 2030," Thomas Frey, 2/3/12, modified 2016.

269 *Id.*

270 https://globaldigitalcitizen.org/the-importance-of-teaching-critical-thinking. "The Importance of Teaching Critical Thinking," Lee Watanabe Crockett, 7/25/15. See also, https://www.theguardian.com/teacher-network/2012/sep/12/critical-thinking-overlooked-in-secondary-education. "Why critical thinking is overlooked by schools and shunned by students," Ben Morse, 9/12/12. "Ben Morse argues that for as long as universities fail to recognise achievements in critical thinking with UCAS points, the subject will continue to be ignored at secondary level."

271 https://www.brainyquote.com/quotes/upton_sinclair_138285.

272 William Egginton, *The Splintering of the American Mind: Identity Politics, Inequality, and Community on Today's College Campuses* (Bloomsbury 2018).

273 https://code.facebook.com/posts/1686672014972296/deal-or-no-deal-training-ai-bots-to-negotiate/. "Deal or no deal? Training AI bots to negotiate," 6/14/17.

274 https://www.theregister.co.uk/2017/06/15/facebook_to_teach_chatbots_negotia-tion/. "Facebook tried teaching bots art of negotiation—so the AI learned to lie: Given train-ing data was real human chatter, this says more about us than anything else," Katyanna Quach, 6/15/17.

275 https://www.theguardian.com/inequality/2017/aug/08/rise-of-the-racist-robots-how-ai-is-learning-all-our-worst-impulses. "Rise of the racist robots—how AI is learning all our worst impulses," Stephen Buranyi 8/8/17. Buranyi writes: "There is a saying in computer science: garbage in, garbage out. When we feed machines data that reflects our prejudices, they mimic them—from antisemitic chatbots to racially biased software. Does a horrifying future await people forced to live at the mercy of algorithms?"

276 https://www.theguardian.com/commentisfree/2016/aug/18/google-re"wir-ing-your-mind-memory-journal-plato. Steven Poole, "Does it matter if Google is rewiring our minds? Ask Plato," 8/18/16.

277 *Id.*

278 *Id.*

279 *Id.*

280 See also: https://www.washingtonpost.com/local/how-millions-of-kids-are-being-shaped-by-know-it-all-voice-assistants/2017/03/01/c0a644c4-ef1c-11e6-b4ff-ac2cf509efe5_story.html?utm_term=.9114bfa62b7a. "How millions of kids are being shaped by know-it-all voice assistants," Michael S. Rosenwald, 3/2/17.

> Kids adore their new robot siblings. Many parents have been startled and intrigued by the way these disembodied, know-it-all voices—Am-azon's Alexa, Google Home, Microsoft's Cortana—are impacting their kids' behavior, making them more curious but also, at times, far less po-lite. ... But psychologists, technologists and linguists are only beginning to ponder the possible perils of surrounding kids with artificial intelli-gence, particularly as they traverse important stages of social and lan-guage development.

281 http://theweek.com/articles/689527/hightech-cities-future-utterly-lonely. "Will the high-tech cities of the future be utterly lonely?," Jessica Brown, 4/24/17.

> Humans are inherently social animals, and our health suffers if we're cut off from social ties. So it's no wonder the so-called loneliness "ep-idemic" is being called a public health crisis. But as we sit on the cusp of massive technological advances, the near future could exacerbate this growing problem. ... *In Britain, more than one in eight people say they don't consider anyone a close friend, and the number of Americans who say they have no close friends has roughly tripled in recent decades. A large proportion of the lonely are young; ... One pervasive source of our loneliness is technology.* While it offers an easy way to keep in contact with friends— and meet new people through dating and friendship apps—technology's omnipresence encourages shallow conversations that can distract us from meaningful, real-life, interactions.

282 Paul Ehrlich and Anne Ehrlich, *The Population Bomb* (1968).

283 https://blog-imfdirect.imf.org/2016/08/17/the-euro-area-workforce-is-aging-cost-ing-growth/. Shekhar Aiyar, Christian Ebeke, and Xiaobo Shao, "The Euro Area Workforce is Aging, Costing Growth," 8/17/16.

284 http://www.nytimes.com/2015/01/07/opinion/an-aging-europes-decline.html?_r=0. Arthur C. Brooks, *"An Aging Europe in Decline,"* 1/6/15.

285 *Id.*

286 *Id.*

287 *Id.*

288 http://www.cnbc.com/2017/07/06/elon-musk-the-worlds-population-is-accelerating-toward-collapse-and-nobody-cares.html. "Elon Musk: The world's population is accelerating toward collapse and nobody cares." 7/6/17.

289 http://www.economist.com/blogs/democracyinamerica/2017/04/not-going-gentle. "Not going gentle: Political polarisation has grown most among the old, so don't blame social media, argues a new study," 4/20/17.

290 "Elon Musk," *supra* note 288.

291 *CNS News,* Terence P. Jeffrey, 5/7/18.

292 *USA Today,* Adam Shell, 1/10/18.

293 *CNBC,* Jessica Dickler, 3/6/18.

294 *Wall Street Journal,* Heather Gillers, 5/8/18.

295 *CNS News,* Susan Jones, 5/7/18.

296 *BBC,* Laurence Peter, 4/23/18.

297 *USA Today,* Doyle Rice, 2/7/18.

298 *Bloomberg News,* Suzanne Woolley, 5/29/18.

299 *CBS News,* 5/29/18

300 https://www.cbsnews.com/news/american-seniors-are-sicker-than-global-peers/. Steve Vernon, MoneyWatch 11/22/17, "American seniors are sicker than global peers."

301 http://www.foxbusiness.com/markets/2017/08/10/decreasing-life-expectancy-could-benefit-us-businesses.html. "Decreasing life expectancy could benefit US businesses," 8/10/17. https://www.bloomberg.com/news/articles/2017-08-08/americans-are-dying-younger-saving-corporations-billions. "Americans Are Dying Younger, Saving Corporations Billions: Life expectancy gains have stalled. The grim silver lining? Lower pension costs," John Tozzi, 8/8/17."

302 https://www.wsj.com/articles/why-are-states-so-strapped-for-cash-there-are-two-big-reasons-1522255521. "Why Are States So Strapped for Cash? There Are Two Big Reasons: *The proportion of state and local tax revenues dedicated to Medicaid and public pensions is the highest since the 1960s,*" Cezary Podkul and Heather Gillers, 3/28/18.

303 http://www.who.int/mediacentre/news/releases/2016/health-inequalities-persist/en/.

304 http://www.bloomberg.com/news/articles/2016-08-11/more-old-than-young-a-population-plague-spreads-around-the-globe. Sunny Oh, "More Old Than Young: A Demographic Shock Sweeps the Globe," *Bloomberg,* 8/11/16.

305 https://www.cnbc.com/2018/03/06/42-percent-of-americans-are-at-risk-of-retiring-broke.html. "42% of Americans are at risk of retiring broke: Nearly half of Americans have less than $10,000 stashed away for retirement, according to a report by GoBankingRates," Jessica Dickler, *CNBC,* 3/6/18.

306 http://www.cbsnews.com/news/slashed-pensions-another-blow-for-heartland-workers/. "Another blow for heartland workers: Slashed pensions," Ed Leefeldt, *MoneyWatch,* 7/20/17.

307 https://www.theguardian.com/money/2016/jul/03/retirement-pensions-money. "Let's make retirement great again—by bringing back a pension system: As Americans look to build their future on 401k plans, they find themselves perched atop nest eggs that are far too small," Suzanne McGee, 7/3/16.

308 http://www.cnn.com/2016/06/22/opinions/pensions-a-financial-time-bomb-andrew-scott/index.html. "What happens if we all live to 100?," Andrew Scott, 9/3/16.

309 http://papers.ssrn.com/sol3/papers.cfm?abstract_id=2800220 (posted 6/28/16). Anne Alstott, "Raising the Retirement Age, Fairly," Chapters 6 and 7 of *A New Deal for Old Age* (Harvard University Press, 2016), Yale Law School, *Public Law Research Paper No. 566,* 3/1/2016.

310 http://www.zerohedge.com/news/2017-04-01/moscow-and-beijing-join-forces-by-pass-us-dollar-global-markets-shift-gold-standard. "Moscow And Beijing Join Forces To Bypass US Dollar In Global Markets, Shift To Gold Trade," Tyler Durden, 4/2/17.

311 http://fortune.com/2015/09/08/germany-migrant-crisis/. Claire Groden, "Here's why Germany is welcoming migrants with open arms," 9/8/15. "Germany has one of the world's most rapidly aging and shrinking populations. With one of the world's lowest birth-rates, the country relies on immigration to plug a growing workforce hole. *According to one expert quoted in* Deutsche Welle *last year, the German economy needs to attract 1.5 million skilled migrants to stabilize the state pension system as more Germans retire.*"

312 http://www.bloomberg.com/news/articles/2016-08-18/aging-danes-hope-robots-will-save-their-welfare-state. Peter Levring, "Aging Danes Hope Robots Will Save Their Welfare State," 8/18/16.

313 "Japan's demography: The incredible shrinking country," 3/25/14. http://www.economist.com/blogs/banyan/2014/03/japans-demography. *See also,* http://www.japantimes.co.jp/news/2015/11/25/national/social-issues/public-pensions-health-care-stretch-japans-population-ages/.

314 "Japan's demography: The incredible shrinking country," 3/25/14, http://www.economist.com/blogs/banyan/2014/03/japans-demography.

315 http://www.latimes.com/world/asia/la-fg-japan-population-snap-story.html.   "As Japan's population shrinks, bears and boars roam where schools and shrines once thrived," Julie Makinen, 7/10/16.

All across Japan, aging villages such as Hara-izumi have been quietly hollowing out for years, even as urban areas have continued to grow modestly. But like a creaky wooden roller coaster that slows at the top of the climb before plunging into a terrifying, steep descent, Japan's population crested around 2010 with 128 million people and has since lost about 900,000 residents, last year's census confirmed. *Now, the country has begun a white-knuckle ride in which it will shed about one-third of its population—40 million people—by 2060, experts predict. In 30 years, 39% of Japan's population will be 65 or older.*"

316 https://www.reuters.com/article/us-japan-ageing-robots-widerimage/aging-japan-robots-may-have-role-in-future-of-elder-care-idUSKBN1H33AB. "Aging Japan: Robots may have role in future of elder care," 3/27/18.

317 *Id.*

318 https://www.theguardian.com/technology/2016/may/20/silicon-assassins-condemn-humans-life-useless-artificial-intelligence. "AI will create 'useless class' of human, predicts bestselling historian: Smarter artificial intelligence is one of 21st century's most dire threats, writes Yuval Noah Harari in follow-up to Sapiens." https://www.penguin.co.uk/books/1111302/homo-deus/. *See also*: http://www.theguardian.com/books/2014/sep/11/sapiens-brief-history-humankind-yuval-noah-harari-review.

319 Juval Noah Harari, *Homo Deus: A Brief History of Tomorrow,* (2016).

320 http://www.breitbart.com/big-government/2016/05/09/cbo-nearly-1-6-young-men-u-s-jobless-incarcerated/. "CBO: Nearly 1 in 6 Young Men in U.S. Jobless or Incarcerated."

According to the Congressional Budget Office (CBO), out of the 38 million young men in the U.S. in 2014, 16 percent were jobless (5 million or 13 percent) or incarcerated (1 million or 3 percent). The share of young men without a job or in prison has increased substantially since 1980, when just 11 percent of young men fit into either category. CBO highlights that the level of joblessness and incarceration varies based on young men's educational attainment. The less they have, the more likely they are to be jobless or incarcerated. The rates also varied among racial and ethnic groups.

321 http://www.atimes.com/article/does-china-really-dominate-southeast-asia/, "Does China really dominate Southeast Asia?—Widespread reports of China's hegemony over the neighboring region miss the nuance of fast-shifting political and strategic dynamics," David Hutt, 8/23/18.

322 "China's Warning About 'Dictatorship' Chant Chills Hong Kong Vigil," David Tweed, 6/3/18.   https://www.bloomberg.com/news/articles/2018-06-03/china-warning-on-dictatorship-chant-chills-hong-kong-vigil.

323 https://www.realclearpolitics.com/articles/2017/11/23/chinas_new_greater_east_asia_co-prosperity_sphere_135602.html. "China's New Greater East Asia Co-Prosperity Sphere," Victor Davis Hanson, 11/23/17.

Chinese President Xi Jinping offered a Soviet-style five-year plan for China's progress at the Communist Party congress in Beijing. Despite his talk of global cooperation, the themes were familiar socialist boilerplate about Chinese economic and military superiority to come. Implicit in the 205-minute harangue were echoes of the themes of the 1930s: A rising new Asian power would protect the region and replace declining Western influence. President Xi promised that the Chinese patronage offered a new option for his neighbors "to speed up their development while preserving their independence." Sound familiar? In the 1930s, Imperial Japan tried to square the same circle of importing Western technology while deriding the West. It deplored Western influence in Asia while claiming its own influence in the region was more authentic. ... Depressed by the superior technology and wealth of Western visitors, late-19th-century Japan entered a breakneck race to create entire new industries—mining, energy, steel—out of nothing.

324 https://www.theguardian.com/science/2018/feb/18/china-great-leap-forward-science-research-innovation-investment-5g-genetics-quantum-internet. "China's great leap forward in science: Chinese investment is paying off with serious advances in biotech, computing and space. Are they edging ahead of the west?," Philip Ball, 2/18/18. Ball writes:

325 http://www.newsweek.com/davos-2017-xi-jinping-economy-globalization-protectionism-donald-trump-543993. "Xi Jinping's Davos Speech Showed the World Has Turned Upside Down," Bessma Momani, 1/18/17. "Where the United States once claimed itself to be the center of research and development, technological innovation and intellectual property, Xi noted how China is encouraging domestic consumption and savings, growing its service sector to diversify from manufacturing, and is investing in and promoting green technologies."

326 *Australia News,* Debra Killalea.

327 *The Guardian,* 10/18/17.

328 *Bloomberg News,* Mark Bergen and David Ramli, 8/14/17.

329 *CNET,* Stephen Shankland, 11/13/17.

330 *McClatchy News,* Tim Johnson, 10/23/17.

331 https://www.theguardian.com/world/2016/dec/09/china-universities-must-become-communist-party-strongholds-says-xi-jinping, "China universities must become Communist party 'strongholds,' says Xi Jinping: All teachers must be 'staunch supporters' of party governance, says president in what experts called an effort to reassert control," Tom Phillips, 12/19/16.

332 *Id.*

333 *Id.*

334 https://www.theguardian.com/world/2018/feb/16/admiral-warns-us-must-prepare-for-possibility-of-war-with-china. "Admiral warns US must prepare for possibility of war with China," Ben Doherty, 2/16/18.

335 https://www.wsj.com/articles/mattis-says-u-s-remains-committed-to-allies-in-asia-1527906395. "Jim Mattis Warns of Consequences If Beijing Keeps Militarizing the South China Sea: U.S. strategy in region rooted in 'principled realism' and shared interests, defense secretary says," Nancy A. Youssef, 6/2/18.

336 http://www.nytimes.com/2005/07/15/washington/world/chinese-general-threatens-use-of-abombs-if-us-intrudes.html. "Chinese General Threatens Use of A-Bombs if U.S. Intrudes," Joseph Kahn, 7/15/2005. *"China should use nuclear weapons against the United States if the American military intervenes in any conflict over Taiwan, a senior Chinese military official said Thursday."* See also, https://www.cnbc.com/2017/01/29/us-china-war-increasingly-a-reality-chinese-army-official-says.html. "US-China war increasingly a 'reality,' Chinese army official says," Evelyn Cheng, 1/29/17. "China is preparing for a potential military clash with the United States, according to an article on the Chinese army's website. "The possibility of war increases" as tensions around North Korea and the South China Sea heat up, Liu Guoshun, a member of the national defense mobilization unit of China's Central Military Commission, wrote on Jan. 20."

337 https://www.washingtontimes.com/news/2012/jun/27/inside-china-pla-says-war-us-imminent/Inside China: PLA says war with U.S. imminent, By Miles Yu, 6/27/12. A Chinese general recently offered an alarming assessment that a future conflict with the United States is coming as a result of U.S. "containment" policies. The release last week of a transcribed speech by People's Liberation Army (PLA) Maj. Gen. Peng Guangqian revealed the harsh words toward the United States and those in China he regards as muddle-headed peacenik intellectuals. Gen. Peng, a well-known PLA strategist, has a hawkish reputation and a large following in China. ... "The United States has been exhausting all its resources to establish a strategic containment system specifically targeting China," Gen. Peng said." The contradictions between China and the United States are structural, not to be changed by any individual, whether it is G.H.W. Bush, G.W. Bush or Barack Obama, it will not make a difference to these contradictions."

338 *Free Beacon,* Bill Gertz, 5/11/18.

339 *Washington Examiner,* Joel Gehrke, 3/26/18.

340 *SCMP,* Stephen Chen, 2/4/18.

341 *SCMP,* 5/28/18.

342 Ellen Nakashima and Paul Sonne, *Washington Post,* 6/8/18.

343 See, e.g., http://harvardpolitics.com/world/chinas-investment-in-africa-the-new-colonialism/. "China's Investment in Africa: The New Colonialism?," Elizabeth Manero, 2/3/17. http://www.coha.org/the-dragon-in-uncle-sams-backyard-china-in-latin-america/. "The Dragon in Uncle Sam's Backyard: China in Latin America," 6/6/14.

344 https://www.census.gov/foreign-trade/balance/c5700.html. U.S. Census Bureau, "Trade in Goods with China" (2017 data).

345 https://www.wsj.com/articles/currency-manipulation-isnt-among-chinas-trade-sins-1539645175, *Currency Manipulation Isn't Among China's Trade Sins:* The yuan is sliding against the dollar, but mostly as a result of U.S. policies like high spending and tariffs," Jason Furman, 10/15/18; https://www.cnbc.com/2018/06/27/larry-summers-china-does-not-need-to-steal-us-technology.html, Larry Summers praises China's state investment in tech, saying it doesn't need to steal from US," Mathew J. Belvedere, 6/27/18. "Chinese companies' leadership in some technologies are not the result of theft from the U.S., the former Clinton Treasury secretary says."

346 https://www.wsj.com/articles/china-offers-to-buy-nearly-70-billion-of-u-s-farm-and-energy-products-1528208835. "China Offers to Buy Nearly $70 Billion of U.S. Products

to Fend Off Trade Tariffs: Beijing says purchase hinges on whether Trump imposes threatened tariffs," Lingling Wei and Bob Davis, 6/5/18.

347 https://www.reuters.com/article/us-global-economy/america-first-is-hurting-chinas-business-idUSKCN1MF1W0. Jonathan Cable, "America First is hurting China's business," 10/5/18.

348 https://www.express.co.uk/news/world/1049793/China-news-robots-artificial-intelligence-donald-trump-trade. Laura O'Callaghan, 11/24/18. "China's AI Revolution: Intelligent robots to power factories—risking US fury: Robots powered by artificial intelligence are set to replace Chinese factory workers in a move aimed at boosting the manufacturing industry which has been hit hard by a rise in wages"

349 Id.

350 https://www.thesun.co.uk/tech/4067800/china-vows-to-become-artificial-intelligence-world-leader-by-2030-but-will-it-spark-a-killer-computer-arms-race/. "China to become artificial intelligence 'world leader' by 2030—but will it spark a killer computer arms race? Experts fear the race to develop a super-smart machine mind could end up creating the digital destroyer which wipes out humanity," Jasper Hamill, 7/21/17.

351 https://www.thesun.co.uk/tech/4067800/china-vows-to-become-artificial-intelligence-world-leader-by-2030-but-will-it-spark-a-killer-computer-arms-race/.

352 https://www.express.co.uk/news/world/868333/Chinas-Xi-Jinping-communist-party-congress-beijing-30-year-plan-global-dominance. "China's 30-year Roadmap: Overtake USA and Dominate world with military 'built for war': Chinese leader Xi Jinping has laid out his plans for world domination with a 30-year plan to transform the country and surpass the US to become the biggest global superpower." Simon Osborne, 10/19/17.

353 http://www.orfonline.org/expert-speaks/lethal-autonomous-weapons-dragon-china-approach-artificial-intelligence/. "Lethal Autonomous Dragon: China's approach to artificial intelligence weapons," Bedavyasa Mohanty, 11/15/17.

354 http://www.mcclatchydc.com/news/nation-world/national/national-security/article179971861.html. "China speeds ahead of U.S. as quantum race escalates, worrying scientists," Tim Johnson, 10/23/17.

355 https://www.theguardian.com/business/2016/may/26/chinas-feud-over-economic-reform-reveals-depth-of-xi-jinpings-secret-state.

356 http://www.theguardian.com/business/2016/may/01/chinas-factories-grow-less-than-expected-raising-recovery-doubts.   http://www.theguardian.com/world/2016/may/08/chinese-economy-exports-fall-by-2-and-imports-by-11-in-april. http://money.cnn.com/2016/04/24/news/economy/china-us-trade/index.html.

357 http://www.theguardian.com/business/2016/feb/29/china-to-cut-jobs-in-coal-and-steel-sectors.   https://www.theguardian.com/business/2016/jun/05/excess-capacity-in-chinese-economy-distorting-world-markets-steel. https://www.theguardian.com/business/2016/jun/16/chinas-debt-is-250-of-gdp-and-could-be-fatal-says-government-expert.

358 The Week, Jeff Spross, 5/7/18

359 National Interest, Peiyuan Lan, 10/20/17.

360 The Guardian, Larry Elliott, 12/6/17.

361 Bloomberg News, 9/21/17.

362 The Guardian, Richard Partington, 8/18/17.

363 Reuters, Benjamin Kang Lim, Matthew Miller, David Stanway, 3/1/16.

364 Daily Mail, 8/16/17

365 The Guardian, Philip Ball, 2/18/18.

366 Bloomberg News, Robert Fenner, January 15, 2018

367 New York Times, Claire Cain Miller, 12/21/16.

368 Daily Star, Anders Anglesey, 5/2/18.

369 North West Indiana Times, Joseph S. Pete, 10/11/15.

370 https://www.theguardian.com/commentisfree/2018/mar/04/apple-users-icloud-services-personal-data-china-cybersecurity-law-privacy. "What price privacy when Apple gets into bed with China?," John Naughton, 3/4/18. He writes: "That guff about improving "the speed and reliability of our iCloud services" is the usual corporate cant designed to conceal a harsh reality—which is that henceforth everything that Chinese Apple users store in the cloud will be accessible to the Chinese state. And although the data is encrypted, *Apple will, apparently, have to store the encryption keys in China—which means that its joint venture will have to comply with the cybersecurity law and provide them to the Chinese authorities if required.* As Amnesty International points out, "Chinese police enjoy sweeping discretion and use broad and ambiguously constructed laws and regulations to silence dissent, restrict or censor information and harass and prosecute human rights defenders and others in the name of 'national security' and other purported criminal offences."

371 https://nypost.com/2016/05/17/why-us-companies-have-started-fleeing-china/. "Why US companies have started fleeing China, Noah Smith, 5/17/16."

372 http://www.euractiv.com/section/innovation-industry/news/eu-berlin-oppose-chinese-bid-for-strategic-german-robotics-maker/. "EU, Berlin oppose Chinese bid for 'strategic' German robotics maker," 5/31/16.

373 http://www.wsj.com/articles/china-is-largest-fastest-growing-market-world-wide-for-industrial-robots-1463584169. "Chinese home-appliance maker Midea Group's planned $5 billion takeover bid for Kuka AG of Germany is the latest example of China's voracious appetite for industrial robots. China is the largest and fastest-growing market world-wide for industrial robotics, accounting for $8.5 billion of a global market of $32 billion in 2014."

374 South China Morning Post, SCMP, 12/13/17.

375 *Wall Street Journal,* Alyssa Abkowitz and Liza Lin, 9/4/17.

376 Manufacturing.net, Mike Collins, [Author, *Saving American Manufacturing*], 01/18/2012.

377 *CNN,* Jethro Mullen, 8/14/17.

378 *Wall Street Journal,* Preetika Rana, Amy Dockser Marcus and We*nxin Fan, 1/21/18.

379 *Politico,* Cory Bennett & Bryan Bender, 5/22/18.

380 http://www.foxnews.com/opinion/2018/02/19/why-chinas-takeover-chicago-stock-exchange-would-have-been-very-bad-thing.html. "Why China's takeover of the Chicago Stock Exchange would have been a very bad thing," Gordon G. Chang, 2/19/18.

381 *Id.*

382 *See supra* note 373. "German robotics makers, including Kuka and Siemens, have opened new production plants in China to more easily access the growing market, said Patrick Schwarzkopf, the managing director of the VDMA. Görg's Mr. Wolf said he expected Chinese bids for German robotics makers to increase. '*The interest [in China] is enormous in anything that has to do with automation,*' he said."

383 https://nypost.com/2016/05/17/why-us-companies-have-started-fleeing-china/. "Why US companies have started fleeing China, Noah Smith, 5/17/16." "*Not long after multinationals showed up in China, they were made to hand over much of their technology to native competitors (almost all of which are directly or indirectly owned by the Chinese government).* This was happening as early as 2006, as the Harvard Business Review reported: "These rules limit investment by foreign companies as well as their access to China's markets, stipulate a high degree of local content in equipment produced in the country, and force the transfer of proprietary technologies from foreign companies to their joint ventures with China's state-owned enterprises."

384 http://foreignpolicy.com/2014/07/11/the-debate-over-confucius-institutes-in-the-united-states/. "The Debate Over Confucius Institutes in the United States," Stephen I. Levine, Matteo Mecacci, Michael Hill, Zha Daojiong, Stephen E. Hanson, Mary Gallagher, 7/11/14. One university critic of the Confucius Institutes observes: "*the problem with CIs cannot be*

remedied by transparency and good governance. No democratic country can ignore their insidiousness, active or potential. CIs should respect the universal value of freedom of expression. If universities instead degrade these values to suit the CI, then universities should be forced to find another way to teach Chinese language and culture."

385 https://www.insidehighered.com/news/2018/02/15/fbi-director-testifies-chinese-students-and-intelligence-threats. "FBI director Christopher Wray tells Senate panel that American academe is naïve about the intelligence risks posed by Chinese students and scholars. Some worry his testimony risks tarring a big group of students as a security threat." Elizabeth Redden, 2/15/18.

386 http://www.telegraph.co.uk/news/world/china-watch/society/more-students-to-study-overseas/. "More Chinese students set to study overseas," Luo Wangshu, 3/21/17. "China is the leading source of international students for foreign universities and colleges, and even more of its young people are preparing to go abroad to study over the next five years, according to an education policy adviser. Yu Minhong, founder and chief executive of the New Oriental Education and Technology Group and a member of the Chinese People's Political Consultative Conference's National Committee, estimates that the number of Chinese studying abroad each year will peak at between 700,000 and 800,000."

387 http://money.cnn.com/2017/12/13/technology/google-ai-research-center-china/index.html. "Google is opening an artificial intelligence center in China," Sherisse Pham, 12/13/17.

388 http://www.computerworld.com/article/3085483/high-performance-computing/china-builds-world-s-fastest-supercomputer-without-u-s-chips.html. "China builds world's fastest supercomputer without U.S. chips: China's massive system runs real applications and is 'not just a stunt machine,' says top U.S. supercomputing researcher," Patrick Thibodeau, 6/20/16.

389 jhamill, 7/27/16, "SPACE RACE 2.0 China to build secret 'orbital internet' using experimental satellite network," https://www.thesun.co.uk/news/1510111/china-building-secret-orbital-internet-using-experimental-satellite-network/.

390 https://www.cnet.com/news/china-surpasses-us-in-supercomputer-usage-on-top-500-list/. "China's supercomputers race past US to world dominance: China doesn't just have the single fastest supercomputer in the world. It now dominates the list of the 500 fastest," Stephen Shankland, 11/1/17.

391 https://www.newscientist.com/article/2133188-ai-will-be-able-to-beat-us-at-everything-by-2060-say-experts/, Timothy Revell, "AI will be able to beat us at everything by 2060, say experts," 5/31/17.

392 https://www.thesun.co.uk/news/3344112/incredible-video-shows-army-of-orange-self-charging-robots-which-sort-200000-packages-a-day-in-a-chinese-warehouse/. "Who Needs Humans? Incredible video shows army of orange self-charging robots which sort 200,000 packages a day in a Chinese warehouse. Tiny machines can work eight hours nonstop and get through 20,000 parcels every hour," Paul Harper, 4/16/17. Harper reports:

> In the STO Express building in Liny, Shandong Province, the 300 robots can get through 20,000 parcels every hour. *The self-charging workers have saved the company, which has 300,000 employees, a staggering 70 per cent of manpower.* ... Chinese manufacturers have been increasingly replacing human workers with machines, the output of industrial robots grew 30.4 per cent last year, reports CNBC. *Apple's supplier FoxConn last year replaced 60,000 factory workers with robots.*

393 http://atimes.com/2016/06/chinese-espionage-and-intelligence-activities-at-all-time-high-experts-say/.

394 http://www.businessinsider.com/only-solution-to-chinas-labor-shortage-2016-2. "This is the only solution to China's labor shortage," John Mauldin, 2/11/16. "So what can China do? *In the short term, the choices are to either import new workers or build better robots to replace people. The latter is already happening. The former is unlikely.* One long-term solution works for everybody: Solve the aging problem and increase "health spans." Help people live longer lives that are more productive. Life extension biotechnologies are developing rapidly. China, whose culture holds deep reverence for older people, could be the first to adopt them aggressively. With few other options, life extension may be China's best answer to labor shortages."

395 http://www.scmp.com/news/china/economy/article/1949918/rise-robots-60000-workers-culled-just-one-factory-chinas. "Rise of the robots: *60,000 workers culled from just one factory as China's struggling electronics hub turns to artificial intelligence:* Kunshan, in Jiangsu province, undergoes makeover as 600 companies look to trim their headcount," 5/21/16. http://www.scmp.com/tech/china-tech/article/1860154/more-half-manufacturers-jiangsu-electronics-hub-planning-switch. "More than half of manufacturers at Jiangsu electronics hub planning switch to robot workforce, survey shows," He Huifeng, 9/21/15.

> A total of 600 major industrial enterprises in Kunshan, Jiangsu province, are going to replace human labour with robots within the next five years, according to a survey conducted by the city's commission of economy and information. Out of 100 manufacturers surveyed at one of China's largest manufacturing hubs for the electronics industry, over 50 per cent of them said they were preparing robot production lines.

396 http://www.wsj.com/articles/as-chinas-workforce-dwindles-the-world-scrambles-for-alternatives-1448293942.   http://www.manzellareport.com/index.php/special/the-real-cause-and-impact-of-china-s-labor-shortage. "The Real Cause and Impact of China's Labor Shortage."

> China continues to suffer a labor shortage in its key coastal manufacturing regions. … But the labor shortage is not due to a lack of available workers. Instead, it is prompted by Chinese government policies, as well as prevailing work and living conditions in affected regions. … During the last two decades of China's development, rural workers migrating to urban manufacturing regions have been the chief source of labor in coastal cities. According to the Chinese government's own statistics, migrant workers have increased to more than 250 million from just over 60 million in the last 20 years.

397 https://www.thesun.co.uk/news/3344112/incredible-video-shows-army-of-orange-self-charging-robots-which-sort-200000-packages-a-day-in-a-chinese-warehouse/, *supra* note 392.

398 https://www.axios.com/the-big-layoff-in-china-2511630146.html. "The big layoff in China," Steve Levine, 11/22/17.

399 https://www.theguardian.com/education/2017/aug/18/cambridge-university-press-blocks-readers-china-quarterly. "Cambridge University Press blocks readers in China from articles: Academics and contributors dismayed after hundreds of CUP articles in China Quarterly become inaccessible in country," Richard Adams, 8/18/17.

400 "China builds world's fastest computer," *supra* note 388.

> China's government last week said it plans to build an exascale system by 2020. The U.S. has targeted 2023. China now has more supercomputers in the Top500 list than the U.S., said Dongarra. "China has 167 systems on the June 2016 Top500 list compared to 165 systems in the U.S.," he said, in an email. Ten years ago, China had 10 systems on the list.

401 The Chinese may have erred too far on the side of Internet control and repression. *See,* e.g., https://www.theguardian.com/world/2017/mar/04/chinese-official-slams-inter-net-censorship. "Chinese official calls for easing of internet censorship: 'Great Firewall' that blocks western websites is too restrictive and hinders economic progress while discouraging foreign investors, says Luo Fuhe," Benjamin Haas, 3/3/17.

402 http://www.bloomberg.com/news/articles/2016-07-25/china-slaps-ban-on-inter-net-news-reporting-as-crackdown-tightens. It is reported that: "China's top internet regulator ordered major online companies including Sina Corp. and Tencent Holdings Ltd. to stop orig-inal news reporting, the latest effort by the government to tighten its grip over the country's web and information industries."

403 *Breitbart,* John Hayward, 4/2/18.

404 *Reuters,* 4/21/18.

405 *CBS Local New York,* Ben Tracy, 4/24/18.

406 *The Guardian,* John Naughton, 5/27/18.

407 *Wall Street Journal,* Eva Dou, July 18, 2017.

408 *The Guardian,* Tom Phillips, 5/20/16.

409 *The Guardian,* John Naughton, 3/4/18.

410 *Fast Company.com,* 3/28/18.

411 *Express UK,* Sean Martin, 7/24/17.

412 https://www.thesun.co.uk/tech/4213358/china-working-on-repression-network-which-lets-speed-cameras-track-and-identify-cars-and-human-motorists/. "China working on 'repression network' which lets cameras identify cars—and humans—with unprecedented accuracy: Peking University reveals results of research aimed at developing next generation of surveillance systems," Jasper Hamill, 8/10/17.

413 http://www.nextgov.com/emerging-tech/2018/02/nvidia-makes-facial-recogni-tion-ai-surveillance/146064/. "Nvidia Making Facial Recognition AI for Smart City Surveil-lance: Do you ever feel like somebody's watching you?," Caitlin Fairchild, 2/20/18. "Tech company Nvidia announced Thursday that it has partnered with AI developer AnyVision to bring a new type of surveillance technology to smart cities. ... According to Anyvision, the technology can continuously scan for faces 24/7, and automatically identify and track individ-uals with 99% accuracy. Then the systems algorithms, with the help of human monitors, will compare identified faces with criminal databases."

414 https://www.wsj.com/articles/chinese-police-go-robocop-with-facial-recogni-tion-glasses-1518004353. "Chinese Police Add Facial-Recognition Glasses to Surveillance Ar-senal" Josh Chin 2/7/18. Chin reports: "As hundreds of millions of Chinese begin traveling for the Lunar New Year holiday, police are showing off a new addition to their crowd-surveillance toolbox: mobile facial-recognition units mounted on eyeglasses."

415 Julie Makinen, 8/10/16, "China's crackdown on dissent is described as the harshest in decades," http://www.latimes.com/world/asia/la-fg-china-crackdown-snap-story.html.

416 https://www.wsj.com/articles/the-all-seeing-surveillance-state-feared-in-the-west-is-a-reality-in-china-1498493020. "China's All-Seeing Surveillance State Is Reading Its Cit-izens' Faces: In vast social-engineering experiment, facial-recognition systems crunch data from ubiquitous cameras to monitor citizens," Josh Chin and Liza Lin, 6/26/17.

China is rushing to deploy new technologies to monitor its people in ways that would spook many in the U.S. and the West. *Unfettered by privacy concerns or public debate, Beijing's authoritarian leaders are in-stalling iris scanners at security checkpoints in troubled regions and using sophisticated software to monitor ramblings on social media.* By 2020, the government hopes to implement a national "social credit" system that would assign every citizen a rating based how they behave at work, in public venues and in their financial dealings.

417 "Xi's China: Smothering dissent," https://next.ft.com/content/ccd94b46-4db5-11e6-88c5-db83e98a590a.

418 "Xi's China: Smothering dissent," https://next.ft.com/content/ccd94b46-4db5-11e6-88c5-db83e98a590a.

419 https://www.uscc.gov/sites/default/files/Sarah%20Cook%20May%204th%20 2017%20USCC%20testimony.pdf. "Chinese Government Influence on the U.S. Media Landscape," Sarah Cook, Senior Research Analyst for East Asia, Freedom House. Testimony before the U.S.-China Economic and Security Review Commission Hearing on China's Information Controls, Global Media Influence, and Cyber Warfare Strategy" 5/4/17.

420 Dan Gillmor, "Obama's NSA phone-record law ignores the other (big) data we're giving away: We are no longer merely creatures of metadata. We are now bystanders to the demise of privacy. Will anyone protect us?," 3/26/14, theguardian.com. http://www.theguardian.com/commentisfree/2014/mar/26/obama-nsa-phone-record-law-big-data-giving-away.

421 https://qz.com/671211/chinas-propaganda-outlets-have-leaped-the-top-of-facebook-even-though-it-banned-at-home/. "China's propaganda news outlets are absolutely crushing it on Facebook," Heather Timmons and Josh Horwitz, 5/6/16. *"China's government-controlled media has embraced Western social media, following top-down orders from president Xi Jinping to "tell China's story" to the world. Facebook, Twitter, and YouTube are now rife with news from China's Communist Party, in the form of English-language articles and videos especially geared to a foreign audience."*

422 http://www.infowars.com/eu-proposes-government-id-to-use-internet/. http://www.marketwatch.com/story/facial-recognition-will-soon-end-your-anonymity-2016-06-02.

423 https://www.chinausfocus.com/culture-history/rethinking-the-foreign-media-as-china-coverage-is-it-biased-. "Rethinking the Foreign Media's China Coverage: Is it Biased?," Ivy Yu, 1/19/18.

424 http://www.bloomberg.com/news/articles/2016-05-03/china-said-to-explore-taking-stakes-in-some-news-websites-apps.   http://bigstory.ap.org/article/608d7cb071684f-328041056c269a93b5/china-proposes-new-web-rules-could-enhance-censorship.   http://www.independent.co.uk/news/world/asia/china-set-to-ban-all-foreign-media-from-publishing-online-a6883366.html.   https://www.theguardian.com/business/2016/may/16/chinese-pour-110bn-into-us-real-estate-says-study.

425 https://www.usatoday.com/story/tech/2018/08/28/facebook-employees-hint-conservative-intolerance/1127988002/. "Facebook 'mobs' attack conservative views within company, some employees say," Jessica Guynn, 8/28/18.

426 https://www.breitbart.com/tech/2018/11/08/google-ceo-sundar-pichai-technology-doesnt-solve-humanitys-problems/, Charlie Nash, 11/8/18. "Google CEO Sundar Pichai: 'Technology Doesn't Solve Humanity's Problems.'" [Analyzing an interview in the New York Times]

427 http://www.foxnews.com/politics/2018/03/24/senior-nsa-official-says-technology-is-great-but-can-bring-great-risk-in-cybersecurity.html. "Senior NSA official says technology is great, but can bring great risk in cybersecurity," Catherine Herridge, 3/24/18. Herridge writes:

> The United States' top four nation-state cyber adversaries are Russia, China, Iran and North Korea, and nothing is off limits, including the nuclear arsenal, a senior NSA official told Fox News in an exclusive interview. "Technology brings great opportunity but also brings great risk," the National Security Agency's Marianne Bailey told Fox News. Bailey is charged with guarding national security systems across the Defense Department and U.S. government—with hundreds of thousands of probing attacks daily. "Cybersecurity is a leveler, so you don't have to be a na-

tion-state to have really advanced tools and to invoke damage," she said. Bailey, whose official title is deputy national manager for national security systems, said that they have seen very active cyber-activity in relation to the four nation-state adversaries and the Islamic State terror group.

428 https://www.theguardian.com/technology/2016/jun/17/self-driving-trucks-impact-on-drivers-jobs-us.
https://www.theguardian.com/technology/2016/feb/13/artificial-intelligence-ai-unemployment-jobs-moshe-vardi.

429 http://www.marketwatch.com/story/hello-self-driving-cars-goodbye-41-million-jobs-2016-09-15, "Hello, self-driving cars, and goodbye to 4.1 million jobs?," Shawn Langlois, 9/17/16. See also, http://www.latimes.com/opinion/op-ed/la-oe-greenhouse-driverless-job-loss-20160922-snap-story.html, "Autonomous vehicles could cost America 5 million jobs. What should we do about it?," Steven Greenhouse, 9/22/16.

430 See various links at Bureau of Labor Statistics under truck, delivery, bus and taxi driver data.

431 http://www.cnet.com/uk/news/you-can-now-sign-up-to-take-a-driverless-car-for-a-spin-around-london/. https://www.theguardian.com/technology/2016/jul/02/elon-musk-self-driving-tesla-autopilot-joshua-brown-risks. https://www.theguardian.com/technology/2016/jul/01/bmw-intel-mobileye-develop-self-driving-cars.

432 http://www.nytimes.com/2016/05/07/business/fiat-chrysler-chief-sees-self-driving-technology-in-five-years.html?emc=edit_th_20160507&nl=todaysheadlines&nlid=60842176. http://www.cnet.com/roadshow/news/ubers-first-self-driving-car-takes-to-the-streets-of-pittsburgh/. https://next.ft.com/content/3af5aa62-33ac-11e6-ad39-3fee5f-fe5b5b. [Rolls Royce self driving "vision" car].

433 Julia Carrie Wong, "'We're just rentals': Uber drivers ask where they fit in a self-driving future," 8/19/16.

434 Id.

435 http://www.sfchronicle.com/news/article/DMV-Humans-soon-no-longer-required-in-10993072.php. "State DMV backs allowing self-driving cars with no human on board," David R. Baker and Carolyn Said, 5/10/17. "Self-driving cars with no human behind the wheel—or, for that matter, any steering wheel at all—may soon appear on California's public roads, under regulations state officials proposed Friday." See also, https://www.yahoo.com/news/self-driving-bus-no-back-driver-nears-california-031247012—finance.html. "Self-driving bus with no back-up driver nears California street," Reuters, 3/6/17.

436 https://apnews.com/b248a02690604d36b79719cc227d2ba3/Autonomous-cars-(no-human-backup)-may-hit-the-road-next-year, "Autonomous cars (no human backup) may hit the road next year," Tom Krisher and Dee-Ann Durbin, 6/8/17."

437 https://www.theguardian.com/business/2016/may/06/lyft-driverless-cars-uber-taxis-us-roads-chevrolet-bolt. http://www.mirror.co.uk/tech/self-driving-robots-deliver-food-7647386.

438 http://bigstory.ap.org/article/81b31f2909c943b1a824a3ecb80fa40d/startup-wants-put-self-driving-big-rigs-us-highways.

439 As to fears that the AI driving systems on cars, trucks and busses could be hacked or remotely controlled for various reasons, including terrorist bombings or the spread of toxic and biological materials, consider: https://www.yahoo.com/tech/cia-mission-cars-shows-concern-next-generation-vehicles-035020504—finance.html. "CIA 'mission' on cars shows concern about next-generation vehicles," Alexandria Sage, Reuters, 3/8/17.

440 The CIA has good reason to be concerned since it has now been demonstrated that the super-secret spy and intelligence gathering agency is unable to protect its own systems against hacking and data theft. See, e.g., http://www.spiegel.de/international/world/wikileaks-data-dump-on-cia-spying-vault-7-a-1137740.html. "New WikiLeaks Revelations CIA Spies

May Also Operate in Frankfurt." Michael Sontheimer, 3/7/17. "WikiLeaks has published thousands of documents pertaining to CIA efforts at global surveillance, including a tool to transform smart-TVs into a powerful spying device."

441 http://thehill.com/policy/defense/342659-top-us-general-warns-against-rogue-killer-robots. "Top US general warns against rogue killer robots," John Bowden, 7/18/17.

442 https://www.theguardian.com/commentisfree/2017/aug/22/killer-robots-international-arms-traders. "We can't ban killer robots—it's already too late: Telling international arms traders they can't make killer robots is like telling soft-drinks makers that they can't make orangeade," Philip Ball, 8/21/17. See also, http://www.foxnews.com/tech/2017/08/21/elon-musk-joins-other-experts-in-call-for-global-ban-on-killer-robots.html. "Elon Musk joins other experts in call for global ban on killer robots," 8/21/17. https://www.thesun.co.uk/tech/4262775/scientists-create-terminator-style-immortal-robot-with-self-healing-flesh/. "Scientists create Terminator-style robot with self-healing 'flesh': In a terrifying new advance for machine-kind, robots are now able to heal themselves," Margi Murphy, 8/17/17. "https://www.thesun.co.uk/news/2992038/human-troops-will-be-battling-terminator-style-killer-robots-within-10-years-experts-warn/. "Human troops could be battling 'Terminator-style' killer robots within 10 years, experts warn: Cyber warfare experts say death machines will soon be locked in mortal combat with soldiers," Mark Moore, 3/2/17. "Peter Singer, a strategist for the New America Foundation, said armies need to brace for battle against killer machines."

443 https://www.independent.co.uk/life-style/gadgets-and-tech/news/killer-robots-un-meeting-autonomous-weapons-systems-campaigners-dismayed-a8519511.html. https://www.independent.co.uk/life-style/gadgets-and-tech/news/killer-robots-un-meeting-autonomous-weapons-systems-campaigners-dismayed-a8519511.html. "'Killer Robots' ban blocked by US and Russia at UN Meeting: Campaigners want to ban the 'morally reprehensible weapons,'" Mattha Busby and Anthony Cuthbertson, 9/3/18.

444 https://www.aol.com/article/news/2017/03/08/stephen-hawking-warns-that-human-aggression-may-destroy-us-all/21876064/?ncid=txtlnkusaolp00000058&. "Stephen Hawking warns that human aggression 'may destroy us all,'" 3/8/17.

445 Margi Murphy, "Scientists create Terminator-style robot with self-healing 'flesh': In a terrifying new advance for machine-kind, robots are now able to heal themselves," supra note 112.

446 The Guardian, Ian Sample, 11/13/17.

447 The Guardian, Daniel Boffey, 9/27/17.

448 Wall Street Journal, Paul Scharre, 4/11/18.

449 Scout.com, Kris Osborn, 11/25/17.

450 Yahoo News, 11/27/17.

451 Military.com Daily News, Martin Egnash, 4/8/18.

452 Free Beacon, Bill Gertz, 1/30/18.

453 Defense One, Patrick Tucker, 10/10/17.

454 Defense One, Patrick Tucker, 11/8/17.

455 http://www.vocativ.com/324304/russia-robot-army/. See also, http://www.dailymail.co.uk/news/article-4457892/Russian-army-puts-new-remote-controlled-robot-tank-test.html. "Russian army puts its enormous new remote-controlled robot tank boasting a 30mm automatic gun and six missiles through its paces," Simon Holmes.

456 http://www.houstonchronicle.com/business/technology/article/Putin-Leader-in-artificial-intelligence-will-12166704.php. "Putin: Leader in artificial intelligence will rule world," 9/1/17.

457 http://bgr.com/2017/11/08/ai-weapons-systems-military-destruction-apocalypse/. "When AI rules, one rogue programmer could end the human race," Mike Wehner, 11/8/17.

458 See, http://news.softpedia.com/news/Skynet-a-Real-Worry-US-Military-Researching-Robotic-Morality-441438.shtml. "Skynet a Real Worry, US Military Researching Robotic Morality," Sebastian Pop, 5/9/14.

459 http://www.techworld.com/social-media/sir-tim-berners-lee-lays-out-nightmare-scenario-where-ai-runs-world-economy-3657280/. "Sir Tim Berners-Lee lays out nightmare scenario where AI runs the financial world: The architect of the world wide web laid out a scenario where AI could become the new masters of the universe by creating and running multitudes of companies better and faster than humans," Scott Carey, 4/10/17.

460 http://www.bloomberg.com/news/articles/2016-06-08/wall-street-has-hit-peak-human-and-an-algorithm-wants-your-job. Hugh Son, 6/18/16.

461 Id.

462 Bloomberg News, Niklas Magnusson & Hanna Hoikkala, 5/15/18.

463 The Guardian, 5/14/18.

464 CNBC, Abigail Hess, 11/8/17.

465 Bloomberg News, Tasneem Hanfi Brogger, 12/10/17.

466 Bloomberg News, Silla Brush, 11/1/17.

467 Wall Street Journal.

468 "Ford's factory robots make coffee and give fist bumps," Duncan Geere http://www.techradar.com/news/car-tech/ford-s-factory-robots-make-coffee-and-give-fist-bumps-1324925.

469 https://www.theguardian.com/technology/2017/jul/29/foxconn-china-apple-wisconsin-trump. "Foxconn's $10bn move to the US is not a reason to celebrate: The company doesn't have a great track record of keeping its job-creation promises, for one. Then there's the issue of worker conditions in China," Zoe Sullivan, 7/29/17.

470 Associated Press, Yuri Kageyama, 4/23/18.

471 Digital Trends, Dyllan Furness, 9/25/17.

472 Bloomberg News, Kyle Stock, 1/30/18.

473 Daily Star UK.

474 The Guardian, Finn Murphy, 11/17/17.

475 Associated Press, Colleen Barry & Charlene Pele, 4/2/18.

476 The Guardian, 1/22/18.

477 Axios, 12/20/17

478 The Guardian, Dominic Rushe, 1/13/18.

479 Fox News, Kathleen Joyce, 1/10/18.

480 The Guardian, Julia Carrie Wong, 8/16/17.

481 Daily Mail UK, Tracy You, 8/2/17.

482 Bloomberg News, Jing Cao, 11/6/17.

483 The Guardian, Dan Hernandez, 6/2/18

484 See, e.g., http://www.lawsitesblog.com/2016/12/10-important-legal-technology-developments-2016.html. Robert Ambrogi, 12/20/16, "The 10 Most Important Legal Technology Developments of 2016."

485 https://www.theguardian.com/commentisfree/2016/jul/07/middle-class-struggle-technology-overtaking-jobs-security-cost-of-living. Laura Donnelly, 7/27/16. http://www.telegraph.co.uk/news/2016/07/27/robots-as-good-as-human-surgeons-study-finds/. "Surgery performed by robots is just as successful as operations carried out by surgeons, a major trial has found. The study of prostate cancer patients found those whose gland was removed by a machine were doing as well after three months as those who went under the knife in the traditional way. They experienced less pain doing day to day activities a week later, and reported better overall physical quality of life after six weeks, but this levelled out over time. Those undergoing robot surgery also lost far less blood and spent less time in hospital."

486 Nadya Sayej, "Robot Customer Service Will Dominate Travel in the Future," 8/4/16. http://motherboard.vice.com/read/future-of-travel-robots-chihira.

487 Telegraph UK, Henry Bodkin, 9/11/17.

488 New York Post, Lauren Tousignant, 8/31/17.

489 *South China Morning Post [SCMP]*, 9/21/17.

490 *The Guardian*, Hannah Devlin, 5/21/18.

491 *Yahoo News*, AFP, 5/28/18.

492 *The Guardian*, Damien Gayle, 2/6/17.

493 http://money.cnn.com/2017/01/10/technology/jack-ma-trump-us-jobs-claim/index.html?iid=EL. "Alibaba's 1 million American jobs promise isn't realistic," Sherisse Pham, 1/11/17.

494 http://www.dailymail.co.uk/news/article-4754078/China-s-largest-smart-warehouse-manned-60-robots.html. "Wifi-equipped robots triple work efficiency at the warehouse of the world's largest online retailer: China's largest 'smart warehouse' is manned by 60 cutting-edge robots," Tracy You, 8/2/17.

495 https://www.nytimes.com/2018/04/01/technology/retailer-stores-automation-amazon.html. "Retailers Race Against Amazon to Automate Stores," Nick Wingfield, Paul Mozur and Michael Corkery, 4/1/18.

496 http://www.latimes.com/local/lanow/la-me-ln-anaheim-homeless-emergency-20170913-story.html. "Anaheim's emergency declaration sets stage for removal of homeless encampment," Anh Do, 9/14/17.

497 http://www.sfgate.com/news/article/Homeless-explosion-on-West-Coast-pushing-cities-12334291.php. "Homelessness soars on West Coast as cities struggle to cope," Gillian Flaccus and Geoff Mulvihill, 11/6/17.

498 *The Guardian*, Alastair Gee, 12/5/17.

499 *The Guardian*, Andrew Gumbel, 3/16/18.

500 *Los Angeles Times*, Gale Holland, 4/11/18.

501 *Fox News*, Tori Richards, 2/26/18.

502 *Seattle Times*, Vernal Coleman, 12/30/17.

503 *East Bay Times*, Louis Hansen, 12/17/17.

504 *Fox News*, Travis Fedschun, 11/27/17.

505 *Fox News*, Tori Richards, 11/22/17.

506 *The Guardian*, Charlotte Simmonds, 12/12/17.

507 *San Francisco Chronicle*, Kevin Fagan and Alison Graham, 9/8/17.

508 *WNYC Report*, Mirela Iverac, 12/6/17

509 http://www.foxnews.com/politics/2017/11/22/homeless-people-defecating-on-la-streets-fuels-horror-hepatitis-outbreak-as-city-faulted.html. "Homeless people defecating on LA streets fuels horror hepatitis outbreak, as city faulted," Tori Richards, 11/22/17. Richards details the challenge.

510 http://www.foxnews.com/us/2018/06/11/voluntarily-vagrant-homeless-youth-crusty-urban-challenge.html. "Voluntarily vagrant, homeless youth a 'crusty' urban challenge," Andrew O'Reilly, 6/11/18.

511 http://www.rawstory.com/2016/01/older-and-sicker-how-americas-homeless-population-has-changed/.

512 https://www.csoonline.com/article/3192519/security/cyber-infrastructure-too-big-to-fail-and-failing.html. "Cyber infrastructure: Too big to fail, and failing: The cybersecurity industry isn't keeping up with cyber threats, the Atlantic Council's Joshua Corman told a Boston audience. And things are about to get even worse," Taylor Armerding, 4/26/17.

513 *Free Beacon*, Adam Kredo, 3/27/18.

514 https://www.brookings.edu/blog/order-from-chaos/2018/03/22/the-next-russian-attack-will-be-far-worse-than-bots-and-trolls/. "Order from Chaos: The next Russian attack will be far worse than bots and trolls," Alina Polyakova, 3/22/18.

515 https://www.domesticpreparedness.com/resilience/cascading-consequences-electrical-grid-critical-infrastructure-vulnerability/. "Cascading Consequences: Electrical Grid Critical Infrastructure Vulnerability," George H. Baker & Stephen Voland, 5/9/18.

516 https://www.eenews.net/stories/1060086303. "Coal plants' vulnerabilities are largely unknown to feds," Blake Sobczak, Energywire: 6/25/18.

517 *The Hill,* Morgan Chalfant, 5/31/18

518 https://www.theguardian.com/business/2018/nov/09/bank-of-england-stages-war-games-combat-cyber-attacks-data-breaches, Angela Monaghan, 11/9/18. "Bank of England stages day of war games to combat cyber-attacks: Spate of data breaches in financial sector prompts voluntary exercise to test resilience."

519 https://www.cnbc.com/2018/06/01/the-next-911-will-be-a-cyberattack-security-expert-warns.html. Natasha Turak, 6/1/18, "The next 9/11 will be a cyberattack, security expert warns."

520 https://www.brookings.edu/blog/order-from-chaos/2017/06/14/cyber-threats-and-how-the-united-states-should-prepare/. Michael E. O'Hanlon, 6/14/17. "Order from Chaos: Cyber threats and how the United States should prepare."

521 https://www.theguardian.com/technology/2018/feb/21/ai-security-threats-cyber-crime-political-disruption-physical-attacks-report. Alex Hern, 2/21/18, "Growth of AI could boost cybercrime and security threats, report warns: Experts say action must be taken to control artificial intelligence tech."

522 https://www.wired.com/story/russian-hackers-attack-ukraine/. Andy Greenberg, 6/20/17. "How an Entire Nation became Russia's Test Lab for Cyberwar."

523 *The Guardian,* Alex Hern, 9/6/17.

524 https://www.infowars.com/black-sky-event-feds-preparing-for-widespread-power-outages-across-u-s/. "Black Sky Event": Feds Preparing For Widespread Power Outages Across U.S.: Experts gear up for catastrophe that would 'bring society to its knees,'" Paul Joseph Watson, 8/9/17.

525 https://www.theguardian.com/technology/2017/jun/13/industroyer-malware-virus-bring-down-power-networks-infrastructure-wannacry-ransomware-nhs. "'Industroyer' virus could bring down power networks, researchers warn: Discovery of new malware shows vulnerability of critical infrastructure, just months after the WannaCry ransomware took out NHS computers," Alex Hern, 6/13/17.

526 "Total Chaos": Cyber Attack Feared As Multiple Cities Hit With Simultaneous Power Grid Failures," Tyler Durden, 4/21/17.

527 https://www.yahoo.com/news/britain-apos-four-meals-away-060000661.html. "Britain 'four meals away from anarchy' if cyber attack takes out power grid," Ben Farmer, *The Telegraph,* 3/17/18.

528 See, e.g., http://www.foxnews.com/tech/2017/06/27/huge-ransomware-attack-hits-europe-sparks-mass-disruption.html. "Huge 'Petya' ransomware attack hits Europe, sparks mass disruption," James Rogers, 6/27/17.

529 http://www.businessinsider.com/warren-buffett-cybersecurity-berkshire-hathaway-meeting-2017-5?utm_source=newsletter&utm_medium=email&utm_campaign=newsletter_axiosam. "Buffett: This is 'the number one problem with mankind,'" Akin Oyedele, 5/6/17.

530 https://www.wired.com/2010/12/ff-ai-flashtrading/. "Algorithms Take Control of Wall Street." Felix Salmon and Jon Stokes, 12/27/10. Since 2010, however, HFT has declined substantially in terms of overall volume for U.S. stock markets as exchanges adapted to the new technology but has also moved into other markets where high-speed algorithm-based trading maintains advantages over traditional traders.

531 Robert J. Kauffman, Yuzhou Hu & Dan Ma, "Will high-frequency trading practices transform the financial markets in the Asia Pacific Region?," *Financial Innovation* (2015) 1:4, DOI 10.1186/s40854-015-003-8.

532 https://www.nytimes.com/2018/01/12/business/ai-investing-humans-dominating.html. Conrad De Aenlle, *The New York Times,* 1/12/2018.

533 http://www.zerohedge.com/news/2017-04-21/total-chaos-cyber-at-tack-feared-multiple-cities-hit-simultaneous-power-grid-failures. "Total Chaos"—Cyber Attack Feared As Multiple Cities Hit With Simultaneous Power Grid Failures," Tyler Durden, 4/21/17.

> [T]he entire national power grid has been mapped by adversaries of the United States and it is believed that sleep trojans or malware may exist within the computer systems that maintain the grid. … *[T]here is also the potential for attacks on individual power substations in the vast network of decentralized and largely unguarded power grid chain.* A U.S. government study established that there would be "major, extended blackouts if more than three key substations were destroyed.

534 https://www.domesticpreparedness.com/resilience/cascading-consequences-elec-trical-grid-critical-infrastructure-vulnerability/. "Cascading Consequences: Electrical Grid Critical Infrastructure Vulnerability," George H. Baker and Stephen Voland, 5/9/18.

535 *Id.*

536 https://www.nytimes.com/2018/08/14/us/puerto-rico-electricity-power.html. "Puerto Rico Spent 11 Months Turning the Power Back On." 8/14/18.

537 Baker and Voland, *supra* note 534.

538 http://dailycaller.com/2017/09/27/fbi-director-terrorist-drones-coming-here-im-minently-video/. "FBI Director: Terrorist Drones 'Coming Here Imminently,'" Chuck Ross, 9/27/17.

539 https://www.technologyreview.com/s/603337/a-100-drone-swarm-dropped-from-jets-plans-its-own-moves/. "A 100-Drone Swarm, Dropped from Jets, Plans Its Own Moves: Once launched, the swarm can decide for itself how best to execute a mission," Jamie Condliffe, 1/10/17.

540 http://www.defenseone.com/threats/2017/07/trumps-special-ops-pick-says-terror-drones-might-soon-reach-us-africa-how-worried-should-we-be/139642/?oref=d-topstory. "Trump's Special Ops Pick Says Terror Drones Might Soon Reach the US from Africa. How Worried Should We Be?," Caroline Houck, 7/21/17.

541 See, e.g., https://www.theguardian.com/world/2018/feb/20/north-korea-cyber-war-spying-study-fire-eye?utm_source=esp&utm_medium=Email&utm_campaign=GU+To-day+USA+-+Collections+2017&utm_term=264822&subid=15825848&CMP=GT_US_col-lection.

"Study reveals North Korean cyber-espionage has reached new heights: Spying unit is widening its operations into aerospace and defence industries, according to US security firm," David Taylor, 2/20/18.

542 https://www.theverge.com/2017/6/23/15860668/amazon-drone-delivery-pat-ent-city-centers. "Amazon's vision for the future: delivery drone beehives in every city: Welcome to Amazontopia," James Vincent, 6/23/17. "In a patent application published yesterday [6/22/17], Amazon described how "multi-level fulfillment centers for unmanned aerial vehicles" could help put drones where they're needed."

543 http://www.fox5ny.com/news/company-plans-drone-to-carry-400-pound-pay-loads. "Company plans drone to carry 400 pound payloads," 4/2/18.

544 *Free Beacon,* Bill Gertz, 5/11/18.

545 https://ca.news.yahoo.com/u-officials-warn-congress-risks-drones-seek-pow-ers-101645306.html. "U.S. officials warn Congress on risks of drones, seek new powers," David Shepardson, *Reuters,* 6/6/18.

546 *Sun UK,* Patrick Knox, 7/21/17.

547 *The Guardian,* Alyssa Sims, 1/19/18.

548 *ABC News,* Geneva Sands, 11/9/17.

549 *McClatchy News,* Tim Johnson, 9/7/17.

550 *Technology Review,* Jamie Condliffe, 1/10/17.

551 Shepardson, *supra* note 545.

552 See, e.g., https://www.thesun.co.uk/news/4072863/plague-disease-biological-weapon-terrorists-isis-spread-from-air/. "Terrorists could weaponise deadly plague disease by releasing it as a cloud above cities killing thousands, experts warn: Warning comes amid fears imploding ISIS have been developing chemical and biological weapons," Patrick Knox, 7/21/17.

553 https://www.theguardian.com/business/2018/apr/07/global-inequality-tipping-point-2030. "Richest 1% on target to own two-thirds of all wealth by 2030: World leaders urged to act as anger over inequality reaches a 'tipping point,'" Michael Savage, 4/7/18.

"World leaders are being warned that the continued accumulation of wealth at the top will fuel growing distrust and anger over the coming decade unless action is taken to restore the balance. ... [I]f trends seen since the 2008 financial crash were to continue, then the top 1% will hold 64% of the world's wealth by 2030."

554 http://globaleconomicanalysis.blogspot.com/2016/01/multiple-jobholders-artificially-boost.html. "Multiple Jobholders Artificially Boost 'Full-Time' Employment: Does the Sum of the Parts Equal the Whole?," 1/8/16. *See also,* http://www.adamtownsend.me/jobs-in-the-gig-economy/. "Are There Any Jobs in the Gig Economy?"

555 https://www.theguardian.com/sustainable-business/2017/may/04/we-need-to-track-more-than-gdp-to-understand-how-automation-is-transforming-work. "We need to track more than GDP to understand how automation is transforming work: *Governments and business don't have the right information to understand what the future of work really looks like,*" Tim Dunlop, 5/4/17.

556 *Id.*

557 http://taxprof.typepad.com/taxprof_blog/2018/07/outstanding-student-loan-debt-hits-15-trillion-women-hold-most-of-it.html. Paul Caron, 7/6/18, "Outstanding Student Loan Debt Hits $1.5 Trillion, Women Hold Most Of It."

558 http://www.theoccidentalobserver.net/2011/07/discrimination-against-whites-in-federal-employment/. Kevin MacDonald, "Discrimination against Whites in Federal Employment," 7/14/11.

559 See,   e.g.,   https://www.afp.com/en/news/2266/despairing-young-italians-seek-greener-pastures-abroad. "Despairing young Italians seek greener pastures abroad," 7/21/17. *"Italy gets worse every time you look at it and offers less and less to young people,"* says Antonio Davide D'Elia, just one of thousands of young Italians who have decided to try their luck abroad."

560 https://www.yahoo.com/news/why-swiss-voted-no-guaranteed-basic-income-163626337.html?ref=gs.   https://ca.news.yahoo.com/swiss-voters-decide-guaranteed-monthly-income-plan-103534182—business.html. "Swiss reject free income plan after worker vs. robot debate," Silke Koltrowitz and Marina Depetris, Reuters, 6/5/16.

561 https://www.lmtonline.com/business/article/A-record-number-of-folks-age-85-and-older-are-13051373.php. Andrew Van Dam, *Washington Post,* 7/5/18. "A record number of folks age 85 and older are working." "Seventy may be the new sixty, eighty may be the new seventy, but 85 is still pretty old to work in America. Yet, in some ways, it is the era of the very-old-worker in America. Overall, 255,000 Americans, 85-years-old and over, were working over the past 12 months. That's 4.4 percent of Americans that age, up from 2.6 percent in 2006, before the recession. It's the highest number on record."

562 https://www.imf.org/external/pubs/ft/wp/2016/wp16238.pdf. INF Working Paper, "The Impact of Workforce Aging on European Productivity," Shekhar Aiyar, Christian Ebeke and Xiaobo Shao, December 2016.

563 http://www.bloomberg.com/news/articles/2016-07-26/golden-years-redefined-as-

older-americans-buck-trend-and-work. "A rising share of Americans is holding jobs into their golden years, bucking the overall trend of people leaving the labor force that is concerning Federal Reserve policy makers trying to boost growth."

564 https://www.theguardian.com/business/us-money-blog/2016/jul/07/fix-us-jobs-report-gig-economy-unemployment-data. "How to fix the jobs report: stop responding to it like Pavlov's dog: *We can't trust the monthly employment data as the labor department often misses the mark and more Americans are working in the gig economy*," Suzanne McGee, 7/7/16. "Zen Payroll has found that in some cities—Los Angeles and Austin among them—the '1099 economy' now makes up 20% or more of the local workforce; the percentage of workers who fall into this category has almost doubled in Seattle."

565 http://apps.npr.org/unfit-for-work/. Chana Joffe-Walt, "Unfit for Work: The startling rise of disability in America." "In other words, people on disability don't show up in any of the places we usually look to see how the economy is doing. ... *It's the story not only of an aging workforce, but also of a hidden, increasingly expensive safety net*."

566 *Id.*

567 http://www.zerohedge.com/news/2016-06-03/americans-not-labor-force-soar-record-947-million-surge-664000-one-month. https://www.theguardian.com/business/2016/jun/03/jobs-report-may-unemployment-rate-economy. "US economy adds paltry 38,000 jobs in May for weakest growth since 2010," Jana Kasperkevic, 6/3/16.

568 https://www.ushmm.org/information/exhibitions/online-exhibitions/special-focus/nazi-persecution-of-the-disabled. United States Holocaust Museum, Nazi Persecution of the Diasabled." We must always keep in mind:

> On July 14, 1933, the German government instituted the "Law for the Prevention of Progeny with Hereditary Diseases." This law called for the sterilization of all persons who suffered from diseases considered hereditary, including mental illness, learning disabilities, physical deformity, epilepsy, blindness, deafness, and severe alcoholism. With the law's passage the Third Reich also stepped up its propaganda against the disabled, regularly labeling them "life unworthy of life" or "useless eaters" and highlighting their burden upon society.

569 http://www.zerohedge.com/news/2017-07-19/social-security-will-be-paying-out-more-it-receives-just-five-years. "*Social Security Will Be Paying Out More Than It Receives In Just Five Years*," Tyler Durden, 7/19/17. (Authored by Mac Slavo via SHTFplan.com), "When social security was first implemented in the 1930s, America was a very different country. Especially in regards to demographics. The average life expectancy was roughly 18 years younger than it is now, and birth rates were a bit higher than they are now. By the 1950s, the fertility rate was twice as high as it is in the 21st century."

570 https://www.washingtonpost.com/news/the-switch/wp/2016/06/02/everything-you-think-you-know-about-ai-is-wrong/. Brian Fung, "Everything you think you know about AI is wrong," 6/2/16.

571 *Id.*

572 Jeff Hawkins & Donna Dubinsky, "What is Machine Intelligence vs. Machine Learning vs. Deep Learning vs. Artificial Intelligence (AI)?," 1/11/16. http://numenta.com/blog/machine-intelligence-machine-learning-deep-learning-artificial-intelligence.html. "We use the term 'machine intelligence' to refer to machines that learn but are aligned with the Biological Neural Network approach. Although there still is much work ahead of us, we believe the Biological Neural Network approach is the fastest and most direct path to truly intelligent machines."

573 Richard Waters, "Investor rush to artificial intelligence is real deal," *Financial Times*, 1/4/15. http://www.ft.com/cms/s/2/019b3702-92a2-11e4-a1fd-00144feabdc0.html#ixzz3Nxmii03Q.

574 http://money.cnn.com/2016/02/22/technology/google-brain-artificial-intelligence-quoc-le/index.html?iid=EL. "AI can solve world's biggest problems: Google brain engineer." Sarah Ashley O'Brien, 2/22/16.

575 http://www.dailymail.co.uk/sciencetech/article-4382162/Scientists-create-AI-LEARNS-like-human-mind.html. "The birth of intelligent machines? Scientists create an artificial brain connection that learns like the human mind," Harry Pettit, 4/5/17.

576 Machine learning, deep learning and algorithms are explained in greater depth at the following source. https://deeplearning4j.org/neuralnet-overview.html.

577 "AI program gets really good at navigation by developing a brain-like GPS system: DeepMind's neural networks mimic the grid cells found in human brains that help us know where we are." *MIT Technology Review*, May 2018. Will Knight, May 9, 2018.

578 https://www.thesun.co.uk/tech/3306890/humanity-is-already-losing-control-of-artificial-intelligence-and-it-could-spell-disaster-for-our-species/. "Humanity is already losing control of artificial intelligence and it could spell disaster for our species: Researchers highlight the 'dark side' of AI and question whether humanity can ever truly understand its most advanced creations," Margi Murphy, 4/11/17.

579 http://theweek.com/articles/689359/how-humans-lose-control-artificial-intelligence. "How humans will lose control of artificial intelligence," 4/2/17. "That's how profoundly things could change. *But we can't really predict what might happen next because superintelligent A.I. may not just think faster than humans, but in ways that are completely different. It may have motivations—feelings, even—that we cannot fathom."*

580 *Id.*

581 https://home.ohumanity.org/breaking-down-superintelligence-890e86c59564. "Breaking Down Superintelligence." This is a cogent analysis of Nick Bostrom's book, *Superintelligence: Paths, Dangers, Strategies.*

582 http://numenta.com/blog/machine-intelligence-machine-learning-deep-learning-artificial-intelligence.html. Jeff Hawkins & Donna Dubinsky, "What is Machine Intelligence vs. Machine Learning vs. Deep Learning vs. Artificial Intelligence (AI)?," 1/11/16.

583 https://www.wsj.com/articles/tiny-hard-drive-uses-single-atoms-to-store-data-1468854001. "Tiny Hard Drive Uses Single Atoms to Store Data: It packs hundreds of times more information per square inch than best currently available technologies, study says." Daniela Hernandez, 7/18/16.

584 For some interesting background on this issue, see: http://www.defenseone.com/technology/2017/09/can-us-military-re-invent-microchip-ai-era/141065/. "Can the US Military Re-Invent the Microchip for the AI Era?," Patrick Tucker, 9/17/17.

585 https://iq.intel.co.uk/exascale-supercomputer-race/. "Who's winning the race to build an exascale supercomputer?," *That Media Thing* Writer.

586 *New Scientist,* Mark Kim, 9/28/17

587 *Technology Review,* Will Knight, 11/10/17.

588 *New Statesman,* Philip Ball, 12/17.

589 *McClatchy News,* Tim Johnson, 10/23/17

590 *Live Science,* Robert Coolman, 9/26/14.

591 *IBM Research,* Dario Gill.

592 *Engadget,* Andrew Tarantola, 2/23/18.

593 http://www.techworld.com/picture-gallery/big-data/9-tech-giants-investing-in-artificial-intelligence-3629737/. http://startuphook.com/tech/startups-leading-the-artificial-intelligence-revolution/918/. http://www.businessinsider.com/10-british-ai-companies-to-look-out-for-in-2016-2015-12?r=UK&IR=T.

594 https://www.ft.com/content/3d2c2f12-99e9-11e4-93c1-00144feabdc0, "Scientists and investors warn on AI: Greater focus needed on safety and social benefits, says open letter," Tim Bradshaw, 1/11/15.

595 *Id.*

596 https://www.nbcnews.com/mach/technology/godlike-homo-deus-could-replace-humans-tech-evolves-n757971, "Godlike 'Homo Deus' Could Replace Humans as Tech Evolves: What happens when the twin worlds of biotechnology and artificial intelligence merge, allowing us to re-design our species to meet our whims and desires?," 5/31/17.

597 https://www.theguardian.com/technology/2017/feb/23/wikipedia-bot-editing-war-study. "Study reveals bot-on-bot editing wars raging on Wikipedia's pages," Ian Sample, 2/23/17. *"Over time, the encyclopedia's software robots can become locked in combat, undoing each other's edits and changing links, say researchers."*

598 https://www.theguardian.com/commentisfree/2017/jul/24/robots-ethics-shakespeare-austen-literature-classics. "We need robots to have morals. Could Shakespeare and Austen help? Using great literature to teach ethics to machines is a dangerous game. The classics are a moral minefield," John Mullan, 7/24/17.

599 https://www.theguardian.com/technology/2016/jun/12/nick-bostrom-artificial-intelligence-machine. "Artificial intelligence: 'We're like children playing with a bomb': Sentient machines are a greater threat to humanity than climate change, according to Oxford philosopher Nick Bostrom," Tim Adams, 6/12/16.

600 https://www.recode.net/2017/1/10/14226564/linkedin-ebay-founders-donate-20-million-artificial-intelligence-ai-reid-hoffman-pierre-omidyar, "LinkedIn's and eBay's founders are donating $20 million to protect us from artificial intelligence: It's part of a $27 million fund being managed by MIT and Harvard," April Glaser, 1/10/17. https://www.wired.com/2015/01/elon-musk-ai-safety/, Davey Alba, "Elon Musk Donates $10M to Keep AI From Turning Evil," 1/15/15.

601 https://www.thesun.co.uk/news/techandscience/1287163/mark-zuckerberg-says-well-be-plugged-into-the-matrix-within-50-years/. "Mark Zuckerberg says we'll be plugged into 'The Matrix' within 50 years: Tech titan claims computers will soon be able to read our minds and beam our thoughts straight onto Facebook," Jasper Hamill.

602 For the interchange see, http://www.cnbc.com/2017/07/25/elon-musk-mark-zuckerberg-ai-knowledge-limited.html. Elon Musk: Facebook CEO Mark Zuckerberg's knowledge of A.I.'s future is 'limited,'" Arjun Kharpal, 7/25/7.http://www.cnbc.com/2017/07/24/mark-zuckerberg-elon-musks-doomsday-ai-predictions-are-irresponsible.html. "Facebook CEO Mark Zuckerberg: Elon Musk's doomsday AI predictions are 'pretty irresponsible,'" Catherine Clifford, 7/24/17.

603 https://www.cnbc.com/2017/09/21/head-of-google-a-i-slams-fear-mongering-about-the-future-of-a-i.html. "Head of A.I. at Google slams the kind of 'A.I. apocalypse' fear-mongering Elon Musk has been doing," Catherine Clifford, 9/21/17.

604 All this may sound like SciFi but really isn't. *See*, e.g., a recent report on a DARPA project. https://futurism.com/darpa-is-planning-to-hack-the-human-brain-to-let-us-upload-skills/. "DARPA Is Planning to Hack the Human Brain to Let Us 'Upload' Skills."

605 https://www.forbes.com/sites/gregsatell/2016/06/03/3-reasons-to-believe-the-singularity-is-near/#62471f817b39. "3 Reasons To Believe The Singularity Is Near," Greg Satell, 6/3/16.

606 https://www.technologyreview.com/s/528656/ray-kurzweil-says-hes-breathing-intelligence-into-google-search/. Tom Simonite, "Ray Kurzweil Says He's Breathing Intelligence into Google Search," *MIT Technology Review*, 6/26/14.

607 "Disrupters bring destruction and opportunity: FT Series: Who has been wreaking havoc on traditional business models in 2014," https://next.ft.com/content/b9677026-8b6d-11e4-ae73-00144feabdc0#slide0.

608 http://www.telegraph.co.uk/technology/2016/03/25/we-must-teach-ai-machines-to-play-nice-and-police-themselves/. "Microsoft's racist bot shows we must teach AI to play nice and police themselves," Madhumita Murgia, 3/29/16.

609 http://www.theguardian.com/technology/2016/mar/03/artificial-intelligence-hackers-security-autonomous-learning. "These engineers are developing artificially intelligent hackers: In a sign of the autonomous security of the future, a $2m contest wants teams to build a system that can exploit rivals' vulnerabilities while fixing its own," Olivia Solon, 3/3/16.

610 http://www.nytimes.com/2016/05/14/science/synthetic-human-genome.html. "Scientists Talk Privately About Creating a Synthetic Human Genome," Andrew Pollack, 5/13/16.

611 http://www.nytimes.com/2016/03/07/technology/taking-baby-steps-toward-software-that-reasons-like-humans.html?_r=0. "Taking Baby Steps Toward Software That Reasons Like Humans," John Markoff, 3/6/16. "The field of artificial intelligence has largely stumbled in giving computers the ability to reason in ways that mimic human thought. Now a variety of machine intelligence software approaches known as 'deep learning' or 'deep neural nets' are taking baby steps toward solving problems like a human."

612 https://www.theguardian.com/books/2016/jun/15/the-age-of-em-work-love-and-life-when-robots-rule-the-earth-robin-hanson-review. "The Age of Em review—the horrific future when robots rule the Earth," Steven Poole, 6/15/16.

613 https://scout.com/military/warrior/Article/Army-Tests-New-Super-Soldier-Exoskeleton-111085386. "Army Tests New Super-Soldier Exoskeleton: The Army is testing an exoskeleton technology which uses AI to analyze and replicate individual walk patterns, provide additional torque, power and mobility," Kris Osborn, 11/25/17.

614 https://www.yahoo.com/news/artificial-muscles-superpower-robots-204049571.html. "Artificial muscles give 'superpower' to robots," 11/27/17.

615 https://www.theladders.com/p/25316/future-of-work-elite-super-workers, "The future of work is medically enhanced 'elite super-workers,' report says," Monica Torres, 8/9/17. https://www.theguardian.com/commentisfree/2017/aug/04/editing-human-genome-consumer-eugenics-designer-babies, "Editing the human genome brings us one step closer to consumer eugenics: Hijacked by the free market, human gene editing will lead to greater social inequality by heading where the money is: designer babies," David King, 8/4/17.

616 http://www.cnbc.com/2017/02/10/cramer-on-ai-the-replacement-of-us-we-dont-need-us-with-nvidia.html. "Cramer on AI: 'This is the replacement of us. We don't need us with Nvidia.'" Berkeley Lovelace Jr., Feb. 10, 2017.

Founder and CEO of Nvidia, Jen-Hsun Huang, said on Thursday that deep learning on the company's graphic processing unit, used in A.I., is helping to tackle challenges such as self-driving cars, early cancer detection and weather prediction. "We can now see that GPU-based deep learning will revolutionize major industries, from consumer internet and transportation to health care and manufacturing. The era of [AI] is upon us," he said.

617 http://fortune.com/2018/06/07/mit-psychopath-ai-norman/. "MIT Scientists Create 'Psychopath' AI Named Norman," Fortune, Carson Kessler, 6/7/18.

618 http://norman-ai.mit.edu/. "NORMAN: World's First Psychopath AI."

619 https://www.theguardian.com/science/2017/jul/19/give-robots-an-ethical-black-box-to-track-and-explain-decisions-say-scientists. "Give robots an 'ethical black box' to track and explain decisions, say scientists: As robots start to enter public spaces and work alongside humans, the need for safety measures has become more pressing, argue academics," Ian Sample, 7/19/17.

620 Although reports such as the following may well be overstated, the fact is that there is ongoing research into how to "join" humans with computerized capabilities through connections and implants. See, http://www.dailymail.co.uk/sciencetech/article-4683264/US-military-reveals-funding-Matrix-projects.html. "US military reveals $65m funding for 'Matrix' projects to plug human brains directly into a computer: System could be used to give soldiers 'supersenses' and boost brainpower," Mark Prigg, 7/10/17.

621 Abigail Beall, Mailonline, 7/8/16, "Scientists are closer to creating a computer with emotions."

622 https://www.livescience.com/62239-elon-musk-immortal-artificial-intelligence-dictator.html, Brandon Specktor, 4/6/18.

623 Michael S. Rosenwald, "Serious reading takes a hit from online scanning and skimming, researchers say," *Washington Post*, 4/6/14. http://www.washingtonpost.com/local/serious-reading-takes-a-hit-from-online-scanning-and-skimming-researchers-say/2014/04/06/088028d2-b5d2-11e3-b899-20667de76985_story.html.

624 http://www.cnbc.com/2017/02/13/elon-musk-humans-merge-machines-cyborg-artificial-intelligence-robots.html. "Elon Musk: Humans must merge with machines or become irrelevant in AI age," Arjun Kharpal, 2/13/17.

625 Jacques Ellul, *The Technological Society* (Alfred A. Knopf 1964).

626 Rosenwald, "Serious reading takes a hit from online scanning and skimming, researchers say," *supra* note 369.

627 https://www.theguardian.com/technology/2017/nov/14/social-media-influence-election-countries-armies-of-opinion-shapers-manipulate-democracy-fake-news, "Thirty countries use 'armies of opinion shapers' to manipulate democracy—report: Governments in Venezuela, the Philippines, Turkey and elsewhere use social media to influence elections, drive agendas and counter critics," Alex Hern, 11/14/17.

628 http://miami.cbslocal.com/2017/03/03/doc-claims-too-much-screen-time-turns-kids-into-digital-junkies/. "Doc Claims Too Much Screen Time Turns Kids Into Digital Junkies," 3/3/17, Lauren Pastrana. "The debate over 'how much screen time is too much' is nothing new but there's one doctor who says the problem is so severe, he's dubbed it 'digital heroin.' In his book, *Glow Kids: How Screen Addiction is Hijacking Our Kids and How to Break the Trance*, author Dr. Nicholas Kardaras likens the effects of excessive screen exposure to the neurological damage caused by drug addiction."

629 http://theweek.com/articles/677922/5-new-brain-disorders-that-born-digital-age. "5 new brain disorders that were born out of the digital age," Tammy Kennon, 2/28/17.

630 http://miami.cbslocal.com/2017/03/03/doc-claims-too-much-screen-time-turns-kids-into-digital-junkies/, *supra* note 628. See also, https://www.theatlantic.com/magazine/archive/2017/09/has-the-smartphone-destroyed-a-generation/534198/. "Have Smartphones Destroyed a Generation?," Jean M. Twenge, *The Atlantic*, September 2017 Issue.

631 http://www.psychguides.com/guides/internet-and-computer-addiction-treatment-program-options/. "Internet and Computer Addiction Treatment Program Options." *"According to the National Institutes of Health, it is the rewarding nature of the Internet that could be responsible for the development of Internet addiction. It has long been established that addictions are responsible for activating various areas of the brain that are linked to pleasure.* Collectively, these sites are known as the pleasure pathway. When these sites are activated, the brain increases production of dopamine and various other neurochemicals."

632 http://www.psychguides.com/guides/computerinternet-addiction-symptoms-causes-and-effects/. Signs and Symptoms of Internet or Computer Addiction. "An Internet or computer addiction is the excessive use of the former or the latter. The latest edition of the *Diagnostic and Statistical Manual of Mental Disorders* (*DSM-V*) actually includes it as a disorder that needs further study and research."

633 http://www.virtualreality-news.net/news/2016/jun/28/can-virtual-reality-really-be-addictive/. "Feature: Can virtual reality really be addictive?," 6/28/16.

634 https://www.theguardian.com/society/2017/nov/30/more-than-half-of-american-children-set-to-be-obese-by-age-35-study-finds. "More than half of American children set to be obese by age 35, study finds: Harvard researchers predict 57% of children will grow up obese," Jessica Glenza, 11/30/17.

635 https://www.forbes.com/sites/reenitadas/2017/07/17/goodbye-loneliness-hel--

lo-sexbots-how-can-robots-transform-human-sex/2/#3b85a22962e3. "Goodbye Loneliness, Hello Sexbots! How Can Robots Transform Human Sex?," Reenita Das, 7/17/17.

636 http://www.wbur.org/onpoint/2015/10/06/fda-oxycontin-heroin-opioid-addiction-crisis. Tom Ashbrook, "American Opioid Addiction Keeps Growing: "American addiction. From prescription painkillers to heroin. The numbers are staggering. Why?," 10/6/15.

637 Henry David Thoreau, *Walden* (1854).

638 https://www.theguardian.com/technology/2016/dec/29/oculus-touch-control-future-vr. "Why the future of VR is all down to touch control: The new controllers from Oculus represent a glimpse of a virtual reality people can really lose themselves in," Samuel Gibbs, 12/29/16.

639 Robert Dahl, *Dilemmas of Pluralist Democracy: Autonomy vs. Control*, 44-45 (1982).

640 *Id.*

641 See *CBO Report* at http://taxprof.typepad.com/taxprof_blog/2016/07/cboin-30-years-us-will-have-highest-debt-to-gdp-ratio-in-our-history141.html.

642 http://www.theinformact.org/. "The Intergenerational Financial Obligations Reform Act." *"This generation of Americans is very likely to be the first generation in our history as a nation to leave a worse economy and a worse fiscal position than the one they inherited. THE INFORM ACT is a step in the right direction toward informing Americans of the magnitude of this problem."* James Heckman, Nobel Laureate in Economics.

643 usdebtclock.org. The US Debt Clock site offers a wide and fascinating range of data on the US economy that graphically demonstrates where our money goes, the number of people in or out of work, official and actual unemployment and employment statistics, retirement and disability information, and much more.

644 https://www.cnbc.com/2018/04/09/the-gop-tax-plan-means-short-term-gains-for-the-economy-but-federal-debt-is-primed-to-explode-cbo-analysis-says.html. "The GOP tax plan means short-term gains for the economy, but federal debt is primed to explode, CBO analysis says," Ylan Mui, 4/9/18.

645 https://www.thebalance.com/current-u-s-federal-government-tax-revenue-3305762. "Current U.S. Federal Government Tax Revenue: Who Really Pays Uncle Sam's Bills?," Kimberly Amadeo, 2/16/18.

646 https://www.thebalance.com/current-u-s-federal-government-tax-revenue-3305762. "Current U.S. Federal Government Tax Revenue: Who Really Pays Uncle Sam's Bills?," Kimberly Amadeo, 2/16/18.

647 *Brookings*, Ron Haskins, 4/8/15.

648 *CNS News*, Terence P. Jeffrey, 4/17/18.

649 *CNS News*, Terence P. Jeffrey, 4/11/18.

650 *Bloomberg News*, Saleha Mohsin and Randy Woods, 4/30/18.

651 *CNBC*, Ylan Mui, 4/19/18.

652 *PJ Media*, Rick Moran, 4/8/18.

653 *CNBC*, Javier E. David, 2/19/18.

654 *Wall Street Journal*, Heather Gillers, 5/8/18.

655 *CNS News*, Terence P. Jeffrey, 5/7/18.

656 *The Guardian*, Graeme Wearden and Larry Elliott, 1/24/18.

657 *Bloomberg News*, Ben Steverman, 4/2/18.

658 *CNN Money*, Katie Lobosco, 6/5/18.

659 *Wall Street Journal*, David Harrison, 6/5/18.

660 https://www.wsj.com/articles/u-s-on-a-course-to-spend-more-on-debt-than-defense-1541937600. Kate Davidson & Daniel Kruger, 11/11/18. "U.S. on a Course to Spend More on Debt Than Defense: Rising interest costs could crowd out other government spending priorities and rattle markets."

661 *Id.*

662 https://www.cbo.gov/publication/45555. Julie Topoleski, "Growing Deficits Over the Long Term Would Cause Federal Debt to Exceed 100 Percent of GDP by 2039," 7/21/14.

663 *Id.*

664 *Id.*

665 http://www.washingtonexaminer.com/debt-to-reach-highest-level-since-1950/ article/2599983. Joseph Lawler, "Debt to reach highest level since 1950 this year," 8/23/16. *"Over the next 10 years, the office sees the debt rising from 77 percent of GDP to 86 percent.* Beyond that, it's supposed to keep rising as interest costs on the debt mount, along with payments for Social Security, Medicare, and other mandatory programs."

666 http://www.washingtonexaminer.com/debt-to-reach-highest-level-since-1950/ article/2599983. Joseph Lawler, "Debt to reach highest level since 1950 this year," 8/23/16.

667 https://thehill.com/blogs/blog-briefing-room/news/259476-ex-gao-head-us-debt-is-three-times-more-than-you-think. "Ex-GAO head: US debt is three times more than you think," Bradford Richardson, 11/7/15.

668 http://www.forbes.com/sites/kotlikoff/2013/06/14/why-governments-need-to-budget-over-the-infinite-horizon/#50e02ca1fa8e. Laurence Kotlikoff, "Why Governments Need to Budget Over the Infinite Horizon," 6/14/13.

Ignoring the distant future when our kids' welfare is at stake is morally repugnant. But it's also forbidden by economic theory. Here's why. Economic theory doesn't tell us whether to call any given dollar the government takes from us "taxes" or or "borrowing." Nor does it tell us whether to call any given dollar the government hands to us a "transfer payment" or "repayment of principal plus interest." *Social Security's $23.1 trillion fiscal gap is off-the-books because politicians from both parties chose to call our FICA contributions "taxes," rather than "borrowing," even though they simultaneously made promises to repay our "taxes" with future "benefits" that could just as well be called "repayment of principal plus interest."* This is the real reason that politicians and their "trustees" budget over short-term horizons. Doing so let's them label their policies in ways that leave most obligations outside the budgeting horizon.

But, here's the catch. Any internally consistent set of labels will produce the same infinite horizon fiscal gap. But each will produce a different fiscal gap over any finite horizon, including 75 years. In short, then, the 75-year fiscal gap of Social Security is not pinned down by economic theory. It can be any size anyone wants to report based on her own internally consistent labeling choice. ... For the U.S. government as a whole, the infinite horizon fiscal gap is a whopping $222 trillion! *Its elimination requires not a 32 percent immediate and permanent tax hike in Social Security FICA taxes or a 22 percent immediate and permanent cut in Social Security benefits, but either a 64 percent immediate and permanent tax hike in all federal taxes or a 40 percent immediate and permanent cut in all expenditures apart from servicing official debt.* So, Social Security's enormous fiscal problem is just a molehill in front of a mountain of horrendous obligations our politicians and their "trustees" are ignoring with their careful choice of words and their finite budgeting horizons.

669 http://www.forbes.com/sites/kotlikoff/2013/06/14/why-governments-need-to-budget-over-the-infinite-horizon/#50e02ca1fa8e. Laurence Kotlikoff, "Why Governments Need to Budget Over the Infinite Horizon," 6/14/13.

670 https://www.forbes.com/sites/johntharvey/2012/09/10/impossible-to-default/#20af1cf11180, "It Is Impossible For The US To Default," John T. Harvey, 9/10/12.

We could choose to do so, just as a person trapped in a warehouse full of food could choose to starve, but we could never be forced to. This is not a theory or conjecture, it is cold, hard fact. The reason the US could never be forced to default is that every single bit of the debt is owed in the currency that we and only we can issue: dollars. Unlike Greece, we don't have to try to earn foreign exchange via exports or beg for better terms. There is simply no level of debt we could not repay with a keystroke.

671 http://www.forbes.com/sites/kotlikoff/2013/06/14/why-governments-need-to-budget-over-the-infinite-horizon/#50e02ca1fa8e. Laurence Kotlikoff, "Why Governments Need to Budget Over the Infinite Horizon," 6/14/13.

672 http://www.brookings.edu/research/opinions/2015/04/08-federal-debt-worse-than-you-think-haskins. "The federal debt is worse than you think," Ron Haskins, 4/8/15. Haskins writes: "CBO estimates that the debt will be well over 100 percent of GDP by 2039 under conservative assumptions about spending and revenue. *When CBO incorporates its estimates of the impact of the continuing large federal deficits on the nation's economy, it estimates that the accumulated debt held by the public will reach an astounding 180 percent of GDP by 2039.*"

673 http://money.cnn.com/2016/06/22/pf/social-security-medicare/index.html. "Social Security trust fund projected to run dry by 2034," Jeanne Sahadi, 6/22/2016. *See also,* "What Happened to the $2.6 Trillion Social Security Trust Fund?," Merrill Mathews, 7/13/11.

674 https://www.cbo.gov/sites/default/files/114th-congress-2015-2016/reports/51580-LTBO.pdf.

In CBO's projections, deficits rise during the next three decades because the government's spending grows more quickly than its revenues do (see Summary Figure 1). In particular, spending grows for Social Security, the major health care programs (primarily Medicare), and interest on the government's debt. Much of the spending growth for Social Security and the major health care programs results from the aging of the population: As members of the baby-boom generation age and as life expectancy continues to increase, the percentage of the population age 65 or older is anticipated to grow sharply, boosting the number of beneficiaries of those programs. *By 2046, projected spending for those programs for people 65 or older accounts for about half of all federal noninterest spending. The remainder of the projected growth in spending for Social Security and the major health care programs is driven by health care costs per beneficiary, which are projected to increase more quickly than GDP per person (after the effects of aging and other demographic changes are removed).* CBO projects that those health care costs will rise—though more slowly than in the past—in part because of the effects of new medical technologies and rising personal income.

675 http://www.newsweek.com/trump-tax-cuts-debt-china-907763. "US Debt Is Growing and Foreigners Are Buying Less: Here's Why That Could Be Disastrous for the Economy," Newsweek, Nicole Goodkind, 5/2/18. "America is taking on record amounts of debt to pay for tax cuts and spending increases, but foreign investors, who currently hold about 43 percent of government debt, are getting skittish about purchasing it."

677 https://www.cnbc.com/2018/06/18/russia-cuts-treasury-holdings-in-half-as-foreigners-start-losing-appetite-for-us-debt.html. "Russia cuts Treasury holdings in half as foreigners start losing appetite for US debt: The U.S. government needs buyers of its debt as the Fed continues to reduce its holdings and the budget deficit is projected to surge in coming years." Jeff Cox, 6/18/18.

678 See, e.g., https://www.theguardian.com/technology/2017/sep/26/tinder-personal-data-dating-app-messages-hacked-sold. "I asked Tinder for my data. It sent me 800 pages of my deepest, darkest secrets: The dating app knows me better than I do, but these reams of intimate information are just the tip of the iceberg. What if my data is hacked—or sold?," Judith Duportail, 9/26/17.

679 http://www.mcclatchydc.com/news/nation-world/national/national-security/article166488597.html. "Is Alexa spying on us? We're too busy to care—and we might regret that," Tim Johnson, 8/10/17.

680 https://ca.news.yahoo.com/lebanese-tourist-referred-criminal-trial-insulting-egypt-facebook-190518538.html. "Lebanese tourist referred to criminal trial for insulting Egypt on Facebook," *Reuters*, 6/3/18.

681 See, e.g., the concerns voiced by a leading private sector AI executive. https://www.theguardian.com/technology/2017/mar/13/artificial-intelligence-ai-abuses-fascism-donald-trump. "Artificial intelligence is ripe for abuse, tech executive warns: 'a fascist's dream': Microsoft's Kate Crawford tells SXSW that society must prepare for authoritarian movements to test the 'power without accountability' of artificial intelligence," Olivia Solon, 3/13/17. Solon writes:

> As artificial intelligence becomes more powerful, people need to make sure it's not used by authoritarian regimes to centralize power and target certain populations, Microsoft Research's Kate Crawford warned on Sunday. *In her SXSW session, titled* Dark Days: AI and the Rise of Fascism, *Crawford, who studies the social impact of machine learning and large-scale data systems, explained ways that automated systems and their encoded biases can be misused, particularly when they fall into the wrong hands."* *"Just as we are seeing a step function increase in the spread of AI, something else is happening: the rise of ultra-nationalism, rightwing authoritarianism and fascism," she said. All of these movements have shared characteristics, including the desire to centralize power, track populations, demonize outsiders and claim authority and neutrality without being accountable.* Machine intelligence can be a powerful part of the power playbook," she said.

682 https://www.goethe.de/en/kul/ges/20440422.html. "Yvonne Hofstetter on Big Data "We Carry The Bugging Device Around With Us In Our Pockets,"" Judith Reker. "Yvonne Hofstetter is herself the managing director of a company that processes and evaluates huge amounts of data. In her book "Sie wissen alles" (They Know All) she calls for a better way of dealing with the digital revolution."

683 https://www.yahoo.com/news/top-experts-warn-against-malicious-ai-014639573.html. "Top experts warn against 'malicious use' of AI," Marlowe Hood, 2/20/18.

> Artificial intelligence could be deployed by dictators, criminals and terrorists to manipulate elections and use drones in terrorist attacks, more than two dozen experts said Wednesday as they sounded the alarm over misuse of the technology. ... "We live in a world fraught with day-to-day hazards from the misuse of AI, and we need to take ownership of the problems." The authors called on policy makers and companies to make robot-operating software unhackable, to impose security restrictions on some research, and to consider expanding laws and regulations governing AI development.

684 https://us.macmillan.com/noplacetohide/glenngreenwald/9781250062581/. Glenn Greenwald, *No Place to Hide: Edward Snowden, the NSA, and the U.S. Surveillance State* (2014). "Fearless and incisive, *No Place to Hide* has already sparked outrage around the globe

and been hailed by voices across the political spectrum as an essential contribution to our understanding of the U.S. surveillance state." *See also*, https://www.theatlantic.com/politics/ archive/2014/05/on-nsa-surveillance-glenn-greenwald-is-not-the-radical/370830/. "*No Place to Hide*: A Conservative Critique of a Radical NSA." *The Atlantic's* analysis argues:

> That totalitarian approach came straight from the top. *Outgoing NSA chief Keith Alexander began using "collect it all" in Iraq at the height of the counterinsurgency. Eventually, he aimed similar tools at hundreds of millions of innocent people living in liberal democracies at peace, not war zones under occupation.* The strongest passages in *No Place to Hide* convey the awesome spying powers amassed by the U.S. government and its surveillance partners; the clear and present danger they pose to privacy; and the ideology of the national-security state. The NSA really is intent on subverting every method a human could use to communicate without the state being able to monitor the conversation. *U.S. officials regard the unprecedented concentration of power that would entail to be less dangerous than the alternative. They can't conceive of serious abuses perpetrated by the federal government, though recent U.S. history offers many examples.*

685   John Kampfner, "As in Russia, the terror threat has become the excuse to curtail our rights" 8/20/13, *The Guardian*, http://www.theguardian.com/commentisfree/2013/aug/20/ russia-terror- excuse-curtail-rights.

686   https://www.telegraph.co.uk/news/2018/12/02/chief-mi6-calls-new-era-spying-using-ai-robots-combat-rogue/. "MI6 chief calls for new era of spying using AI and robots to combat rogue states," Robert Mendick and Dominic Nicholls, 12/2/18.

687   Kampfner, *supra* note 685.

688   See, e.g., http://www.lifezette.com/polizette/silicon-valley-tightens-its-grip-on-free-speech/. "Silicon Valley Tightens Its Grip on Free Speech: Alliance between progressives and tech is killing the unfettered exchange of ideas," Edmund Kozak, 8/14/17. Kozak writes: " *Political totalitarianism is coming to America, and it is being ushered in not by government thugs in jackboots but by progressive activists and their allies in Silicon Valley.*" http://www. mcclatchydc.com/news/nation-world/national/national-security/article166488597.html. "Is Alexa spying on us? We're too busy to care—and we might regret that," Tim Johnson, 8/10/17. https://www.thesun.co.uk/tech/4213358/china-working-on-repression-network-which-lets-speed-cameras-track-and-identify-cars-and-human-motorists/. "China working on 'repression network' which lets cameras identify cars—and humans—with unprecedented accuracy: Peking University reveals results of research aimed at developing next generation of surveillance systems," Jasper Hamill, 8/10/17.

689   https://www.yahoo.com/news/privacy-fears-over-artificial-intelligence-crimestopper-015326163.html. "Privacy fears over artificial intelligence as crimestopper," Rob Lever, 11/11/17.

690   http://www.bloomberg.com/view/articles/2016-07-08/freedom-is-receding-around-the-world. "Freedom Is Receding Around the World: By itself, Brexit isn't a big deal. But it symbolizes the decade-long weakening of the U.S.-led bloc that advanced liberal values," Noah Smith, 7/8/16.

691   http://www.mcclatchydc.com/news/nation-world/national/national-security/article212173259.html. "Big Tech firms march to the beat of Pentagon, CIA despite dissension," *McClatchy News*, Tim Johnson, 6/4/18.

692   Scott Moxley, "FBI Used Best Buy's Geek Squad To Increase Secret Public Surveillance," 3/8/17. http://www.ocweekly.com/news/fbi-used-best-buys-geek-squad-to-increase-secret-public-surveillance-7950030.

693   Kampfner, *supra* note 685.

694 https://www.axios.com/the-growing-antitrust-concerns-about-u-s-tech-giants-2433870013.html. Kim Hart, 6/15/17, "Policing the power of tech giants: The world's largest tech companies—Google, Facebook, Amazon, Microsoft and Apple—have become enormous concentrations of wealth and data, drawing the attention of economists and academics who warn they're growing too powerful." "Platform companies have captured the economy," said Jonathan Taplin, who argues in a new book and a recent NYT op-ed that the dominant platforms are so big that they're undermining competition. https://www.wsj.com/articles/amazon-is-leading-techs-takeover-of-america-1497653164, "Amazon Is Leading Tech's Takeover of America: The tentacles of a handful of tech giants are reaching into industries no one ever expected them to, reshaping our world in their image," Christopher Mims, 6/16/17.

695 http://www.foxnews.com/opinion/2017/03/09/andrew-napolitano-spies-among-us-congress-has-created-monster-that-is-coming-for-us.html. "Spies among us—Congress has created a monster that is coming for us," Andrew P. Napolitano, 3/9/17. *"The NSA has 24/7/365 access to all the mainframe computers of all the telephone and computer service providers in America."*

696 *Daily Star UK*, Rachel O'Donoghue, 12/12/17.

697 *CBS Local NY*, Ben Tracy, 4/24/18.

698 *Washington Post*, 1/7/18.

699 *Forbes*, Thomas Fox-Brewster, 4/16/18.

700 *Loss Prevention Media*, Chris Trlica, 6/19/17.

701 *New York Times*, Nick Wingfield, 5/22/18.

702 *Physics News*, 8/27/17.

703 *Wall Street Journal*, Dan Strumpf and Wenxin Fan, 11/1/17.

704 *Next Government*, Caitlin Fairchild, 2/20/18.

705 *Yahoo News*, Rob Lever, 11/11/17.

706 *Ars Technica*, Dan Goodin, 11/20/2017.

707 https://arstechnica.com/tech-policy/2017/11/an-alarming-number-of-sites-employ-privacy-invading-session-replay-scripts/. "No, you're not being paranoid. Sites really *are* watching your every move: Sites log your keystrokes and mouse movements in real time, before you click submit," Dan Goodin, 11/20/2017. Goodin explains: "If you have the uncomfortable sense someone is looking over your shoulder as you surf the Web, you're not being paranoid. *A new study finds hundreds of sites—including microsoft.com, adobe.com, and godaddy.com—employ scripts that record visitors' keystrokes, mouse movements, and scrolling behavior in real time, even before the input is submitted or is later deleted."*

708 http://www.cnn.com/2017/03/08/politics/james-comey-privacy-cybersecurity/. "Comey: 'There is no such thing as absolute privacy in America,'" Mary Kay Mallonee and Eugene Scott, 3/9/17.

709 https://www.theguardian.com/technology/2017/apr/27/facebook-report-government-propaganda. *"Facebook admits: governments exploited us to spread propaganda: Company will step up security to clamp down on 'information operations,'"* Olivia Solon, 4/27/17. *"Facebook has publicly acknowledged that its platform has been exploited by governments seeking to manipulate public opinion in other countries—including during the presidential elections in the US and France—and pledged to clamp down on such 'information operations.'"*

710 See, e.g., http://www.foxnews.com/opinion/2017/03/09/andrew-napolitano-spies-among-us-congress-has-created-monster-that-is-coming-for-us.html. "Spies among us—Congress has created a monster that is coming for us," Andrew P. Napolitano, 3/9/17.

711 http://www.foxnews.com/tech/2016/10/04/yahoo-built-email-spying-software-for-intelligence-agencies-report-says.html. "Yahoo built email spying software for intelligence agencies, report says," October 04, 2016. See also, https://www.washingtonpost.com/local/public-safety/us-courts-electronic-surveillance-up-500-percent-in-dc-area-since-2011-al-

most-all-sealed-cases/2016/10/22/48693ffa-8f10-11e6-9c52-0b10449e33c4_story.html. Spencer S. Hsu and Rachel Weiner, 10/24/16. "U.S. courts: Electronic surveillance up 500 percent in D.C.-area since 2011, almost all sealed cases."

712   http://www.nextgov.com/cloud-computing/2017/11/amazon-web-services-announces-secret-cloud-region-cia/142662/. "Amazon Web Services Announces Secret Cloud Region For CIA," Frank Konkel, 11/20/17.

713   https://www.yahoo.com/tech/us-nsa-spy-agency-halts-controversial-email-sweep-215107654.html. "US NSA spy agency halts controversial email sweep." AFP, 4/28/17.

714   *See*, e.g., Jane Black, "Some TIPS for John Ashcroft: Mr. Attorney General, forget your plan for a system to promote Americans spying on Americans. It won't work—and is un-American," *BusinessWeek Online*, 7/25/2002. http://www.businessweek.com/bwdaily/dnflash/jul2002/nf20020725_8083.htm [visited 8/1/05]. Lawrence Donegan, "Pentagon creates a Big Brother so Uncle Sam can keep his eye on us," 11/172002, http://observer.guardian.co.uk/international/story/0,6903,841731,00.html. *The Guardian* reports that:

> [T]he IAO has begun work on a global computer surveillance network which will allow unfettered access to personal details currently held in government and commercial databases around the world. Contracts worth millions of dollars have been awarded to companies to develop technology which will enable the Pentagon to store billions of pieces of electronic personal information—from records of internet use to travel documentation, lending library records and bank transactions—and then access this information without a search warrant. The system would also use video technology to identify people at a distance. "Total Information Awareness," or TIA, was proposed to the Pentagon by Admiral John Poindexter after the terrorist attacks of September 2001. A former official in the Reagan administration who was convicted for his leading role in the Iran-Contra scandal, Poindexter was appointed head of the IAO in February.

715   https://www.theguardian.com/technology/2017/nov/14/social-media-influence-election-countries-armies-of-opinion-shapers-manipulate-democracy-fake-news, "Thirty countries use 'armies of opinion shapers' to manipulate democracy—report: Governments in Venezuela, the Philippines, Turkey and elsewhere use social media to influence elections, drive agendas and counter critics, says report," Alex Hern, 11/14/17. Hern reports:

> The governments of 30 countries around the globe are using armies of so called opinion shapers to meddle in elections, advance anti-democratic agendas and repress their citizens, a new report shows. Unlike widely reported Russian attempts to influence foreign elections, most of the offending countries use the internet to manipulate opinion domestically, says US NGO Freedom House. "Manipulation and disinformation tactics played an important role in elections in at least 17 other countries over the past year, damaging citizens' ability to choose their leaders based on factual news and authentic debate," the US government-funded charity said. "Although some governments sought to support their interests and expand their influence abroad, as with Russia's disinformation campaigns in the United States and Europe, in most cases they used these methods inside their own borders to maintain their hold on power."

716   https://www.theguardian.com/technology/2017/jan/23/china-vpn-cleanup-great-firewall-censorship. "China cracks down on VPNs, making it harder to circumvent Great Firewall: A 14-month government 'cleanup' of internet access services will make it harder for users to access websites that are usually censored or restricted," Olivia Solon, 1/23/17.

717 https://www.theguardian.com/world/2017/mar/04/chinese-official-slams-internet-censorship. "Chinese official calls for easing of internet censorship: 'Great Firewall' that blocks western websites is too restrictive and hinders economic progress while discouraging foreign investors, says Luo Fuhe," Benjamin Haas, 3/3/17.

718 https://www.nytimes.com/2018/03/04/world/europe/turkey-erdogan-internet-law-restrictions.html. "Erdogan's Next Target as He Restricts Turkey's Democracy: The Internet," Carlotta Gall, 3/4/18.

719 https://www.independent.co.uk/news/world/europe/french-court-convicts-three-over-homophobic-tweets-in-case-hailed-as-a-significant-victory-by-lgbt-9996878.html, "French court convicts three over homophobic tweets, in case hailed as a 'significant victory' by LGBT rights campaigners," Kashmira Gander, 1/22/15. Staff and agencies, "Bardot weeps over racism charges," 5/7/2004, The Guardian [online]. See also, Polly Toynbee, "Get off your knees: Afraid of being labelled Islamophobic, the left has fallen into an embarrassed silence on religion. We must speak up," 6/11/2004, The Guardian [online]; Jon Henley, "Jail sentence for sexist insults under new French law," 6/24/2004, The Guardian [online]. http://www.worldmagblog.com/blog/archives/015541.html. "How civilizations die," Posted by Veith, 6/23/2005.

720 http://www.theverge.com/2016/11/3/13507126/iot-drone-hack. Thomas Ricker, "Watch a drone hack a room full of smart lightbulbs from outside the window," 11/3/16. Ricker writes:

721 See, e.g., http://www.mcclatchydc.com/news/nation-world/national/national-security/article166488597.html. "Is Alexa spying on us? We're too busy to care—and we might regret that," Tim Johnson, 8/10/17.

722 https://www.theguardian.com/technology/2017/mar/08/wikileaks-cia-leak-apple-vault-7-documents. "Apple to 'rapidly address' any security holes as companies respond to CIA leak: Company says it already fixed many exploits described in 'Vault 7' documents released by WikiLeaks, as CIA and Trump administration refuse to comment." Alex Hern, 3/8/17. https://www.theguardian.com/world/2017/mar/08/fbi-james-comey-privacy-wikileaks-cia-hack-espionage. Julian Borger, 3/8/17. "FBI's James Comey: 'There is no such thing as absolute privacy in America'—FBI director's assessment deepened privacy concerns raised by the details of CIA tools to hack consumer electronics for espionage published by WikiLeaks."

723 Softbank has an enormous stake in Alibaba, although it has begun selling off a significant portion of its shares to raise new money. See, http://fortune.com/2016/06/03/softbank-alibaba-shares-sale/. "SoftBank Bulks Up Alibaba Deal by Selling $1.1 Billion More Shares," Reuters, 6/3/16.

724 https://www.theguardian.com/business/2016/jul/18/arm-holdings-to-be-sold-to-japans-softbank-for-234bn-reports-say. "ARM Holdings to be sold to Japan's SoftBank for £24bn: Chancellor says sale of country's most successful technology company shows 'Britain has lost none of its allure to international investors,'" Sean Farrell, 7/18/16.

725 https://www.newscientist.com/article/2094629-beware-the-brexit-bots-the-twitter-spam-out-to-swing-your-vote/. 6/21/16, "Beware the Brexit bots: The Twitter spam out to swing your vote," Chris Baraniuk.

726 See, e.g., https://www.theguardian.com/technology/2017/apr/27/race-to-build-world-first-sex-robot. "The race to build the world's first sex robot: The $30bn sex tech industry is about to unveil its biggest blockbuster: a $15,000 robot companion that talks, learns, and never says no," Jenny Kleeman, 4/27/17.

727 We don't have far to look for examples. Spencer Ackerman, "NSA under renewed fire after report finds it violated its own privacy rules," 8/16/13. The Guardian, http://www.theguardian.com/world/2013/aug/16/nsa-violated-privacy-rules-audit.

> The NSA serially violated its own restrictions on bulk surveillance, according to a report that puts further pressure on beleaguered intelligence chief James Clapper and strengthens claims by a leading Senate critic that

a "culture of misinformation" exists at the agency. The Washington Post reported, with information provided by whistleblower Edward Snowden, that internal NSA audits found thousands of instances where the powerful surveillance agency collected, stored and possibly searched through vast swaths of information it is not permitted to acquire. *The revelations contradict repeated assurances this summer from senior Obama administration and intelligence officials that the NSA's programs to collect Americans' phone records and foreigners' communications in bulk contain adequate privacy protections.*

728   Peter Drucker, *The New Realities* 76 (Harper & Row 1989).

729   https://theintercept.com/2016/11/26/washington-post-disgracefully-promotes-a-mccarthyite-blacklist-from-a-new-hidden-and-very-shady-group/. Ben Norton & Glenn Greenwald, 11/26/16. "Washington Post Disgracefully Promotes a McCarthyite Blacklist From a New, Hidden, and Very Shady Group."

730   https://www.theguardian.com/technology/2017/mar/11/tim-berners-lee-online-political-advertising-regulation. "Tim Berners-Lee calls for tighter regulation of online political advertising: Inventor of the worldwide web described in an open letter how it has become a sophisticated and targeted industry, drawing on huge pools of personal data," Olivia Solon, 3/11/17. See also, http://www.usatoday.com/story/tech/news/2017/03/11/world-wide-webs-inventor-warns-s-peril/99005906/. "The World Wide Web's inventor warns it's in peril on 28th anniversary," Jon Swartz, *USA Today*, 3/11/17.

731   *Id.*

732   *Id.*

733   https://www.yahoo.com/news/sir-tim-berners-lee-launches-214716734.html, "Sir Tim Berners-Lee launches 'Magna Carta for the web' to save internet from abuse," Laurence Dodds, *The Telegraph*, 11/5/18.

734   *Id.*

735   Philip Hensher, "The bigger a community gets, the easier and more virulent anonymity becomes," *The Guardian*, 8/23/13; http://www.theguardian.com/commentisfree/2013/aug/23/bigger-community-easier-virulent-anonymity. He adds: "We are now much more anonymous than we used to be. We are less and less likely to know even our most immediate neighbours—one survey found that over 50% of us don't even known their names. Robert D. Putnam, in his celebrated 2000 study, *Bowling Alone*, found that everyday personal interaction had been on the decline in North America since 1950."

736   https://www.theguardian.com/world/2017/nov/06/workplace-surveillance-big-brother-technology. "Big Brother isn't just watching: workplace surveillance can track your every move: Employers are using a range of technologies to monitor their staff's web-browsing patterns, keystrokes, social media posts and even private messaging apps," Olivia Solon, 11/6/17.

737   For just a taste of the development see, http://www.worldpolicy.org/blog/2016/09/06/brink-artificial-intelligence-arms-race. "On the Brink of an Artificial Intelligence Arms Race." 9/6/16.   http://www.bbc.com/future/story/20150715-killer-robots-the-soldiers-that-never-sleep. Simon Parkin, 7/16/15.

738   http://www.dailymail.co.uk/sciencetech/article-3613443/The-device-eavesdrops-voices-head-Mind-reaching-machine-soon-turn-secret-thoughts-speech.html.   http://mashable.com/2016/05/19/perching-robot-bee/#Ohrhp7CxSgqX.

739   https://www.theguardian.com/technology/2017/jun/26/google-will-stop-scanning-content-of-personal-emails. "Google will stop scanning content of personal emails: Company did read emails in personal Gmail accounts to target users with tailored adverts but said it would stop," Alex Hern, 6/26/17.

740 Dan Gillmor, "Obama's NSA phone-record law ignores the other (big) data we're giving away: We are no longer merely creatures of metadata. We are now bystanders to the demise of privacy. Will anyone protect us?," 3/26/14, theguardian.com. http://www.theguardian.com/commentisfree/2014/mar/26/obama-nsa-phone-record-law-big-data-giving-away.

741 https://www.nytimes.com/2018/06/05/technology/facebook-device-partnerships-china.html. "Facebook Gave Data Access to Chinese Firm Flagged by U.S. Intelligence," Michael LaForgia and Gabriel J.X. Dance, 6/5/18. The report indicates: *"Facebook has data-sharing partnerships with at least four Chinese electronics companies, including a manufacturing giant that has a close relationship with China's government,* the social media company said on Tuesday. The agreements, which date to at least 2010, gave private access to some user data to Huawei, a telecommunications equipment company that has been flagged by American intelligence officials as a national security threat, as well as to Lenovo, Oppo and TCL.

742 *Axios,* Steve LeVine.

743 *The Guardian,* Mark Sweney, 1/13/18.

744 *The Guardian,* Jonathan Freedland, 3/23/18

745 https://www.theguardian.com/technology/2018/mar/11/tim-berners-lee-tech-companies-regulations. "Tim Berners-Lee: we must regulate tech firms to prevent 'weaponised' web: The inventor of the world wide web warns over concentration of power among a few companies 'controlling which ideas are shared,'" Olivia Solon, 3/11/18.

746 *New York Post,* Kevin Carty, 2/3/18.

747 *The Verge,* Vlad Savov, 5/17/18.

748 *The Guardian,* Dylan Curran, 5/19/18.

749 https://www.bloomberg.com/news/articles/2017-11-13/top-tech-stocks-1-7-trillion-gain-eclipses-canada-s-economy. "World's Top Tech Giants Amass $1.7 Trillion in Monster Year," Sofia Horta E Costa, 11/12/17. "That's more than Canada's entire economy, and exceeds the worth of Germany's biggest 30 companies put together. The eight tech giants— Facebook Inc., Amazon Inc., Apple Inc., Netflix Inc. and Google parent Alphabet Inc., as well as their Asian peers Baidu Inc., Alibaba Group Holding Ltd. and Tencent Holdings Ltd.—have amassed as much money in 2017 as Pacific Investment Management Co., one of the world's biggest fund managers, has done in about 46 years."

750 https://www.bloomberg.com/news/articles/2018-02-21/the-rise-of-amazon-facebook-may-be-bad-news-for-the-economy. "The Rise of Tech Giants May Be Bad News for the Economy: The dominance of a few firms risks harming productivity and growth, study finds," Alessandro Speciale, 2/21/18.

751 Solon, *supra* note 745.

752 *Hollywood Reporter,* Paul Bond, 12/3/2017

753 *The Guardian,* Larry Elliott, 4/19/18.

754 *The Guardian,* Nick Srnicek, 8/30/17.

755 *Wall Street Journal,* Mark Epstein, 12/18/17.

756 *Bloomberg News,* Alessandro Speciale, 2/21/18.

757 *The Guardian,* Chris Hughes, 4/27/18.

758 *Fox News,* James Rogers, 3/30/18. [Or, it's not our fault if people use us to do bad things.]

759 *The Guardian,* John Naughton, 10/8/17.

760 Solon, *supra* note 736.

761 *See,* e.g., http://www.seattletimes.com/business/amazon-isnt-technically-dominant-but-it-pervades-our-lives/. "Amazon isn't technically dominant, but it pervades our lives," Anick Jesdanun, 7/19/17.

762 https://www.independent.co.uk/news/world/europe/facebook-google-too-big-french-president-emmanuel-macron-ai-artificial-intelligence-regulate-govern-a8283726.html. "Facebook and Google are becoming too big to be governed, French

president Macron warns: 'At a point of time, your government, your people, may say, 'wake up',' Jane Dalton, 4/1/18.

763 *The Guardian,* Nicola Perrin and Danil Mikhailov, 11/3/17

764 *Washington Post,* Michael Birnbaum, 5/12/18.

765 *The Hill,* Harper Neidig, 4/1/18.

766 *The Guardian,* 5/21/18.

767 *Yahoo News,* Rob Lever, 5/25/18.

768 *The Guardian,* Alex Hern, 4/19/18.

769 https://wjla.com/news/nation-world/does-the-government-have-an-antitrust-case-against-amazon-google-and-facebook, "Does the government have an antitrust case against Amazon, Google and Facebook?," Leandra Bernstein, 9/10/18.

770 https://www.theringer.com/tech/2018/6/7/17436870/apple-amazon-google-facebook-break-up-monopoly-trump, "Monopoly Money: How to Break Up the Biggest Companies in Tech," Victor Luckerson, 6/7/18.

771 See e.g., http://www.businessinsider.com/peter-thiel-google-monopoly-2014-9; "Peter Thiel: Google Has Insane Perks Because It's A Monopoly," Drake Baer, 9/16/14. http://theweek.com/articles/693488/google-monopoly-crushing-internet, "Google is a monopoly—and it's crushing the internet," Ryan Cooper, 4/21/17.

772 http://www.nextgov.com/cloud-computing/2017/09/amazon-web-services-can-now-host-defense-departments-most-sensitive-data/140973/. "Amazon Web Services Can Now Host the Defense Department's Most Sensitive Data," Frank Konkel, 9/13/17.

773 http://observer.com/2016/08/tech-companies-apple-twitter-google-and-instagram-collude-to-defeat-trump/. Liz Crokin, "Tech Companies Apple, Twitter, Google, and Instagram Collude to Defeat Trump," 8/12/2016.

774 http://www.news.com.au/technology/online/social/leaked-document-reveals-facebook-conducted-research-to-target-emotionally-vulnerable-and-insecure-youth/news-story/d256f850be6b1c8a21aec6e32dae16fd. Nick Whigham, "Leaked document reveals Facebook conducted research to target emotionally vulnerable and insecure youth: A secret document shows in scary detail how Facebook can exploit the insecurities of teenagers using the platform."

775 http://www.cnbc.com/2016/05/27/mark-zuckerberg-is-dictator-of-facebook-nation-the-pirate-bay-founder.html. "Mark Zuckerberg is 'dictator' of Facebook 'nation': The Pirate Bay founder," Arjun Kharpal, 5/27/16.

776 http://www.thedailybeast.com/articles/2016/08/11/today-s-tech-oligarchs-are-worse-than-the-robber-barons.html. Joel Kotkin, "Today's Tech Oligarchs Are Worse Than the Robber Barons," 8/11/16.

777 https://www.theguardian.com/us-news/2017/sep/26/tech-industry-washington-google-amazon-apple-facebook. "'From heroes to villains': tech industry faces bipartisan backlash in Washington: In an effort uniting such disparate figures as Steve Bannon and Elizabeth Warren, leaders are calling for a clampdown on what some see as unchecked power," Sabrina Siddiqui, 9/26/17.

778 https://www.theguardian.com/us-news/2017/sep/26/tech-industry-washington-google-amazon-apple-facebook. "'From heroes to villains': tech industry faces bipartisan backlash in Washington: In an effort uniting such disparate figures as Steve Bannon and Elizabeth Warren, leaders are calling for a clampdown on what some see as unchecked power," Sabrina Siddiqui, 9/26/17.

779 http://www.politico.com/agenda/story/2017/09/17/open-markets-google-antitrust-barry-lynn-000523. "Inside the new battle against Google: Barry Lynn and his team think monopoly is the next great Democratic political cause. But what happens when they aim for the tech giants?," Danny Wink, 9/17/17. https://www.bloomberg.com/news/articles/2017-09-15/the-silicon-valley-backlash-is-heating-up. "The Silicon Valley Backlash is Heating Up," Eric Newcomer, 9/15/17.

780 https://www.theguardian.com/world/2017/aug/23/silicon-valley-big-data-extraction-amazon-whole-foods-facebook. "Silicon Valley siphons our data like oil. But the deepest drilling has just begun: Personal data is to the tech world what oil is to the fossil fuel industry. That's why companies like Amazon and Facebook plan to dig deeper than we ever imagined," Ben Tarnoff, 8/23/17.

781 http://www.telegraph.co.uk/technology/2017/11/11/eu-closes-google-prepares-second-antitrust-fine/. "EU closes in on Google as it prepares second antitrust fine," James Titcomb, 11/11/17.

782 https://www.theguardian.com/commentisfree/2017/aug/30/nationalise-google-facebook-amazon-data-monopoly-platform-public-interest. "We need to nationalise Google, Facebook and Amazon. Here's why," Nick Srnicek, 8/30/17.

783 http://www.independent.co.uk/life-style/gadgets-and-tech/news/google-my-activity-shows-everything-that-company-knows-about-its-users-and-there-s-a-lot-a7109256.html.   https://www.theguardian.com/technology/2016/jun/29/facebook-privacy-secret-profile-exposed.

784 https://www.axios.com/artificial-intelligence-pioneer-calls-for-the-breakup-of-big-tech-2487483705.html. Steve LeVine, 9/21/17. "Artificial intelligence pioneer calls for the breakup of Big Tech: Yoshua Bengio, the artificial intelligence pioneer, says the centralization of wealth, power and capability in Big Tech is "dangerous for democracy" and that the companies should be broken up." "Says Bengio: "Concentration of wealth leads to concentration of power. That's one reason why monopoly is dangerous. It's dangerous for democracy."

785 Tom Z. Spencer, "Police investigating reports of peeping drones spying inside NH homes," NH1.com, 7/30/16. http://www.nh1.com/news/police-investigating-reports-of-peeping-drones-spying-inside-nh-homes/.

786 http://www.nbcnews.com/tech/tech-news/biometric-scanning-use-grows-so-do-security-risks-n593161.

787 Id.

788 http://www.breitbart.com/london/2016/06/01/claim-corrupt-google-suppressing-eurosceptic-website-says-founder/.

789 http://www.telegraph.co.uk/news/2016/06/01/twitter-suspends-popular-anti-putin-parody-accounts/.   http://www.theguardian.com/world/2016/jun/02/twitter-unblocks-darthputinkgba-spoof-russia.

790 http://www.breitbart.com/tech/2016/05/18/facebook-censoring-content-critical-immigration/.http://www.independent.co.uk/voices/facebook-is-censoring-our-views-and-this-is-feeding-extremism-a7029251.html.https://www.washingtonpost.com/news/the-switch/wp/2016/03/28/mass-surveillance-silences-minority-opinions-according-to-study/.

791 Id.

792 https://ca.news.yahoo.com/edward-snowden-warns-against-relying-092615819.html. Mary Pascaline, "Edward Snowden Warns Against Relying On Facebook For News," 11/17/16.

793 http://fortune.com/2017/02/16/mark-zuckerberg-facebook-globalism/. "Mark Zuckerberg Warns Against Threats to Globalism and Says Facebook Is Here to Help," 2/17/17. "Facebook chief executive Mark Zuckerberg laid out a vision on Thursday of his company serving as a bulwark against rising isolationism, writing in a letter to users that the company's platform could be the 'social infrastructure' for the globe."

794 Yahoo News/Reuters, 4/21/18.

795 Breitbart, John Hayward, 4/2/18.

796 The Guardian, John Naughton, 3/4/18.

797 The Guardian, Alex Hern, 6/19/17.

798 Technology Review, Will Knight, 3/27/18.

799 The Guardian, Alex Hern, 11/14/17.

800 https://www.theguardian.com/commentisfree/2017/jun/12/general-election-so-cial-media-facebook-twitter. "Facebook needs to be more open about its effect on democracy: Social media plays a huge role in elections. But while Twitter allows access to its data, Facebook's secrecy means the extent of its influence may never be known," John Gallacher and Monica Kaminska, 6/12/17.

801 See, e.g., http://www.cnbc.com/2017/06/26/mark-zuckerberg-compares-facebook-to-church-little-league.html. "Mark Zuckerberg: Facebook can play a role that churches and Little League once filled: Mark Zuckerberg wants Facebook groups to be as important to people's lives as their local, community-support groups. Facebook's AI software led to a 50% rise in people signing up for online groups. Zuckerberg praised the role played in society by Little League coaches and leader of local religious congregations," John Shinal, 6/26/17.

802 https://www.washingtonpost.com/technology/2018/10/11/face-book-purged-over-accounts-pages-pushing-political-messages-profit/?utm_ter-m=.260d701d6d9c, "Facebook purged over 800 U.S. accounts and pages for pushing political spam," Elizabeth Dwoskin and Tony Romm, 10/11/18.

803 https://www.wsj.com/articles/amazon-is-leading-techs-takeover-of-ameri-ca-1497653164. "Amazon Is Leading Tech's Takeover of America: The tentacles of a handful of tech giants are reaching into industries no one ever expected them to, reshaping our world in their image," Christopher Mims, 6/16/17.

804 https://pjmedia.com/trending/google-reveals-plans-to-monitor-our-moods-our-movements-and-our-childrens-behavior/. "Google Reveals Plans to Monitor Our Moods, Our Movements, and Our Children's Behavior at Home," Phil Baker, 11/24/18.

805 https://www.usatoday.com/story/tech/2018/08/28/facebook-employees-hint-con-servative-intolerance/1127988002/. "Facebook 'mobs' attack conservative views within company, some employees say," Jessica Guynn, 8/28/18.

806 https://www.usatoday.com/story/opinion/2018/09/10/trump-google-you-tube-search-results-biased-against-republicans-conservatives-column/1248099002/. "Trump is right: More than Facebook & Twitter, Google threatens democracy, online freedom," Brad Parscale, 9/10/18.

807 Ewen MacAskill, "NSA paid millions to cover Prism compliance costs for tech companies," The Guardian, 8/23/13, http://www.theguardian.com/world/2013/aug/23/nsa-prism-costs-tech- companies-paid.

808 Ackerman, "US should re-evaluate surveillance laws, ex-NSA chief acknowledges," see supra note 727. Brian Bergstein, "In this data-mining society, privacy advocates shudder," Seattle Post-Intelligencer, 1/2/04; Kim Zetter, "GAO: Fed Data Mining Extensive," Wired News, 5/27/04; "Large Volume of F.B.I. Files Alarms U.S. Activist Groups," NYT 7/18/05; John Markoff, "Marrying Maps to Data for a New Web Service," NYT 7/18/05; Jeremiah Marquez, "LAPD Recruits Computer to Stop Rogue Cops," 7/24/05.

809 Forbes, Thomas Fox-Brewster, 4/16/18.

810 CBS Local New York, Ben Tracy, 4/24/18.

811 Quartz/QZ.com, Hannah Kozlowska, 12/19/17.

812 CNET, Sean Hollister, 4/13/18.

813 The Hill, Ron Yokubaitis, 01/19/18.

814 Daily Mail, Tim Collins, Matt Leclere and Nicole Pierre, 3/1/18.

815 http://blogs.wsj.com/economics/2016/07/21/a-shrinking-world-spurs-calls-to-re-write-the-tax-guidebook/. "A Shrinking World Spurs Calls to Rewrite the Tax Guidebook: The argument against taxing capital income relatively more than wages is losing its force," Adam Creighton, 7/21/16.

816 https://www.theguardian.com/world/2016/aug/06/two-resign-from-panama-pa-pers-commission-over-publicity-of-report. "Stiglitz resigns from Panama Papers commission," 8/5/16.

"We can only infer that the government is facing pressure from those who are making profits from the current non-transparent financial system in Panama," Stiglitz said. The Panama Papers cover a period from 1977 to December 2015, and show that some companies set up in tax havens may have been used for money laundering, arms and drug deals, as well as tax evasion. In addition to embarrassing leaders worldwide who had interests tied to secretive business concerns, the leak heaped pressure on Panama, a well-known global tax haven, to clean up its act. "I have had a close look at the so called Panama Papers and I must admit that, even as an expert on economic and organised crime, I was amazed to see so much of what we talk about in theory was confirmed in practice," Pieth said. In the paper he said he found evidence of crimes such as money laundering for child prostitution rings.

817 https://www.wsj.com/articles/the-numberof-americans-caught-underpaying-sometaxes-surges-40-1502443801. "Number of Americans Caught Underpaying Some Taxes Surges 40%: People who pay taxes quarterly—such as gig workers, retirees and business owners—are getting their payments wrong," Laura Saunders, 8/11/17.

818 Eric Lipton and Julie Creswell, "Panama Papers Show How Rich United States Clients Hid Millions Abroad," 6/5/16. http://www.nytimes.com/2016/06/06/us/panama-papers.html?emc=edit_na_20160606&nlid=60842176&ref=headline&_r=0. "Federal law allows United States citizens to transfer money overseas, but these foreign holdings must be declared to the Treasury Department, and any taxes on capital gains, interest or dividends must be paid—just as if the money had been invested domestically. Federal officials estimate that the government loses between $40 billion and $70 billion a year in unpaid taxes on offshore holdings."

819 https://www.nytimes.com/2017/11/05/world/paradise-papers.html. "Millions of Leaked Files Shine Light on Where the Elite Hide Their Money," Michael Forsythe, 11/5/17.

820 http://equitablegrowth.org/report/taxing-capital/. David Kamin, "Taxing Capital: Paths to a fairer and broader U.S. tax system," 8/10/16.

821 Id.

822 Id.

823 Id.

824 http://www.heritage.org/taxes/report/the-laffer-curve-past-present-and-future. "The Laffer Curve: Past, Present, and Future," Arthur Laffer, The Heritage Foundation, June 1, 2004.

825 Arthur Laffer noted that the Laffer Curve is subject to limitations and qualifications. "Revenue responses to a tax rate change will depend upon the tax system in place, the time period being considered, the ease of movement into underground activities, the level of tax rates already in place, the prevalence of legal and accounting-driven tax loopholes, and the proclivities of the productive factors." Likewise, the effectiveness of any change in tax rates—such as a tax cut—depends upon the size, timing, and location of the change. *See supra* note 824.

826 http://www.telegraph.co.uk/news/worldnews/europe/france/11844532/Actor-Gerard-Depardieu-to-sell-everything-in-France.html. "Actor Gérard Depardieu to 'sell everything' in France."

827 https://www.ft.com/content/19feb16a-1aaf-11e7-a266-12672483791a.   "France's wealth tax riles and divides presidential candidates: Amid cries for tariffs on rich, critics say it drives entrepreneurs away," Harriet Agnew, 4/10/17.

828 https://papers.ssrn.com/sol3/papers.cfm?abstract_id=2912395.
"Defending Worldwide Taxation with a Shareholder Based Definition of Corporate Residence," *Brigham Young University Law Review*, Vol. 2016, No. 6, 2017. Posted: 5 Mar 2017, J. Clifton Fleming Jr.

829 http://fortune.com/2016/12/19/apple-eu-tax-ireland/. "Ireland Says the EU Over-stepped in Its $14 Billion Apple Tax Ruling," 12/19/16.

830 https://www.cnbc.com/2018/06/18/how-amazon-made-jeff-bezos-the-richest-man-alive-worth-141-billion.html.

831 http://www.bloomberg.com/news/articles/2016-08-22/bill-gates-s-net-worth-hits-record-high-of-90-billion-chart. Devon Pendleton, "Bill Gates's Net Worth Hits Record High of $90 Billion," 8/22/16.

832 https://www.socialeurope.eu/2017/05/getting-robots-pay-tax/. "Getting The Robots To Pay Tax," Vincenzo Visco, 5/2/17.

833 Id.

834 Id.

835 Id.

836 http://clsbluesky.law.columbia.edu/2017/06/06/how-tax-policy-favors-robots-over-workers-and-what-to-do-about-it/, "How Tax Policy Favors Robots over Workers and What to Do About it," Ryan Abbott and Bret Bogenschneider, 6/6/17.

837 Id.

838 http://clsbluesky.law.columbia.edu/2017/06/06/how-tax-policy-favors-robots-over-workers-and-what-to-do-about-it/, "How Tax Policy Favors Robots over Workers and What to Do About it," Ryan Abbott and Bret Bogenschneider, 6/6/17.

839 http://www.telegraph.co.uk/technology/2017/08/09/south-korea-introduces-worlds-first-robot-tax/. "South Korea introduces world's first 'robot tax,'" Cara McGoogan, 8/9/17. "Though it is not about a direct tax on robots, it can be interpreted as a similar kind of policy considering that both involve the same issue of industrial automation," an industry source told the *Korea Times*. Korea is the first country to implement a robot tax, but it is not the only one to have proposed a technology levy. Bill Gates has previously called for a tax on robots to balance the Government's income as jobs are lost to automation. He said the levy could help slow down the pace of change and provide money to hire additional employees in sectors that require people, such as health care."

840 https://www.theguardian.com/business/2016/nov/15/joseph-stiglitz-what-the-us-economy-needs-from-donald-trump. "Joseph Stiglitz: what the US economy needs from Donald Trump," Joseph Stiglitz, 11/15/16.

841 Paul Krugman, "Borrow-and-build spree is precisely what U.S. needs," *Plain Dealer*, 8/9/16, at p. A10.

842 http://www.marketwatch.com/story/productivity-declines-for-third-straight-quar-ter-2016-08-09. Greg Robb, 8/9/16, "Productivity declines for third straight quarter."

843 For an intriguing analysis, see: https://www.theguardian.com/commentisfree/2017/jun/09/seven-years-of-pain-austerity-experiment-over-general-election. "After seven years of pain, the austerity experiment is over ," Larry Elliott, 6/9/17.

844 https://www.theguardian.com/business/2016/nov/15/joseph-stiglitz-what-the-us-economy-needs-from-donald-trump. Joseph Stiglitz, 11/15/16, "Joseph Stiglitz: what the US economy needs from Donald Trump."

845 https://www.theguardian.com/business/2018/oct/03/world-economy-at-risk-of-another-financial-crash-says-imf. "World economy at risk of another financial crash, says IMF: Debt is above 2008 level and failure to reform banking system could trigger crisis," Phillip Inman, 10/3/18. "The world economy is at risk of another financial meltdown, following the failure of governments and regulators to push through all the reforms needed to protect the system from reckless behaviour, the International Monetary Fund has warned. With global debt levels well above those at the time of the last crash in 2008, the risk remains that unregulated parts of the financial system could trigger a global panic, the Washington-based lender of last resort said."

846 http://fortune.com/2017/04/05/jobs-automation-artificial-intelligence-robotics/. "The Bright Side of Job-Killing Automation," Barb Darrow, 4/5/17.

847 Catherine Clifford, "Elon Musk: Robots will take your jobs, government will have to pay your wage," 11/4/2016. http://www.cnbc.com/2016/11/04/elon-musk-robots-will-take-your-jobs-government-will-have-to-pay-your-wage.html.

848 *Id.*

849 *Id.*

850 Novelist Kurt Vonnegut, in "Harrison Bergeron," *The Magazine of Fantasy and Science Fiction* (October 1961), envisioned a society that rigidly enforced principles of perfect equality among its members—the strong were forced to wear heavy weights, the intelligent were required to medicate or wear distracting earbuds, the beautiful were disfigured, and so on.

851 *CNN Money,* Tami Luhby, 5/18/18.

852 *CBS MoneyWatch,* Aimee Picchi, 8/24/17.

853 *Pasadena Star,* Kevin Smith, 7/25/17.

854 *CNBC,* Emmie Martin, 3/15/18.

855 *Wall Street Journal,* Laura Sanders, 4/6/18.

856 *Washington Examiner,* Paul Bedard, 10/27/17.

857 *BBC,* Laurence Peter, 4/23/18.

858 http://freebeacon.com/issues/1-5-families-u-s-no-one-works/. "No One Works in I in 5 U.S. Families: In 2015, there were 16,060,000 families with no member employed," Ali Meyer, 4/22/16.

859 http://www.marketwatch.com/story/45-of-americans-pay-no-federal-income-tax-2016-02-24.

860 Paul Vigna, "Bill Gross: What to Do After the Robots Take Our Jobs: Get ready for driverless trucks, universal basic income, and less independent central banks," 5/4/16. http://blogs.wsj.com/moneybeat/2016/05/04/bill-gross-what-to-do-after-the-robots-take-our-jobs/.

861 *Id.*

862 https://ca.news.yahoo.com/swiss-voters-decide-guaranteed-monthly-income-plan-103534182—business.html.

863 See https://finance.yahoo.com/news/universal-basic-free-monthly-income-utopian-switzerland-silicon-valley-finland-canada-122210548.html. "Free income is a great idea—unfortunately, it sort of doesn't work," Melody Hahm, 6/7/16. Hahm writes: "in theory, the idea of having a fall-back income foundation sounds delightful. But what most advocates aren't getting specific about is where exactly this money would come from." See also, http://forums.canadiancontent.net/news/153316-what-we-can-learn-finland.html. "What We Can Learn From Finland's Basic Income Experiment." The report explains:

> Olli Kangas, Kela's coordinator for the program, told *The Economist* that it was currently in a state of neglect, comparing politicians' actions to "small boys with toy cars who become bored and move on." ... India, the world's largest democratic country, has endorsed the system—claiming in a report that it is "basically the way forward"—and is now considering the best way to introduce it to its populous [sic]. ... The system is not without its skeptics, however. Experts question who would provide the money to fund such projects, asserting that a universal basic income of $10,000 a year per person could add approximately $3 trillion to national spending in the U.S. Individuals such as Mark Cuban and Robert Gordon, an economist at Northwestern University, have suggested that we should optimize existing benefits systems. Gordon told the MIT Technology Review that his idea is to make "benefits more generous to reach a reasonable minimum, expand the Earned Income Tax Credit, and greatly expand preschool care for children who grow up in poverty."

864 Other discussions of UBI include: https://www.theguardian.com/commentis-free/2017/mar/06/utopian-thinking-poverty-universal-basic-income. "Utopian think-ing: the easy way to eradicate poverty," Rutger Bregman, 3/6/17. http://www.pressherald.com/2017/03/05/to-panhandlers-program-may-offer-welcome-change-jobs/. "To Portland panhandlers, program may offer welcome change: Jobs," Randy Billings, 3/5/17.

865 https://www.bloomberg.com/news/articles/2017-08-21/people-start-hating-their-jobs-at-age-35. "People Start Hating Their Jobs at Age 35: The shiny newness of life in the workforce begins to wear off," Chris Stokel-Walker, 8/21/17. https://www.theguardian.com/us-news/2017/aug/21/missouri-fast-food-workers-better-pay-popeyes-economics. "Fran works six days a week in fast food, and yet she's homeless: 'It's economic slavery.'" Dominic Rushe and Tom Silverstone, 8/21/17.

866 http://money.cnn.com/2017/01/02/news/economy/finland-universal-basic-in-come/. "Finland is giving 2,000 citizens a guaranteed income," Ivana Kottasova, 1/3/17.

> Participants will receive €560 ($587) a month—money that is guar-anteed regardless of income, wealth or employment status. The idea is that a universal income offers workers greater security, especially as techno-logical advances reduce the need for human labor. It will also allow unem-ployed people to pick up odd jobs without losing their benefits. The initial program will run for a period of two years. Participants were randomly selected, but had to be receiving unemployment benefits or an income subsidy. The money they are paid through the program will not be taxed. If the program is successful, it could be expanded to include all adult Finns. The Finnish government thinks the initiative could save money in the long run. The country's welfare system is complex and expensive to run, and simplifying it could reduce costly bureaucracy. The change could also encourage more jobless people to look for work, because they won't have to worry about losing unemployment benefits. Some unemployed workers currently avoid part time jobs because even a small income boost could result in their unemployment benefits being canceled. "Incidental earnings do not reduce the basic income, so working and ... self-employ-ment are worthwhile no matter what," said Marjukka Turunen, the head of the legal unit at Kela, Finland's social insurance agency.

867 See, e.g., Robert Skidelsky, "A basic income could be the best way to tackle inequal-ity," 6/23/16. https://www.theguardian.com/business/2016/jun/23/universal-basic-income-could-be-the-best-way-to-tackle-inequality.

868 Id.

869 http://www.ibtimes.co.uk/alcoholism-epidemic-more-1-8-americans-are-now-al-coholics-1634315. "Alcoholism epidemic: More than 1 in 8 Americans are now alcoholics: Alcoholism has risen 49% in the US in just 11 years, national surveys find," Martha Henriques, 8/9/17.

870 Charles Murray, 6/3/16, http://www.wsj.com/articles/a-guaranteed-income-for-ev-ery-american-1464969586. "A Guaranteed Income for Every American." Murray writes: "[T] he system has to be designed with certain key features. In my version, every American citizen age 21 and older would get a $13,000 annual grant deposited electronically into a bank ac-count in monthly installments. Three thousand dollars must be used for health insurance (a complicated provision I won't try to explain here), leaving every adult with $10,000 in dispos-able annual income for the rest of their lives."

871 Murray, id, explains:

> The UBI is to be financed by getting rid of Social Security, Medi-care, Medicaid, food stamps, Supplemental Security Income, housing

subsidies, welfare for single women and every other kind of welfare and social-services program, as well as agricultural subsidies and corporate welfare. As of 2014, the annual cost of a UBI would have been about $200 billion cheaper than the current system. By 2020, it would be nearly a trillion dollars cheaper.

872 http://www.scholarsstrategynetwork.org/brief/why-us-federal-government-needs-guarantee-jobs-all-willing-workers. "Why the U.S. Federal Government Needs to Guarantee Jobs for All Willing Workers," Mark Paul, William Darity Jr., , Darrick Hamilton.

873 Terence P. Jeffrey, "21,995,000 to 12,329,000: Government Employees Outnumber Manufacturing Employees 1.8 to 1," September 8, 2015. http://cnsnews.com/news/article/terence-p-jeffrey/21955000-12329000-government-employees-outnumber-manufacturing.

874 Charles S. Clark, "Even CBO Is Stumped on the Size of the Contractor Workforce," 3/12/15, http://www.govexec.com/contracting/2015/03/even-cbo-stumped-size-contractor-workforce/107436/.

875 http://blogs.wsj.com/economics/2014/11/07/the-federal-government-now-employs-the-fewest-people-since-1966/.

876 http://www.zerohedge.com/news/2017-02-08/most-government-workers-could-be-replaced-robots-new-study-finds. "Most Government Workers Could Be Replaced By Robots, New Study Finds," Tyler Durden, 2/8/17.

877 Id.

878 http://www.theoccidentalobserver.net/2011/07/discrimination-against-whites-in-federal-employment/. Kevin MacDonald, "Discrimination against Whites in Federal Employment," 7/14/11.

879 http://www.epi.org/publication/bp339-public-sector-jobs-crisis/. David Cooper, Mary Gable and Algernon Austin, "The public-sector jobs crisis: Women and African Americans hit hardest by job losses in state and local governments," 5/2/12. Briefing Paper # 339.

880 Tom Leonard, Daily Mail, Feb. 3, 2017. http://www.dailymail.co.uk/news/article-4190322/Tech-billionaires-building-boltholes-New-Zealand.html. "Apocalypse island: Tech billionaires are building boltholes in New Zealand because they now fear social collapse or nuclear war. So what do they know that we don't?"

881 http://www.foxnews.com/us/2017/07/20/concealed-handgun-permits-surging-blacks-women-lead-growth.html. "Concealed handgun permits surging, blacks, women lead growth," 7/20/17. "Concealed handgun permits in the United States soared by 1.83 million since last July, setting a record for the fourth consecutive year, according to an analysis released Thursday."

882 Yuval Noah Harari, Homo Deus: A Brief History of Tomorrow (2015); Nick Bostrom, Superintelligence: Paths, Dangers, Strategies (2014); James Barrat, Our Final Invention: Artificial Intelligence and the End of the Human Era (2013).

883 http://www.dailymail.co.uk/sciencetech/article-3721365/Can-predict-turn-crime-Minority-Report-computers-soon-mark-children-likely-criminals.html. Paul McGorrery and Dawn Gilmore, "Can we predict who will turn to crime? 'Minority Report' computers may soon mark out children as 'likely criminals,'" The Conversation, 8/9/16. This fear is not without foundation. See, e.g., http://www.foxnews.com/tech/2016/08/22/chicago-police-push-back-on-criticism-crime-prediction-system.html. Stephanie Mlot, "Chicago Police push back on criticism of crime-prediction system," 8/22/16.

884 http://www.sciencemag.org/news/2017/04/even-artificial-intelligence-can-acquire-biases-against-race-and-gender. "Even artificial intelligence can acquire biases against race and gender," Mathew Hutson, 4/13/17. https://www.theguardian.com/technology/2017/jul/16/how-can-we-stop-algorithms-telling-lies. "How can we stop algorithms telling lies? Algorithms can dictate whether you get a mortgage or how much you pay for insurance. But sometimes they're wrong—and sometimes they are designed to deceive," Cathy O'Neil, 7/16/17.

885 https://www.theguardian.com/technology/2016/sep/08/artificial-intelligence-beauty-contest-doesnt-like-black-people, "A beauty contest was judged by AI and the robots didn't like dark skin," Sam Levin, 9/8/16.

886 https://arxiv.org/ftp/arxiv/papers/1607/1607.05402.pdf. See, "Web-based Teleoperation of a Humanoid Robot." The research paper on the ability to use the Internet to control others' devices remotely is found at the above link.

887 David Rotman, "How Technology Is Destroying Jobs," 6/12/13, *MIT Technology Review* magazine July/August 2013, https://www.technologyreview.com/magazine/2013/07/.

888 https://www.bloomberg.com/enterprise/blog/weve-hit-peak-human-and-an-algorithm-wants-your-job-now-what/. "We've hit peak human and an algorithm wants your job. Now what?," Hugh Son, *Bloomberg Markets*, 6/8/16.

889 http://freebeacon.com/issues/1-5-families-u-s-no-one-works/. "No One Works in 1 in 5 U.S. Families: In 2015, there were 16,060,000 families with no member employed," Ali Meyer, April 22, 2016.

890 http://www.marketwatch.com/story/45-of-americans-pay-no-federal-income-tax-2016-02-24.

891 Who carries the US tax burden? http://www.marketwatch.com/story/45-of-americans-pay-no-federal-income-tax-2016-02-24. "45% of Americans pay no federal income tax: 77.5 million households do not pay federal individual income tax," 4/18/16. "The top 1% of taxpayers pay a higher effective income-tax rate than any other group (around 23%, according to a report released by the Tax Policy Center in 2014)—nearly seven times higher than those in the bottom 50%. On average, those in the bottom 40% of the income spectrum end up getting money from the government. Meanwhile, the richest 20% of Americans, by far, pay the most in income taxes, forking over nearly 87% of all the income tax collected by Uncle Sam."

892 http://www.zerohedge.com/news/2017-04-21/total-chaos-cyber-attack-feared-multiple-cities-hit-simultaneous-power-grid-failures. ""Total Chaos"—Cyber Attack Feared As Multiple Cities Hit With Simultaneous Power Grid Failures," Tyler Durden, 4/21/17. "The U.S. power grid appears to have been hit with multiple power outages affecting San Francisco, New York and Los Angeles."

893 http://www.infowars.com/jpmorgan-chase-ceo-economic-crisis-inevitable/, "JPMorgan Chase CEO: Economic Crisis Inevitable." Kit Daniels, *Infowars.com*, 4/11/15. "Another economic crisis like the Great Recession is inevitable, according to JPMorgan Chase CEO Jamie Dimon.

NAMES INDEX

ACKNOWLEDGEMENTS

David would like to acknowledge and thank the following individuals who provided input at various points in the development and completion of *The Artificial Intelligence Contagion.*

My co-author and son, Daniel Barnhizer, in addition to interacting with me on many ideas about the structure, focus, and content of *Contagion*, provided editing and his own important textual additions. Sue Barnhizer not only provided continual moral support but made significant editorial and intellectual contributions. Joshua Barnhizer frequently offered helpful perspectives on the issues discussed in *Contagion*. Brian Barnhizer provided a continuous stream of relevant articles related to the AI/robotics concerns. Barry Barnhizer shared his extensive business experience in ways that enriched our insights, and Bret Barnhizer also contributed his significant and diverse business experience as well as reminded me early on of how critical is the role China is playing in AI/robotics.

In addition to the above individuals, Robert "Bob" Bryson was, as always, a continual font of ideas, critique and suggestions on a number of *Contagion's* issues. My lifelong friend Bill Dornan, an experienced provider of technology to numerous Midwestern US companies, offered examples of the speed and scale at which his customers were shifting production activities to AI/robotics systems and reducing inputs from human workers. Ridge Cooper offered a frequent stream of ideas about artificial intelligence developments, as did Timothy Halverson who recommended various readings, particularly the work of Yuval Noah Harari. Robert Baskette, owner of a successful AI-based consulting company, provided stimulating insights into the steady evolution of research into quantum computing technology. Marc Moretz offered suggestions and introduced me to the work of MIT's Max Tegmark. David Cooper is a New York based business and technology consultant who provided a substantive overview of *Contagion* including the observation that it presented a "sober and compelling" look at our future. I also want to thank my Dixie Roadhouse buddies Greg Nichols, Ron Hanson and Keith Sauerwald, with whom I engaged in regular discussions about the book as it developed.

Finally, I can't say enough to express my admiration for Clarity's Diana Collier who performed a detailed and professional job of stylistic editing and offered very useful substantive and structural suggestions.

Daniel: it's always an amazing opportunity to collaborate on a project with my dad, David, who pushed me to challenge, question, and explore the world and the impact of human action and decisions. This is an important book that tells a frightening story of the world we may be building for our own children, and I'm thankful to have been a part of this project.

ABOUT THE AUTHORS

David Barnhizer is Professor of Law *Emeritus* at the Cleveland State University. He received law degrees from the Ohio State University where he graduated *summa cum laude* and from Harvard University where he was a Ford Foundation Urban Law Fellow, a CLEPR Clinical Teaching Fellow, and earned a Masters of Law degree. He was Articles Editor of the Ohio State Law Journal, and began his legal career as a Reginald Heber Smith Community Lawyer in Colorado with the Colorado Springs Legal Services Office. He has been a Senior Research Fellow at the University of London's Institute for Advanced Legal Studies, a frequent Visiting Professor at the Westminster University School of Law in London, taught human rights and international environmental law in St. Petersburg, Russia in a joint program with St. Petersburg State University, and instructed in Harvard's Intersession program in Trial Advocacy. He has served as Senior Adviser in the International Program of the Natural Resources Defense Council (NRDC). He was a Senior Fellow with Earth Summit Watch, and a board member of the International Shrimp Action Network (ISANET) an international NGO network made up of more than twenty NGOs involved in environment, development and coastal zone management in developing countries. He was Executive Director of the Washington, DC-based Year 2000 Committee, and has consulted extensively with environmental and development organizations. These include the World Resources Institute, the International Institute for Environment and Development, the United Nations Development Program, the President's Council on Environmental Quality, the World Bank, the United Nations Food and Agricultural Organization (FAO), the World Wildlife Fund, the Mongolian government, and the Center for Global Change. In addition, he spent nine years as a member of the board of directors for Performance Capital Management, was General Counsel for a high-tech development company, NanoLogix, and did strategic consulting with Sovonics Solar Systems, a subsidiary of British Petroleum. He also served as *Rapporteur* for the Foresight Capability Workshop of the Energy and Commerce Committee of the US House of Representatives and has worked on projects in conjunction with numerous federal agencies. He has written over fifty law review articles and book chapters, and authored and edited a number of books including *The Warrior Lawyer*, a work developing the strategic principles of Sun Tzu's *Art of War* and Musashi's *A Book of Five Rings*. Other books include *Strategies for*

Sustainable Societies, Environment Cleveland, The Blues of a Revolution, two volumes on *Effective Strategies for Protecting Human Rights,* and *Hypocrisy & Myth: The Hidden Order of the Rule of Law* co-authored with Daniel Barnhizer.

Daniel Barnhizer is the Bradford Stone Faculty Scholar at the Michigan State University School of Law. He received his undergraduate degree *summa cum laude* from Miami University and is a member of Phi Beta Kappa honorary society. He earned Juris Doctor degree, with honors, from Harvard Law School where he was an Editor of the Harvard Environmental Law Review. He teaches and writes in the areas of tax, contract law and theory, conservation law, comparative law, and the jurisprudence associated with the rule of law. He is a coauthor of casebooks in the fields of Contracts and Commercial Transactions. Professor Barnhizer directs the Conservation Law Program at MSU, as well as the MSU College of Law Institute for Comparative Law & Jurisprudence at the University of Bialystok Faculty of Law in Poland. He has regularly taught in Poland and Lithuania and been a Lecturer at the Jinan University Law School in Guangzhou, China. Prior to coming to Michigan State in 2001, Professor Barnhizer has extensive experience in business and worked for the law firms of Hogan & Hartson and Cadwalader, Wickersham & Taft in Washington, D.C. He served as a judicial law clerk for the Honorable Richard L. Nygaard, U.S. Court of Appeals for the 3rd Circuit, and for the Honorable Robert B. Krupansky, U.S. Court of Appeals for the 6th Circuit, sitting by designation on the U.S. District Court for the Northern District of Ohio.